Globular clusters are spherical, densely packed groups of stars found around galaxies. They are thought to have formed at the same time as their host galaxy and thus provide a powerful probe for understanding stellar and galaxy evolution, as well as being studied as objects of interest in their own right. This timely volume presents invited articles by a team of world leaders who gathered at the X Canary Islands Winter School of Astrophysics to review our current understanding of globular clusters. It provides an accessible introduction to the field for graduate students and a comprehensive and up-to-date reference for researchers.

This volume is dedicated to the memory of
Rebecca Elson,
who died tragically on May 19, 1999.

CAMBRIDGE CONTEMPORARY ASTROPHYSICS

Globular Clusters

CAMBRIDGE CONTEMPORARY ASTROPHYSICS

Series editors
José Franco, Steven M. Kahn, Andrew R. King and Barry F. Madore

Titles available in this series
Gravitational Dynamics, *edited by O. Lahav, E. Terlevich
and R. J. Terlevich* (ISBN 0 521 56327 5)

High-sensitivity Radio Astronomy, *edited by N. Jackson and R. J. Davis*
(ISBN 0 521 57350 5)

Relativistic Astrophysics, *edited by B. J. T. Jones and D. Marković*
(ISBN 0 521 62113 5)

Advances in Stellar Evolution, *edited by R. T. Rood and A. Renzini*
(ISBN 0 521 59184 8)

Relativistic Gravitation and Gravitational Radiation,
edited by J.-A. Marck and J.-P. Lasota
(ISBN 0 521 59065 5)

Instrumentation for Large Telescopes,
edited by J. M. Rodríguez Espinosa, A. Herrero and F. Sánchez
(ISBN 0 521 58291 1)

Stellar Astrophysics for the Local Group,
edited by A. Aparicio, A. Herrero and F. Sánchez
(ISBN 0 521 63255 2)

Nuclear and Particle Astrophysics, *edited by J. G. Hirsch and D. Page*
(ISBN 0 521 63010 X)

Theory of Black Hole Accretion Discs,
edited by M. A. Abramowicz, G. Björnsson and J. E. Pringle
(ISBN 0 521 62362 6)

Interstellar Turbulence
edited by J. Franco and A. Carramiñana
(ISBN 0 521 65131 X)

Globular Clusters,
edited by C. Martínez Roger, I. Pérez Fournón and F. Sánchez
(ISBN 0 521 77058 0)

Globular Clusters
X Canary Islands Winter School of Astrophysics

Edited by
C. MARTÍNEZ ROGER
Instituto de Astrofísica de Canarias

I. PEREZ FOURNÓN
Instituto de Astrofísica de Canarias

F. SÁNCHEZ
Instituto de Astrofísica de Canarias

CAMBRIDGE
UNIVERSITY PRESS

PUBLISHED BY THE PRESS SYNDICATE OF THE UNIVERSITY OF CAMBRIDGE
The Pitt Building, Trumpington Street, Cambridge, United Kingdom

CAMBRIDGE UNIVERSITY PRESS
The Edinburgh Building, Cambridge CB2 2RU, UK   www.cup.cam.ac.uk
40 West 20th Street, New York, NY 10011-4211, USA   www.cup.org
10 Stamford Road, Oakleigh, Melbourne 3166, Australia
Ruiz de Alarcón 13, 28014 Madrid, Spain

First published 1999

Printed in the United States of America

Typeset by the author

*A catalog record for this book is available from
the British Library.*

Library of Congress Cataloging-in-Publication Data

Canary Islands Winter School on Astrophysics (10th)
Globular clusters : X Canary Islands Winter School of Astrophysics
/ edited by C. Martinez Roger. I Pérez Fournón. F. Sánchez.
p. cm. — (Cambridge contemporary astrophysics)
ISBN 0-521-77058-0 (hb)
1. Stars — Globular clusters Congresses.   I. Martínez Roger, C.
II. Pérez-Fournon. I.   III. Sánchez, F.   IV. Title.   V. Series.
QB853.5.C36   1999
523.8'55—dc21   99-32417
CIP

ISBN 0 521 77058 0   hardback

Contents

The Observational Approach to Populations in Globular Clusters

I. R. King

Stellar Populations and the Formation of the Milky Way

S. Majewski

Globular Clusters as a Test for Stellar Evolution

V. Castellani

Early Nucleosynthesis and Chemical Abundances of Stars in Globular Clusters

R. Gratton

Stellar Dynamics in Globular Clusters

R. A. W. Elson

Pulsating Stars in Globular Clusters and Their Use

M. W. Feast

X-Ray Sources in Globular Clusters

R. Canal

Globular Clusters Systems: Formation Models and Case Studies

W. E. Harris

Participants

Agudo Rodríguez, Iván	Universidad de Granada (Spain)
Álvarez, Carlos A.	University of Leeds (U.K.)
Andreuzzi, Gloria	Osservatorio Astronomico di Roma (Italy)
Bakos, Gaspar	Eötuös Lòrànd University (Hungary)
Beasley, Michael Andrew	University of Durham (U.K.)
Benkö, Józef	Konkoly Observatory (Hungary)
Butler, Raimond	University of Edinburgh (U.K.)
Canal, Ramón	Universitat de Barcelona (Spain)
Castellani, Marco	Osservatorio Astronomico di Roma (Italy)
Castellani, Vittorio	Università di Pisa (Italy)
Chapelon, Sandra	Université de Provence (France)
Cora, Sofía Alejandra	Universidad Nacional de La Plata (Argentina)
Corte Rodríguez, Natalia B.	Universidad de Oviedo (Spain)
Degl'Innocenti, Scilla	Università di Pisa (Italy)
Deiters, Stefan	Universität Heidelberg (Germany)
Dinescu, Dana	Yale University (U.S.)
Domínguez Tagle, Carlos	Instituto de Astrofísica de Canarias (Spain)
Elson, Rebecca	University of Cambridge (U.K.)
Feast, Michael W.	University of Cape Town (South Africa)
Funato, Yoko	University of Tokio (Japan)
García Pérez, Ana E.	University of Uppsala (Sweden)
Gómez Velarde, Gabriel	Instituto de Astrofísica de Canarias (Spain)
Gozzi, Giacomo	Università di Pisa (Italy)
Gratton, Raffaele	Osservatorio Astronomico di Padova (Italy)
Harris, William E.	University of McMaster (Canada)
Hilker, Michel	Universidad Católica de Chile (Chile)
Iaria, Rosario	Università di Palermo (Italy)
Ivanova, Borjana	Bulgarian Academy of Sciences (Bulgaria)
Johnson, Winfrey Yeamans	University of Virginia (U.S)
Katajainen, Seppo J.	University of Turku (Finland)
Khalisi, Emil	Universität Heidelberg (Germany)
King, Ivan	University of California (U.S.)
Kopacki, Grzegorz	Wroclaw University (Poland)
Kundu, Arunav	University of Maryland (U.S.)
Kurth, Oliver Michael	Universitary Observatory Goettingen (Germany)
Larsen, Soeren	Copenhagen University Astronomical Obs. (Denmark)
Lejeune, Thibault	Astron. Institut der Universität Basel (Switzerland)
Lyubchyk, Yuriy	Main Astronomical Observatory (Ukraine)
Majewski, Steven R.	University of Virginia (U.S.)
Marconi, Marcella	Osserv. Astronomico di Capodimonte (Italy)
Markov, Haralambi	Bulgarian Academy of Sciences (Bulgaria)
Martínez Delgado, David	Instituto de Astrofísica de Canarias (Spain)
Martínez Roger, Carlos	Instituto de Astrofísica de Canarias (Spain)
Masana Fresno, Eduard	Univetsitat de Barcelona (Spain)
Mathieu, Anne	Sterrenkundig Observatorium (Belgium)
Matute, Israel	Instituto de Astrofísica de Canarias (Spain)
de Morais Barros Fdes., João M.	University of Coimbra (Portugal)
Motta, Veronica	Instituto de Astrofísica de Canarias (Spain)

Parmentier, Geneviève	University of Liege (Belgium)
Pérez Fournon, Ismael	Instituto de Astrofísica de Canarias (Spain)
Pignatelli, Ezio	Osservatorio di Padova (Italy)
Pugliese, Giovanna	Max Planck I. für Radioastronomie (Germany)
Puzia, Thomas Hyazinth	University of Bonn (Germany)
Pych, Wojciech	Warsaw University Observatory (Poland)
Raimondo, Gabriella	Osserv. Astron. "V. Cerulli" di Collurania (Italy)
Recio Blanco, Alejandra	Instituto de Astrofísica de Canarias (Spain)
Rengel Lamus, Miriam Elina	Centro de Investigaciones de Astronomía (Venezuela)
Ripepi, Vincenzo	Osserv. Astronomico di Capodimonte (Italy)
da Rocha, Cristiano	Universidad de Sao Paulo (Brazil)
Rosenberg González, Alfred	Osservatorio Astronomico di Padova (Italy)
Salgado, Raquel J.	Universidad de Extremadura (Spain)
di Salvo, Tiziana	Università di Palermo (Italy)
Saviane, Ivo	Università di Padova (Italy)
Schroder, Linda	University of California (U.S.)
Sosa Brito, Rafael Mario	Max Planck I. für Extraterrestrische Physik (Germany)
Suárez, Olga	Universidad de Vigo (Spain)
Subramaniam, Annapurni	Indian Institute of Astrophysics (India)
Uribe Botero, José Antonio	Universidad Nacional de Colombia (Colombia)
Valle, Giada	Università di Pisa (Italy)
VanDalfsen, Marcel	McMaster University (Canada)
Vivas, Anna Katherina	Yale University (U.S.
Westera, Pieter	Astron. Institut der Universität Basel (Switzerland)
Yasarsoy, Bulent	Ege University Observatory (Turkey)
Zoccali, Manuela	Università di Padova (Italy)
Zivkov, Vesna	Belgrade University (Yugoslavia)

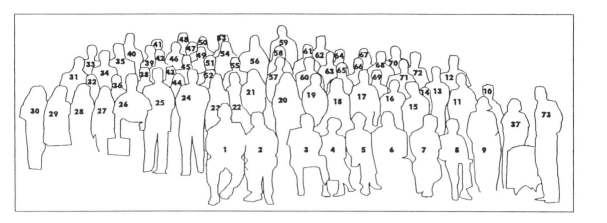

1	William E. Harris	26	Thomas Hyazinth Puzia	51	Rafael Mario Sosa Brito
2	Ramón Canal	27	Marcella Marconi	52	Giovanna Pugliese
3	Vittorio Castellani	28	Giada Valle	53	Thibault Lejeune
4	Rebecca Elson	29	Scilla Degl'Innocenti	54	Oliver Michael Kurth
5	Michael W. Feast	30	Natalia B. Corte Rodríguez	55	Carlos Álvarez
6	Steven R. Majewski	31	Marco Castellani	56	Joao M. de Morais Barros Fdes.
7	Raffaele Gratton	32	Gloria Andreuzzi	57	Yuriy Lyubchyk
8	Ivan King	33	Wojciech Pych	58	Pieter Westera
9	Lourdes González	34	Vincenzo Ripepi	59	Gabriel Gómez Velarde
10	Ismael Pérez Fournón	35	Grzegorz Kopacki	60	Rosario Iaria
11	Ivo Saviane	36	Gabriella Raimondo	61	Raimond Butler
12	Alfred Røsenberg González	37	Nieves Villoslada	62	Iván Agudo Rodríguez
13	Michael Andrew Beasley	38	Annapurni Subramaniam	63	Marcel VanDalfsen
14	Yoko Funato	39	József Benkö	64	Arunav Kundu
15	Ezio Pignatelli	40	Giacomo Gozzi	65	Emil Khalisi
16	Manuela Zoccali	41	Carlos Domínguez Tagle	66	Bulent Yasarsoy
17	José Antonio Uribe Botero	42	Seppo J. Katajainen	67	Dana Dinescu
18	Tiziana Di Salvo	43	Sofía Alejandra Cora	68	Winfrey Yeamans Johnson
19	Geneviève Parmentier	44	Vesna Zivkov	69	Sandra Chapelon
20	Anna Katherina Vivas	45	Eduard Masana Fresno	70	Cristiano da Rocha
21	Miriam Elina Rengel Lamus	46	Alejandra Recio Blanco	71	Stefan Deiters
22	Olga Suárez	47	Veronica Motta	72	Gaspar Bakos
23	Ana E. García Pérez	48	Linda Schroder	73	Carlos Martínez Roger
24	Soeren Larsen	49	David Martínez Delgado		
25	Michel Hilker	50	Israel Matute		

Preface

The study of globular clusters has been and still is essential for furthering our knowledge of such astrophysical phenomena as stellar and galactic evolution, variable and X-ray emission stars, chemical abundances (primordial nucleosynthesis), etc. Globular clusters are ideal laboratories for testing theories of stellar evolution, the chemical evolution of the Universe and the dynamics of N-body systems. They are the oldest known objects whose ages can be independently determined, the closest in proximity to the origin of the Universe and the sole surviving structures of the first stages in the formation of the Galaxy. They provide us with important evidence concerning on the age and formation processes of the Galaxy. Globular Clusters are a fundamental unit of the known Universe, they are also found in all other galaxies within our observational grasp. They are possibly a necessary stage in the formation of galaxies.

Research on Globular Clusters covers a vast amount of territory that was reviewed and collected in the present book. From the photographic plate to the HST most recent results, the field of Globular Clusters was actualised and presented by Ivan R. King, with an interesting Observational Approach to Populations in Globular Clusters, where discusses the observations on which our understanding of globular clusters lies. Steven Majewski, reviews the Stellar Populations and Formation of the Milky Way, with particular emphasis on the role of globular clusters in tracing stellar populations and unravelling the Galactic history. Vittorio Castellani in Globular Clusters as test Stellar evolution, reviews the theoretical predictions concerning the evolution of old, metal-poor stars in galactic globulars, in the light of recent improvements of the input physics. Raffaele Gratton, reviews the Early nucleosynthesis and chemical abundances of stars, where presents a self consistent sketch of current understanding in the topic. Rebecca A.W. Elson, on the Stellar Dynamics, provides an overview of the life of a globular cluster and presents some recent results from a large HST project to study the formation and evolution of rich star clusters in the Large Magellanic cloud. Michael W. Feast, reviews the scenario of the Pulsating stars and their use. Ramón Canal begin his chapter with the formation mechanism of neutron stars and end by discussing about the origin and evolution of x-ray sources and millisecond pulsars in globular clusters. Finally William E. Harris, reviews and actualises the scenario of Globular Clusters Systems, from the Milky way globulars and thorough the study of a few case individual galaxies show the richness and diversity of this field.

As scientific research becomes ever more highly specialised, researchers, particularly those who are now beginning their careers in this atmosphere of intense specialisation, are finding it harder and harder to keep abreast and properly orientated in all the disciplines related to their line of work. The X Winter School of Astrophysics of the Instituto de Astrofísica de Canarias and this book, was planned with a view to offering a thorough review of research on Globular Clusters and is intended to cover all the relevant disciplines with the aid of the best possible international team of specialists, including the theoretical and observational aspects of stellar populations.

This book collects an up-to-date overview of the Globular Cluster field, in order to gain clearer insight into the "big picture", and to help experts link their own field to its surroundings while improving their understanding of their own works. With regard to the young researcher we seek to install in them an awareness of the "great game of investigation" enjoyed by their more experienced peers. This aspect of the book will ensure that it plays a twofold role in the fields of education and training, as well as becoming a vehicle for the dissemination of the latest findings.

Carlos Martínez Roger
Instituto de Astrofsica de Canarias
January 1999

Foreword

For a decade now, the Instituto de Astrofisica de Canrias (IAC) has hosted the Canary Islands Winter School of Astrophysics in which young astrophysicists from all over the world have the opportunity of meeting accredited specialists to study the topics of most active concern in present-day astronomy. During these ten years 80 lecturers and more than 600 students have attended the Winter School, an even higher number not being able to come due to the limited number of places available.

The X Canary Islands Winter School on Astrophysics was dedicated to Globular Clusters, one of the basic sources of our knowledge concerning the lives of the stars and the physics of their evolution.

The School intended to portray a thorough review of research in this field, covering all the relevant disciplines with the aid of the best possible international team of specialists (Canada, Italy, South Africa, Spain, the United Kingdom and the United States), including the theoretical and observational aspects of stellar populations, stellar evolution and chemical abundances, dynamics, variable stars, X-ray sources and the globular clusters of other galaxies.

We take the opportunity to thank local Canarian authorities - Cabildo Insular de La Palma, and Cabildo Insular de Tenerife, as well as the Town Hall of La Laguna, for their continuous support during this and also previous editions of the School.

This tenth Winter School marks a milestone on a long but gratifying journey, in spite of occasional difficulties. I am confident that the next ten Winter Schools will be equally enriching and beneficial for all that take part.

Francisco Sánchez Martínez
Director of the Instituto de Astrofísica de Canarias

Acknowledgements

The Editors should like to express our gratitude to all who have made the X Canary Islands Winter School of Astrophysics possible. It has been the fruit of the good will of many individuals at the IAC, among whom we wish to mention Lourdes González, Nieves Villoslada, Jesús Burgos, Carmen del Puerto, Begoña López Betancor, Mónica Murphy and Campbell Warden. The School had the financial support of the Instituto de Astrofísica de Canarias, the Training and Mobility of Researchers programme of the European Comission, IBERIA airlines of Spain, the Cabildos of Tenerife and La Palma Islands and the Faculty of Physics of the University of La Laguna. Finally, but no less importantly, our thanks are due to the participants themselves: the lecturers and students from over 26 countries, from all the continents, for their interest and the work they have presented from their most recent research, and to the Professors because they have prepared their manuscripts ready for the publication in a very short time. And we want to acknowledge specially the work done by Dr. Gabriel Gómez in preparing the camera ready original for Cambridge University Press.

IVAN R. KING was born in New York in 1927.

He studied at Harvard University, where he obtained his doctorate in 1952.

He then worked as a mathematician for a private firm, as an instructor for Harvard University (1951-52) and, during the following four years was attached on Active Duty to the US Naval Reserve and the Department of Defense (1954-56) as a methods analyst.

From then onwards, his career followed a more academic line: professor of the University of Illinois from 1956 to 1964, and Professor of the University of California Berkeley, where he occupied a chair from 1966 to 1992 and where he is a present Emeritus Professor. King has served as President of the Dynamical Astronomy Division of American Astronomical Society and of the Astronomy Section of the American Association for the Advancement of Science (1973-1974). He was elected member of the American Academy of Arts and Sciences in 1980 and of the National Academy of Science in 1982.

The Observational Approach to Populations in Globular Clusters

By IVAN R. KING

Astronomy Dept., University of California, Berkeley, CA 92720-3411, USA

This introductory chapter discusses the observations on which our understanding of globular clusters lies. Successive sections deal with photometry, chemical abundances, the details of color–magnitude diagrams, the distance scale, luminosity and mass functions, and the lower end of the main sequence. An appendix treats the dynamical role of binaries in globular clusters.

Astronomy aims at an understanding of the facts and phenomena that we see, and the processes by which they came about—and in the best of possible cases, the recognition of why they had to be this way and could not have been otherwise. The first stage in this endeavor is to see what is there, and, to the extent that we can, how it became that way.

Globular clusters can in many ways be considered the crossroads of astronomy. They have played a central role in the unfolding of our astronomical understanding, to which they bring two singular advantages: first, each cluster (with a possible rare exception) is a single and specific stellar population, stars born at the same time, in the same place, out of the same material, and differing only in the rate at which each star has evolved. Such a group is much easier to study than the hodgepodge that makes up the field stars of the Milky Way. Second, globular clusters are made up of nearly the oldest—perhaps the very oldest—stars of the Universe, and as such they give us an unparalleled opportunity to probe the depths of time that are the remotest to reach.

This chapter introduces the facts about globular clusters—occasionally relying also on well-established pieces of understanding to help organize the facts and hold them together, but concentrating on the facts. These facts are presented in the context of the observations, and indeed the specific processes by which the observations have been gathered, and continue, in this vitally living science, to be gathered. It is in that spirit that this chapter is presented.

1. Photometry in Globular Clusters

1.1. *The evolution of techniques*

It has been known for a very long time that globular clusters (GCs) have color–magnitude diagrams (CMDs) that differ in appearance from the traditional diagram that was first drawn by Hertzsprung and by Russell. (It is of course this difference that was central to Baade's [1944] original breakthrough in defining stellar populations.) Even back in the 1930's and 40's there were a few good CMDs of GCs (e.g., Greenstein 1939, Hachenberg 1939), but the creation of CMDs did not become a real industry until the 1950's; it may be said that its start was centered on Sandage's thesis work on M3 (Sandage 1954), which was among the first to show the main-sequence turnoff clearly. This was of course a cornerstone of the physical understanding of stellar evolution that was awakening at that time, and that led so quickly to the level of the Vatican conference of 1957 on stellar populations (O'Connell 1958).

What had held the field back, and what was still slowing it down, was the cumbersome and inaccurate method that was needed in order to derive good stellar magnitudes on photographic plates. The basic problem is that the response of the photographic plate is non-linear and difficult to calibrate, whereas magnitudes require such a calibration. It had helped a great deal when photoelectric photometers were introduced. Their response is linear, and could be used to set up sequences of stars of known magnitude. For decades, however, photoelectric devices had been almost the exclusive province of those few who could build them and keep them working—pre-eminently Stebbins and Whitford (whose work was too voluminous for specific references here). Then, after the end of the Second World War, photomultiplier tubes became available, and it was then possible for the ordinary astronomer to carry out good photographic photometry—but even so, by a technique that to us seems almost incredibly laborious. One would first use the photoelectric photometer, with its linear response, to set up a sequence of standard stars of known relative magnitude. Then one would put each photographic plate, with its notoriously non-linear response, into some sort of device where the stars could be pointed at one by one, to measure the sizes of their images. Relative magnitudes were established by using the photoelectric standards to calibrate the curve of star-image size against magnitude. Finally, on some perfectly clear ("photometric") night, some of the standards would be compared with one of the few internationally recognized magnitude sequences in the sky, so that a zero point could be put on the photographically measured magnitudes. No wonder progress was slow.

What makes photoelectric photometry itself slow and laborious is that it observes only one star at a time. What was needed was an imaging photometric device that had a linear response. A number of such were developed and tried, but the modern era of photometry in globular clusters can be said to have begun with the wide availability of charge-coupled devices (CCDs), after about 1980.

The CCD has three great advantages over the photographic plate. First, obviously, is its linearity, which allows the immediate determination of relative magnitudes of all the stars in the frame, with only a simple zero-point comparison needed in order to make the job complete. Second, the CCD has a much higher quantum efficiency; it registers photons 10 to 100 times as well as a photographic plate. The third advantage is much less obvious; it arises from the fact that the photographic image of a star consists of a very limited number of grains of metallic silver, in the small area covered by the seeing disk of the star. The number of grains of silver limits the S/N ratio of a faint star on top of the sky background; by contrast, a CCD can accumulate a large number of photons, even in this limited area, and thus detect fainter stars. Their greater sensitivity indeed makes CCDs *faster*, but it is this escape from discrete grains that allows them to go fainter.

One other advantage of electronic registration of images is obvious but is worthy of note nevertheless. There is only one original photograph, but a CCD image can easily be copied by co-investigators without loss of fidelity.

In quantitative terms, the quantum efficiency (i.e., the fraction of incident photons that achieve a detectable effect) of the best photographic plates is about 1%, whereas a good CCD can reach a peak QE of 80% at its most sensitive wavelengths.

Only in one respect does the photographic plate retain an advantage: at a well-corrected telescope focus, it can cover a much larger area. This hardly matters for faint stars, where even a small CCD can gather thousands of stars in a single field of a globular cluster; but the smallness of CCD fields has definitely held back one area of globular-cluster research: the study of the stars in the upper part of the CMD. These stars are few in number, and therefore the study of adequate numbers of them requires a wide area—sometimes the whole cluster, which usually covers a considerable fraction of a square degree. While the present-day explorers of the lower main sequence have been glorying in fainter and

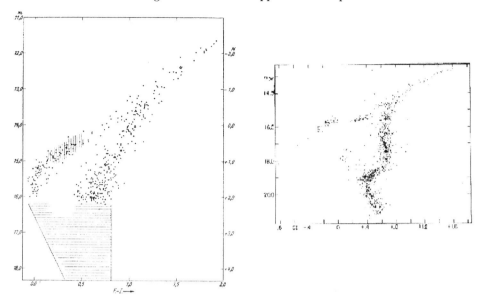

FIGURE 1. CMD of M92 by Hachenberg (1939, left) and of M3 by Sandage (1954, right), showing the progress after photoelectric calibration became relatively easy. The second diagram reaches 5 magnitudes fainter.

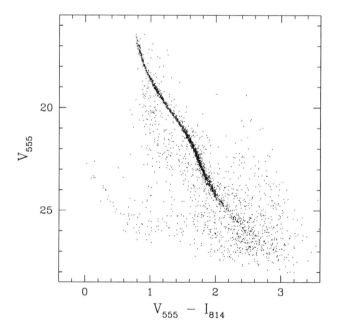

FIGURE 2. A recent CMD made from HST CCD imaging with the WFPC2 camera. The brightest data shown are at the main-sequence turnoff (since stars brighter than that were saturated even on the short exposures). The contrast with the previous figure shows how much difference CCDs and HST have made: another 6 magnitudes of faintness, and higher accuracy. (From Cool, Piotto, & King 1996.)

fainter magnitude limits, many theoreticians have been languishing for data at CCD accuracy for the evolved branches of the CMD. Only recently have a few observers begun to mount CCDs on smaller telescopes (for the larger field that they have) and take the multiple short exposures that are needed to measure the brighter stars over large regions of a cluster.

In specific numbers, the photographic foci of the Kitt Peak and Cerro Tololo 4-meter telescopes take photographs with a usable diameter of 20 cm. The earlier CCDs of the 1980's were postage-stamp size, registering fields a few hundred picture elements (pixels) across, covering only a few minutes of arc. CCDs are now larger, 2048 × 2048 formats being common now (giving fields of 10 or 15 arcmin), with even larger chips, and sometimes arrays of them, coming into use in a few places.

1.2. *The use of CCDs*

I will turn now to a brief description of CCDs themselves, from the point of view of the user, and will move finally to the particular problems of CCD photometry of globular clusters.

A CCD consists of an ultra-thin wafer of silicon, one side covered with a grid of electrodes that define its individual pixels. The thinness itself is a factor, in several ways. First, in most CCDs the light must pass through the silicon, which is not highly transparent at the shorter wavelengths. Hence CCDs achieve their highest quantum efficiency at red and near-infrared wavelengths, while they are less sensitive, or in many cases hopelessly insensitive, in the ultraviolet, or in many cases even in the blue. The second problem of thinned chips is uniformity of thickness, which is most difficult to achieve for the chips that are thinnest. The pixels in regions of the chip that are thinner have less silicon and therefore less sensitivity. Moreover, the transparency problem makes this effect wavelength-dependent. Some thin chips also fail to be completely flat, making the quality of star images vary over the field.

On the whole, however, non-uniformity tends to be less serious than lack of sensitivity at short wavelengths. Since non-uniformity is a problem for other reasons also, it is normal for reduction systems for CCD observations routinely to include procedures for correction of non-uniformity ("flatfielding" the image), whereas limited sensitivity at short wavelengths can be an immutable drawback.

Each pixel of a CCD has an electrode, and it is the arrangement of the electrodes that fixes the pixels, whose boundaries are rather well defined. Except for the "bleeding" effects referred to below, very few of the electrons released by the incident photons are counted into a pixel other than the one into whose discrete boundaries the photon fell.

The CCD accumulates charge in each of its pixels during an exposure that typically lasts 1 to 15 minutes; then it is read out. Readout consists of shifting the batches of accumulated charge down each column from row to row, and reading each successive row (simultaneously in each column) as it reaches the bottom. The readout process is imperfect in several ways, however. First, a pixel that has an outstandingly large number of counts (from a bright star) will affect neighboring rows in its column, "bleeding" into them. Second, a CCD can suffer from inadequate "charge transfer efficiency," in which the readout process loses electrons during the shifting. Finally, the emerging electrons are not counted individually, but go into an analog-to-digital converter, which introduces noise into the counts. This "readout noise" is quoted as an r-m-s value, so that a readout noise level of 5 means, for example, that on the average each pixel has 25 noise electrons added to the output of the readout. (Readout noise is almost always more serious than thermal dark current.) Given this noise level, it would be wasteful of data space to record the exact number of electrons; instead each number of electrons that comes from the A/D converter is divided by a reduction factor called the DN (digital number). In the above case the readout might be set to record an output number that is only 1/10 of the actual number of electrons ("DN = 10").

The use of a DN has the effect of increasing the dynamical range of a CCD. Although the "full-well" capacity of most CCDs is greater than 100,000, the data systems that record their outputs are usually limited to 15 or 16 bits (32,767 or 65,535 counts). Even so, the centers of bright stars saturate, and observers generally include short exposures on which they can be successfully measured.

Stellar magnitudes can be measured on CCD images (i.e., in the digital array that represents the image) by aperture photometry, a procedure in which one simply adds up the light in a small number of pixels around the center of each star image; but more accurate results are usually achieved by use of some procedure that involves fitting to each star image a correctly chosen multiple of the point spread function (PSF), which should be determined from a number of well-exposed stars in the same image. Thus the result of PSF-fitting, for each star, is a position and a brightness factor that fit the star.

Many ready-made routines are available for PSF-fitting. The most commonly used is DAOPHOT (Stetson 1987), which is available as one of the the the many sets of tasks in the massive IRAF package that has been developed by the National Optical Astronomy Observatories, with some important additions from the Space Telescope Science Institute. A number of other packages are available; besides, many users have written their own software for stellar photometry, often to cope with particular situations that they have encountered.

1.3. *The role of HST*

Rather than digress on photometric methods, however, I prefer here to concentrate on the problems that are peculiar to globular-cluster photometry, and in particular to the results that I will be emphasizing later in this chapter, that is, photometry with the Wide Field/Planetary Camera 2 (WFPC2) of the Hubble Space Telescope (HST).

In globular-cluster work, HST has gone several magnitudes fainter than previous ground-based observations, and this has made a big difference to the scientific results. (In principle, the largest ground-based telescopes, Keck and VLT, are capable of going almost equally faint; but for various reasons they have not yet done so.)

HST gets its excellent faintness limit, relative to ground-based telescopes of equal or even greater aperture, mainly from its high resolving power. The faintness limit at which stars can be detected and measured is set not purely by the number of photons collected from the star, but rather by the ratio of the signal from the star to the noise of the sky background in the area that the star image covers. Although HST does in fact see a somewhat darker sky than the ground-based observer does (*much* darker in the infrared, in fact), the more important factor is that the HST star image is so small, compared with a ground-based seeing disk. Thus a star in an HST image competes with a very much smaller area of sky, and we are able to observe fainter stars. The difference between the limiting magnitudes of HST CMDs and those observed from the ground is, in fact, about 4 magnitudes.

There are some important practical problems in WFPC2 photometry of globular clusters, some of them peculiar to HST and others connected more with the nature of the clusters themselves.

It should be noted first that WFPC2 has four 800×800-pixel CCDs, which record adjacent pieces of sky that are re-imaged onto them by transfer optics, from the focal plane of HST. The re-imaging for one chip, PC1, is such that its pixel size is 0.045 arcsec, while the other three, WF2, WF3, and WF4, have a scale of 0.100 arcsec/pixel. The image that HST forms of a star is comparable in size to a PC1 pixel. Since the Nyquist criterion states that we need at least two samples per resolution element, even PC1 is undersampled, and the WFs are grossly undersampled.

In spite of the undersampling, a PSF can be determined reasonably well, by combining measurements of many stars that are situated in different positions with respect to pixel boundaries. But applying this PSF to a single star image is a different matter; here the undersampling really hurts. The best cure for this is dithering—displacements of otherwise identical images by well-chosen fractions of a pixel; the average of photometry from a set of well-dithered images is more accurate than photometry from images that are identically placed.

It is beneficial for dithering to include, at the same time, displacements by a small integral number of pixels in addition to the fractional displacements. The reason for this is that the flat-field corrections for the chips of WFPC2 are known only on a rather smooth scale; individual pixels have sensitivities that may differ by a few per cent (equivalent to a few hundredths of a magnitude). Dithering by a few pixels randomizes these errors and gives a much better average.

In fact, the most accurate globular-cluster photometry that has been published (Cool, Piotto, & King 1996, Rubenstein & Bailyn 1997) relied on well-dithered images. (It is interesting to note that a globular-cluster CMD offers an automatic index of its accuracy: the color width of the main sequence. Intrinsic widths of GC main sequences are too small to be measured; hence the observed width reflects the photometric accuracy. Main sequences have been produced with color widths of only 0.02–0.03 mag.; cf. Fig. 2.)

1.4. *Crowding problems*

If there is one characteristic that most distinguishes globular-cluster photometry from photometry of any other objects, it is crowding, which hinders photometry in several ways. The most obvious effect of crowding is the partial overlapping of star images. There are two basically different ways of dealing with the interaction of neighboring images with each other. DAOPHOT (like some of its competitors) works simultaneously on a group of star images each of which partially overlaps at least one other member of the group. It makes a fit of the PSF to each star in turn, while holding each of the others fixed at a previously determined level; the process is iterated till it converges. A quite different approach is neighbor subtraction (Yanny et al. 1994). In this method, the magnitudes of all the stars are first measured by some method (e.g., DAOPHOT). Then they are all subtracted from the image, by subtracting at the location of the star a replica of the PSF scaled to fit the magnitude of that star. The final measurements are made by adding each star back into the image, temporarily—an operation which has the effect of restoring all of the pixels of that star to their original values—and measuring it alone, in surroundings from which all its neighbors have been removed.

Another effect of crowding is more subtle; it comes not from the visible neighbors but from the invisible ones. Part of the background that surrounds a star comes not from the sky or from scattered light of bright stars, but rather from the combined light of the stars that are below the threshold of detection. Since these are subject to random fluctuations of number, the background around a star is noisier than it would otherwise be. This interference makes the magnitudes of stars in crowded regions, even when a star appears well isolated, less accurate than they would be in regions that are less crowded.

Scattered light of bright stars was mentioned above. In a crowded cluster field there is so much scattered light from the halos of the images of the numerous bright stars that its sum is equivalent to an increase in the sky background, making it difficult or impossible to detect and measure the faint stars. At the dense center of a rich globular cluster, even if the faint stars are resolvable the limiting magnitude may be considerably less faint—sometimes by magnitudes—than the level that can be reached farther out in the cluster.

One other method of coping with crowding is worthy of mention (Anderson 1997). If one's objective is to achieve photometry with the highest level of accuracy (for judging the color width of the main sequence, for example), one can use some objective criterion, such as the distance to the nearest neighbor, to select a subset of stars that are, *a priori*, likely to be measured with higher accuracy than the others.

1.5. *The choice of color bands*

Finally, a word about color systems. For the photography-based study of globular clusters, the color bands that were convenient and natural were the blue B and yellow V for which photographic materials had good sensitivity; and it became standard to study globular clusters in B and V. As already noted, most CCDs have inferior sensitivity at B; quite to the contrary, they have excellent sensitivity in the near infrared, with the result that modern CCD studies are tending more and more to be made in the V and I bands. These questions will be taken up in more detail in Section 3, which goes into the details of the photometric results.

2. Chemical Abundances in Globular Clusters

One of the prominent differences between globular clusters is in the chemical mix of the material of which their stars are made. Early in the history of stellar populations it had become evident that the spectra of the kindred stars of the halo population (as we now call it—in those days they were the "high-velocity stars", because the halo stars that are passing through our neighborhood have velocities that differ so much from our near-solar reference standard)—had noticeable differences from those of stars like the Sun (Lindblad 1922, Roman 1950); the differences were traced to a deficiency in the abundances of the elements heavier than helium (Chamberlain & Aller 1951). Because of the faintness of even the brightest globular-cluster giants, it was not until much later, with much greater effort, that two globular-cluster giants were observed and found indeed to be deficient in heavy elements (Helfer, Wallerstein, & Greenstein 1959).

The elements beyond helium are generally referred to as the "metals," even though the most abundant of them include C, N, O, Ne, Si, etc. Since among the more abundant heavy elements iron is the most prominent in stellar spectra, the practice has generally been to cite the "metallicity,"

$$[\text{Fe/H}] = \log_{10} \frac{\text{Fe/H}}{(\text{Fe/H})_\odot},$$

with the implicit assumption that all the other heavy elements go in lock step with iron. It has now become evident that this assumption was unjustified, but [Fe/H] has been used as a metallicity index for many decades, and the differing behavior of other elements is still insufficiently explored. I will therefore use the index [Fe/H] throughout most of this section, shifting only at the end to the modern picture of greater complexity.

Baade's original population distinction had merely contrasted the color–magnitude diagrams, but the spectroscopic studies just mentioned showed that the globular clusters were metal-poor relative to the Sun. The weakening of metal lines was a strong indication; but, interestingly, the most graphic demonstration of low metallicity came from theory. In their epoch-making first calculation of an evolutionary track, Hoyle & Schwarzschild (1955) found that with solar chemical abundances they produced the red-giant branch of the open cluster M67, while they could match the RGB of the globular cluster M3 only by lowering the abundance of heavy elements by two orders of magnitude.

Differences in metallicity among globular clusters soon became apparent. In Morgan's study of the spectra that Mayall had taken to measure radial velocities of globular clusters (Morgan 1956, 1959), the Great Classifier found that he could distinguish eight different degrees of metal-line strength. (In those papers the distinction between disk and halo globulars was also first made.) Morgan's subjective classes were soon replaced, however, by the quantitative [Fe/H].

Metallicities in globular clusters have been estimated by every method that astronomers have been able to conceive of: line strengths in integrated spectra, various types of color indices or spectral indices, or spectra of individual stars. Only the last of these can provide a direct measure of chemical abundances; the other methods each require calibration onto this true abundance scale.

Even individual spectra must be used with care, because of the complexity of interpreting line strengths in a stellar atmosphere. Low-dispersion spectra are dangerous to use, because their information is contained in the strong spectral lines. These, by the very nature of their strong absorption coefficient, are formed in the upper layers of the star's atmosphere, whereas information about the temperature of the atmosphere, which governs excitation and ionization, comes from the deeper photospheric layers, around unit optical depth. This means that the temperature of the layer in which the strong lines are formed is known only by extrapolation of the photospheric temperature via a model atmosphere, any errors in which will be reflected in abundance errors. The interpretation of strong lines also requires use of a curve of growth, whose characteristics also depend on the model atmosphere; furthermore, it is in the nature of the curve of growth that strong lines are less sensitive to abundance than are weak lines.

The most reliable abundance measurements—perhaps the only reliable ones—come from measurements of weak lines, because, having an absorption coefficient that differs only a little from that of the neighboring continuum, they are formed in the atmospheric layer whose temperature we know the best. The price of this advantage is the cost of obtaining high-dispersion spectra—even higher dispersion than one might think, because of the need to detect even weaker lines, neglect of which might cause the continuum level to be set incorrectly. Getting high-dispersion spectra of stars as faint as those in a globular cluster costs valuable time on large telescopes, but the price is worth it.

Abundance measurements are discussed in more detail by Gratton in Chapter 4.

Values of [Fe/H] in globular clusters range from close to zero down to about -2.3. The best abundance determinations have accuracies of about ± 0.15, or in common parlance, 0.15 dex. Some of the photometric abundance indices are capable of comparable accuracy, but have the danger of systematic errors due to inadequate calibration of the index against high-dispersion spectroscopic measures.

Although globular clusters differ from one another, nearly every cluster contains stars of a uniform chemical abundance, as one would expect from a single event of star formation. An outstanding exception, however, is Omega Centauri, where spectroscopy of red giants and subgiants has revealed an abundance spread of nearly half a dex. The stars of the main sequence of the cluster are too faint for easy spectroscopic study, but it has been recognized for some time that the main sequence has a color width indicative of a range of metallicities, and a recent Berkeley thesis (Anderson 1997) has found an actual split of the middle main sequence into two parallel sequences.

Smaller abundance differences have also been observed in M22 (Anthony-Twarog, Twarog, & Craig 1995, which also gives references to earlier studies).

In recent years it has become evident, however, that it is an oversimplification to describe the chemical composition of a globular cluster, or of any other stellar population, by a single parameter such as [Fe/H]. The complexity first showed up as an overabundance of oxygen relative to iron (stated as [O/Fe]) in some clusters. A similar behavior had been seen in field stars, and there turned out to be a systematic behavior there, in that [O/Fe] was most enhanced in the stars of lowest [Fe/H] (by about 0.5 dex) but gradually decreased with increasing [Fe/H] until it leveled off at zero (i.e., a solar ratio of oxygen to iron) for values of [Fe/H] greater than −1.0. A recent preprint (Chiappini et al. 1998) gives abundant references to this problem, as does the general review of abundances by Wheeler, Sneden, & Truran (1989).

This behavior of the abundances received a ready explanation, which notes that shortly after the birth of a population the only supernovae that appear are of Type II, because their progenitors are upper-main-sequence stars that have rather short lifetimes. These stars produce a lot of α-process elements, and relatively little iron. (Oxygen is an α-process element, and further study has shown that the heavier α-process elements are also enhanced along with oxygen.) It is only after a billion years or so that supernovae of Type I begin to appear, and then the iron abundance goes up. Thus the initial build-up of metallicity emphasizes the α-process elements, while iron also increases slowly; then the larger production of iron takes over.

Unfortunately recent discoveries in clusters near the Galactic bulge have shattered the uniformity of this picture (Barbuy et al. 1999). NGC 6553, with [Fe/H] = −0.5, has a striking level of α-enhancement, whereas 47 Tuc, with a very similar metallicity but in a different part of the Galaxy, does not show this peculiarity at all. (The same also seems to be true of NGC 6528, another bulge cluster [Barbuy, personal communication].) As Bernard Pagel was so kind as to point out to me, in a discussion shortly after the end of the Winter School, what this means is that there was a difference, for the material that was to go into the stars of each of these two clusters, in the length of time that elapsed in building up a level of [Fe/H] that is similar in both clusters. In the material that was to make up NGC 6553 (and presumably the other bulge clusters), element enrichment by ejection from Supernovae II was so rapid that [Fe/H] was already up to −0.4 in less than 1 Gyr, i.e., before any Supernovae of Type Ia could form and decrease the early [α/Fe] level of 0.4–0.5. For 47 Tuc, by contrast—and apparently for other clusters outside the bulge—enrichment to the same level of [Fe/H] took a much larger number of years, so that there the Sne Ia were able to play their role of bringing up the level of iron relative to the α elements.

This difference in the rate of enrichment of pre-cluster material leads to two different possible scenarios. If all enrichment began at the same time, then the bulge clusters are older than those outside the bulge. On the other hand, if we wish to view all globular clusters as having similar ages, we would then have to postulate that element enrichment began earlier outside the bulge than in it—but with a time lapse in the bulge that contrived to be just long enough for cluster formation to begin simultaneously everywhere.

In any case, it is an observational fact that the chemical-abundance history has been different in different locations in the Galaxy, and the specter thus threatens us that α-enhancement may be a quantity that has to be measured in each individual cluster, quite separately from [Fe/H].

Finally, I leave it to others to speculate on the significance of the fact that CMD properties associated with this sort of abundance difference have also been observed in the stars of the M31 halo (see Sect. 3.3).

3. The Morphology of Color–Magnitude Diagrams of Globulars

The vehicle for describing the populations of globular clusters has always been the color–magnitude diagram (CMD).

Note first that the CMD, a plot of the absolute magnitudes of the stars against their colors, is the observational expression of the HR diagram of the theoreticians, whose coordinates are instead $\log L$ and $\log T_e$, where L is the energy output of the star added up over all wavelengths and T_e is its effective temperature, defined by the relation

$$L = 4\pi R^2 \cdot \sigma T_e^4,$$

where σ is the Stefan–Boltzmann constant. The relation between the observational and the theoretical coordinates of a given star is by no means obvious; the determination of such relationships has consumed a large amount of discussion and effort on the part of both observers and theoreticians. Its details will not be discussed here; we merely note that it is an important problem in the comparison of observations with theory, for which the theoreticians need to produce not just an L and a T_e but a complete spectral energy distribution for each star.

In the present section the coordinates of the CMD will be taken to be M_V and $(B-V)_0$, where B and V are two of the bands of the Johnson system, defined in principle by Johnson & Morgan (1953) and operationally by the standard stars of Landolt (1992). The zero subscript on $B - V$ is intended to remind us that all colors must be corrected for interstellar reddening, just as the definition of absolute magnitude implies that it corresponds to a true distance, corrected for interstellar absorption.

Again, it is well to mention in passing that the conversion between observational and theoretical coordinates depends on the details of the bandpass, so that the conversions will not be the same for, e.g., the standard Johnson V and the V defined by the filter and detector in use at a particular observatory at a particular time. It is a common practice of observers to transform their observed magnitudes to a standard system, as best they can; but in fact this can be a damaging step. In comparisons with theory it is essential instead to integrate the theoretical results over the specific passband in which the observations were made.

Note also that observations tend now to be made increasingly in quite different bandpasses, such as the V and I that are increasingly favored by observers who use CCDs, either from the ground or with HST, or the J and K that are appearing in an increasing number of papers by observers in the near infrared. The present section, which discusses CMDs from the point of view of morphology alone, will express everything in the more traditional B and V, with the understanding that the conversion from other bands can be made in principle, although (as just emphasized) detailed comparisons between observation and theory should always be made with the bandpass actually used in the observations.

Figure 3 is a schematic plot of a globular-cluster color–magnitude diagram, showing the sequences (and parts of sequences) that are generally recognized. Other features are often described and referred to, but the sequences labeled in Fig. 3 are the regions of the CMD that will be discussed in this section. The figure is intended only to be schematic; it can indeed be no better than that, because the positions of the sequences differ from cluster to cluster, and some branches are totally or partially lacking in some clusters.

3.1. *The main sequence(MS)*

The main sequence extends from the turnoff at its top (which will be discussed in Subsection 3.1.1) to the limit of hydrogen burning (discussed in Section 6). Aside from the locations of these two limits, clusters differ in the colors of their MSs at a given ab-

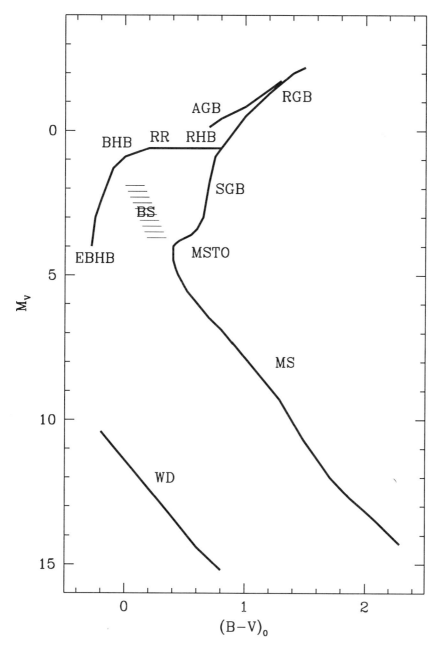

FIGURE 3. A schematic CMD of a globular cluster, with the various sequences labeled. RGB = red giant branch, AGB = asymptotic giant branch, RHB = red horizontal branch, RR = RR Lyrae region, BHB = blue horizontal branch, EBHB = extreme blue horizontal branch, SGB = subgiant branch, BS = blue stragglers, MSTO = main-sequence turnoff, MS = main sequence, WD = white dwarfs.

solute magnitude. This difference arises mainly from blanketing by metal lines in the atmospheres of the stars; the number of metal lines heaps up at the shorter wavelengths, depressing them relative to the longer-wavelength bands, and this makes the star redder. Since a low metallicity means weaker metal lines at a given temperature, the gross effect of blanketing is to make the MS redder for high-metallicity clusters and bluer for low-metallicity clusters. (Note that a similar effect occurs in the subdwarf stars of the field, and that this will figure in our discussion, in Section 4, of the use of subdwarf fitting to get cluster distances.)

A noteworthy characteristic of the main sequence is its bends. The one near $M_V \simeq 9$ will also be referred to as a possible absolute-magnitude criterion, in Section 4.

The reader should be aware, also, that I have somewhat falsified the lowest bend in the MS. In $B - V$ it is actually a downward rather than an upward bend; I have shown the latter instead, because that is the way that it goes in the $V - I$ CMD that is used for the faintest HST CMDs (cf. Figs. 2 and 12). (The difference is due to the fact that the bend is caused largely by the onset of strong molecular absorptions in the V band; they depress it and cause an opposite effect in the $B - V$ and $V - I$ colors.)

3.1.1. *The main-sequence turnoff (MSTO)*

The main-sequence turnoff is the region that is occupied by stars that have just exhausted the hydrogen at their centers and begun to evolve off the MS onto the subgiant branch. It of course occupies a central role in the age-dating of clusters, which is discussed in detail in Chapter 3 of this book. The shape and location of the MSTO depend in such complicated ways on age and details of chemical abundances that no generalizations can be given here.

3.1.2. *The hydrogen-burning limit*

The lower limit of hydrogen burning is discussed in more detail in Section 6. Suffice it to say here that it is not a perfectly defined cutoff in the CMD, but rather a rapid petering out of the luminosity function, and that its location is predicted to be metallicity-dependent, being fainter at higher metallicities.

3.2. *Blue stragglers (BS)*

Above the MSTO the main sequence is not devoid of stars. There is a scattering of stars that lie roughly along an upward extension of the MS. These "blue stragglers" form a sequence that is not narrowly defined in color. A few exist in outer parts of some clusters, but one of the surprise discoveries by HST was that they exist in the center of every cluster that it has observed, and that they are much more concentrated to the cluster center than are the other visible stars.

The physical nature of the BS stars is not understood. The one thing that seems reasonably clear is they can occupy this position in the CMD only by being more massive than the stars of the MSTO, and through not yet having evolved off the MS. Since primeval stars of this mass would long since have left the MS, they must have been formed relatively recently. The two hypotheses that are most commonly suggested both involve a recent merger of two lower-mass stars into a star of greater mass than that of the MSTO. One hypothesis is that they have come from collisional mergers; the other involves binaries spiraling together. (For a discussion see Bailyn & Pinsonneault 1995.)

3.3. *The subgiant branch (SGB) and the red giant branch (RGB)*

After a star leaves the MS, it moves briefly toward the right in the CMD before encountering the Hayashi track, up which it then moves on the so-called subgiant branch. The

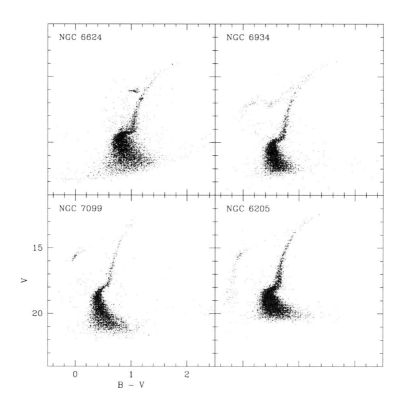

FIGURE 4. Clusters with four very different types of horizontal branch. (Data courtesy of M. Zoccali & G. Piotto [private communication]; source is HST archival images of the centers of globular clusters.)

length and the horizontality of the lead-in to the SGB differ from cluster to cluster, being in general longer and flatter at high metallicity.

The SGB and the RGB are not really distinguished in any way; they are merely names given to the parts of this rising branch that are below and above the horizontal branch, respectively.

There are striking differences between the red giant branches of different clusters, with the height of the RGB depending directly on metallicity. The parameter most commonly used to characterize an RGB is ΔV, which is defined as the difference in magnitude between the horizontal branch and the RGB at $(B - V)_0 = +1.4$. It ranges roughly from 2 for high-metal-abundance clusters to 3 for those of low metal abundance.

Another type of RGB behavior has recently come to light: an extension far to the red, with a downturn at the red end (see, e.g., Ortolani, Barbuy, & Bica 1990). Interestingly, this strange extension of the RGB is also observed in the Galactic bulge (Rich et al. 1998, Rich 1996) and in the halo of M31 (Rich, Mighell, & Neill 1996). Note the probable connection of this CMD peculiarity with the anomalies in α-element abundances discussed at the end of Section 2.

3.4. *The horizontal branch (HB)*

The horizontal branch—so called because its V magnitude is almost independent of color except at its blue end—has long been recognized to be the helium-core-burning stage into which a star moves after it has undergone the helium flash at the tip of the RGB. Its parts are often given separate names, largely on account of the natural marker set by the RR Lyrae region. The RR Lyraes are pulsating stars that occupy the part of the HB that intersects the instability strip of the general HR diagram. The part of the HB to the redward of it is referred to as the red horizontal branch (RHB) and the part to the left as the blue horizontal branch (BHB). In some clusters the HB extends unusually far to the blue, and droops steeply in V magnitude; this region is referred to as the extreme blue horizontal branch (EBHB).

(Two points of clarification: [1] The falling V magnitude on the blue side of the HB is not caused by a lower luminosity for these stars, but rather by a steeply increased bolometric correction with increasing temperature; furthermore, the bend in the HB is accentuated by the fact that for such hot stars the region between the B and V bands is almost on the Rayleigh–Jeans part of the Planck curve, and the color of such stars is therefore insensitive to temperature. [2] Some astronomers have used the term "RR Lyrae gap," which is a complete misnomer; it is not a gap at all, but appears in some globular-cluster CMDs merely because an observer has omitted the photometrically inconvenient RR Lyraes from the published CMD. A good example is shown in the right-hand half of Fig. 1.)

Theory shows clearly that the position of a star on the HB is determined by the amount of envelope that overlies its core. What is not understood is what causes individual stars to lose different amounts of mass, so as to create a spread along the HB.

It is the horizontal branch that exhibits the largest differences among the CMDs of globular clusters. Some clusters have HBs that are predominantly red, and some have HBs that are predominantly blue. This difference can be very strong. Among the clusters that have no RR Lyraes at all, some lack them because their HB is too red to reach the RR Lyrae region and others because their HB is too blue to reach it. Examples of HB types are shown in Figure 4.

To a first approximation, high-metallicity clusters have a red HB—perhaps only a short stub coming off the SGB/RGB—and low-metallicity clusters have a blue HB. But clusters are known that grossly violate this rule, e.g., NGC 362, which has a low metal abundance but a red-stub HB. Since something other than metallicity must be acting here, anomalies such as this are called the "second-parameter effect." Interestingly, behavior of the opposite type—a high-metallicity cluster with only a blue HB, has not yet been found (although cases with both blue *and* red pieces of HB will be shown below). The nature and identity of the second parameter have been subjects of debate and speculation ever since its existence was discovered.

Most clusters have a continuously populated horizontal branch, like those illustrated in Fig. 4. But a few clusters have HBs that consist of distinctly separate pieces. The most striking of these is NGC 2808 (Fig. 5), whose HB has one obvious gap, and two more that are less conspicuous but almost certainly real (at $V \sim 18.7$ and 20).

Another kind of two-piece HB is exhibited by NGC 6388 and 6441, clusters that have the red-stub HB that is normal for their metallicity and also a BHB/EBHB that seems completely out of place (Fig. 7).

3.5. *The asymptotic giant branch (AGB)*

After a star has exhausted the helium fuel at its center, it develops a helium shell source, analogous to the hydrogen shell source in RGB stars, and moves to the asymptotic giant branch, whose lower part good photometry can easily distinguish from the RGB, although the upper parts of the two branches run together. The Mira variables in globular clusters are at the tip of the AGB.

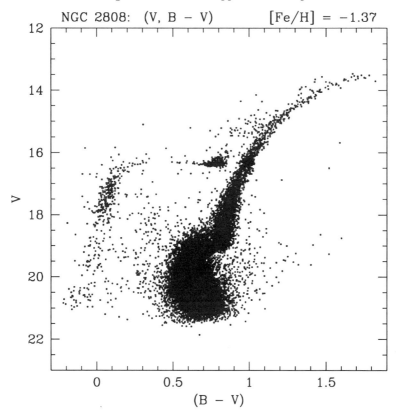

FIGURE 5. NGC 2808 has a horizontal branch that not only extends very far to the blue but has one prominent gap and two other gaps that are statistically significant. (From Sosin et al. 1997.)

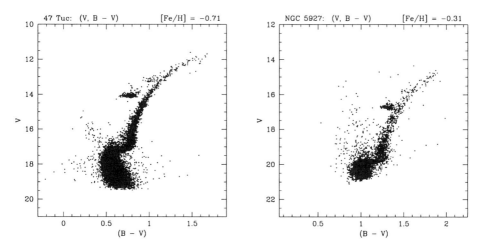

FIGURE 6. These two clusters have the kind of red-stub horizontal branch that is characteristic for high-metal-abundance globular clusters.

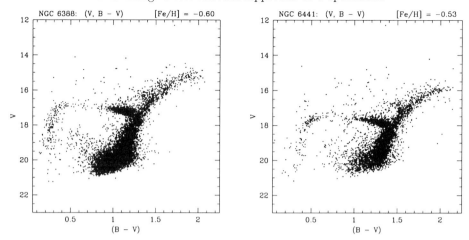

FIGURE 7. In these two clusters there is an additional unexpected EBHB in addition. (This figure and the preceding one from Rich et al. 1997.)

It should be noted that some stars, especially in metal-rich clusters, do not pass through the AGB stage; after their HB lives they pass to a pre-white-dwarf stage. For them Renzini has invented the name "AGB-manqué." Such stars almost certainly exist, but they have not shown up prominently in any cluster.

3.6. *The white dwarfs (WD)*

After a star leaves the AGB, it evolves so rapidly through planetary-nebula and post-AGB phases that such stars do not contribute appreciably to the CMD of a cluster; it finally slows its evolution when it reaches the white dwarf sequence. The WDs in globular clusters are so faint that they were not reliably detected until the HST era, but now they have been seen in several clusters. The colors and magnitudes along the WD sequence can now be used as an observational test of WD models, and the luminosity function along that sequence as a test of theories of the rate of WD cooling. In addition, the use of the WD sequence as a distance indicator will be discussed in Section 4.

4. Estimating the Distances of Globular Clusters

I have chosen the title of this section deliberately to emphasize the uncertainties in all of our methods of getting the distances of globular clusters; "estimating" is certainly a more appropriate word than "measuring." Indeed, the number of methods that I shall discuss should make it immediately evident that none of them is satisfactory by itself.

Nearly all of the methods of estimating GC distances depend on having a knowledge of the absolute magnitude of some feature of the color–magnitude diagram (CMD), either from calibration through measurements of similar field stars or through the assertion that theory can predict the absolute magnitude of that feature. (Notice the hedge-words again: "having a knowledge of" rather than "knowing," and "assertion" rather than "fact.")

4.1. *The horizontal branch*

Historically the most-often-used method of estimating distances of globulars has depended on the absolute magnitude of the horizontal branch (HB), through assigning an

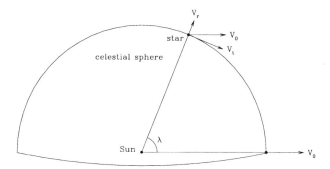

FIGURE 8. Resolution, for a particular star, of the mean group velocity into its contribution to the radial velocity and the proper motion of the star.

absolute magnitude to the HB itself, or more traditionally to the RR Lyrae stars that form a part of the HB in so many clusters.

4.1.1. *RR Lyrae stars*

The RR Lyraes are pulsating variable stars that populate the part of the HB that crosses the instability strip. Thus their number in a given cluster (or better, their *specific number*, normalized to a standard absolute magnitude) depends on the HB morphology. What makes this fact relevant to the present discussion is that some clusters lack RR Lyraes completely, because their HBs are either too red or too blue to reach the intermediate color of the instability region.

Absolute magnitudes of RR Lyraes are estimated in four different ways: statistical parallax, trigonometric parallax, the Baade–Wesselink (BW) method, and by observing them in an external system, such as the Large Magellanic Cloud.

Statistical parallaxes. Since statistical parallax is one of the general methods of distance measurement in astronomy, it is important to understand how it works. In practice solutions are usually carried out by methods of great mathematical sophistication and complexity (see, e.g., Popowski & Gould 1998, whose introductory section gives an admirable discussion of the principles of statistical parallax), but the underlying principles can be easily explained. I will do so here, because I do not know of any explanation in an existing textbook that is clear enough and detailed enough.

First let it be noted (specifically for the RR Lyraes here) that two independent methods both fall into the category of statistical parallax, one based on their mean velocity vector with respect to the Sun and the other based on their velocity dispersion. In both cases, of course, the underlying principle is to compare the angular motions expressed in the proper motions with the linear motions expressed in the radial velocities.

It is the mean-velocity-based method that gives statistical parallax its name, since this method is logically equivalent to finding the parallax that is caused by the Solar motion with respect to the group, by taking a weighted mean over all its members; in this process the dispersion of RR Lyrae velocities acts in just the same way that errors of measurement do in an ordinary mean.

The geometry on which this method is based is illustrated in Figure 8, which shows how the mean velocity vector, V_0, of the group resolves, at the position of a particular star, into a radial and a transverse component, which are reflected, of course, in the radial velocity and the proper motion of the star.

The dispersion-based method is simply a comparison of the manifestations of the

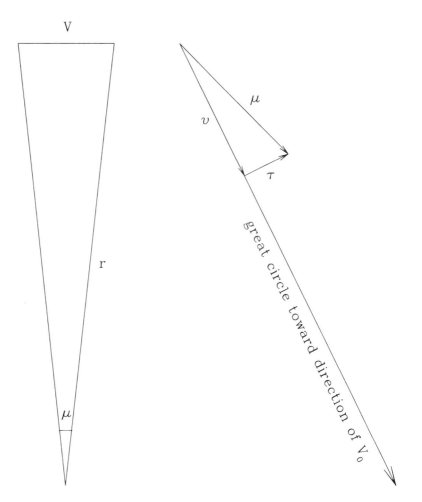

FIGURE 9. (left) The relation between transverse velocity, proper motion, and distance. (right) Resolution of the proper motion of a group member into tau and upsilon components.

velocity-dispersion ellipsoid of the RR Lyraes in their proper motions and in their radial velocities; the distance follows from a comparison of the angular quantity proper motion with the linear quantity radial velocity. It is clear from the figure that

$$V_r = V_0 \cos \lambda \tag{4.1}$$
$$V_t = V_0 \sin \lambda. \tag{4.2}$$

The first of the above equations is of no direct use, since *ab initio* the direction of the mean velocity, and hence the value of λ for each star, is unknown. The observed radial velocities of RR Lyraes all around the sky are instead used in a standard solar-motion solution (see, e.g., Binney & Merrifield 1998, p. 626) to find the value of the mean velocity vector of the RR Lyraes with respect to the Sun, i.e., the magnitude V_0 of the vector and its direction. The value of λ can then be calculated for each star.

We now turn to the proper motions, using the second of the above equations. But first we must have a relationship between the star's transverse linear motion (call it V), its proper motion, and its distance. From the left-hand diagram in Figure 9 it is clear that $V = r\mu$, where μ is in radians and V and r must be expressed in the same linear units. For practical purposes, however, we like to express velocities in km/s, distances in parsecs, and proper motions in arcsec/yr. A little calculation shows that in these units

$$V = 4.74r\mu, \tag{4.3}$$

a relationship that is worth remembering.

Now we come up against the fact that, because of the peculiar motion of the star, on top of the mean group motion, V is not the same as V_t, nor μ the same as the corresponding proper motion. In fact, V is the V_t attributable to the group mean, plus some additional motion peculiar to the star, part of which will be along the direction of V_t and part in the cross direction.

Here is where we can, in making use of the star's observed proper motion μ, take advantage of our knowledge of the direction of the group motion, as found from the solution of the radial velocities. In Fig. 8 the direction of the group motion is represented by the arc curving downward from the star to the bottom of the hemisphere that is shown. If we now draw another diagram, at the position of the star but laid out in the plane of the sky rather than perpendicular to it (right-hand side of Fig. 9), the direction of the proper motion that corresponds to V_t is in a known position angle, as shown. The actual proper motion, μ, is in a somewhat different direction, because of the peculiar motion of the star. So what we do is to resolve μ into a component, υ, that is in the direction of group motion, and a perpendicular component τ. The situation now is that all of the information about the group motion is contained in υ (even though the latter also contains some peculiar motion), while the τ component of proper motion speaks purely to the peculiar motion.

Our strategy now is clear: use the υ components of the proper motions to study the group-mean motion and the τ components to study the velocity dispersions. In both cases, of course, we will derive a distance by comparing the angular size of proper motions with the linear size of the radial velocities.

(An aside here, for clarification regarding that last phrase, "the mean distance of the group of stars." The discussion here has tacitly assumed that all stars are at the same distance. They are of course not; the assumption was made only for simplicity of exposition. Differences in distance are easily allowed for; if all the stars in a group have the same absolute magnitude, they need to be corrected for differences in apparent magnitude, by multiplying the proper motion of each star by a factor $10^{0.2m-\text{const.}}$, and adjusting its weight accordingly; this reduces all stars to an equivalent distance. If the absolute magnitudes are known to be different, an additional correction is accordingly included in the exponent.)

Now for the proper-motion solution itself. Eq. 4.2, combined with Eq. 4.3, and the tau-upsilon decomposition described above, gives

$$\upsilon = \frac{V_0 \sin \lambda}{4.74r}. \tag{4.4}$$

This equation serves as an equation of condition for a least-squares solution for $1/r$, i.e., the mean parallax of the group. This is the solution for statistical parallax from proper motions.

The tau components, by contrast, contain peculiar motion only; thus they are ideal for studying velocity dispersions. In doing so, we must take into account the entire velocity ellipsoid, determining all of its components from the proper motions and also from the radial velocities. Again, comparison of the two gives the statistical parallax. How to do so simultaneously for the 6 independent parameters that describe the velocity ellipsoid is beyond the level of the present discussion of principle. This matter is discussed further in the introductory section of the Popowski & Gould (1998) paper already referred to, as are the problems of terminology discussed in the next paragraph.

Some questions of terminology deserve mention here. Older papers in Galactic astronomy refer to statistical parallaxes determined from the group mean as "secular parallaxes," while for statistical parallaxes from velocity dispersions they reserve the name "mean parallax"; in some cases the term "statistical parallax" was also reserved solely for determinations from velocity dispersions. Indeed, the term "secular parallax" does serve to highlight the basis on which the group-motion approach rests; just as the heliocentric trigonometric parallax is caused by the Earth's motion back and forth around the Sun, the Sun's motion through space with respect to the mean of the group (which is just the negative of the group motion with respect to the Sun) can be considered as establishing a baseline from which we take a parallax. In this case, however, the baseline increases secularly with time; it is this building up of a long baseline that gives statistical parallax its power. Nevertheless, I cannot agree more thoroughly with Popowski & Gould (1998, their fourth paragraph) that "secular parallax and classical statistical parallax are actually two aspects of the same generalized method, now simply called statistical parallax."

A peculiarity of the observations that go into statistical parallaxes is that each star contributes to the radial-velocity and proper-motion parts of the data set in ways that are completely complementary. In the mean-velocity method, the stars that contribute most to the Solar-motion vector, through their radial velocities, are situated in parts of the sky where their proper motions contribute least to the solution, and vice versa. In the dispersion method, it is the contribution of a star to the different components of the velocity-dispersion tensor that is complementary. Because of this complementarity it is important to have an even distribution of stars over the sky.

Some of the particular properties of RR Lyraes must be taken into account in statistical-parallax studies. One of these is the dependence of absolute magnitude on metallicity. What needs to be found is not a single absolute magnitude for all RR Lyraes, but rather the two constants in the equation

$$M_V = a([\text{Fe/H}] + 1.6) + b \qquad (4.5)$$

(where the inclusion of the constant in the parentheses is in recognition of the fact that the mean [Fe/H] of the sample is likely to be ~ -1.6, in which case the symmetry around that value of [Fe/H] will make the values of a and b nearly independent statistically). Some treatments, however, recognizing that considerable information about the value of a comes from other sources, have preferred to adopt a value for a and concentrate all of the weight of the solution on b.

Interesting attempts have been made to use the globular clusters in M31, which are all at essentially the same distance from us, to compare the absolute magnitudes of the horizontal branches in clusters of different metallicities. These studies (Fusi Pecci et al. 1996, Ajhar et al. 1996) are photometrically difficult, but have given a weak indication that the value of a is somewhere near 0.1.

On the whole, however, the better determinations of the value of a have come from simultaneous solutions for both a and b, made by either the statistical-parallax method or the Baade–Wesselink method (see below). They put the value of a in the range 0.2–0.3.

Another property of the RR Lyraes is much more important to treat with great care, however. This is their division into two kinematic groups, probably corresponding to members of the thick-disk and halo populations. There is a sharp break in the kinematic properties of RR Lyraes at a ΔS value (the metallicity index most commonly used for RR

Lyraes) between 3 and 5. To mix stars of the two groups in a single statistical-parallax solution can lead only to chaos. Separate solutions must be made for the two groups, and I would even advocate giving up the statistical contribution that might be made by stars near the dividing line in metallicity, rather than running the risk of contaminating either sample.

Reduced proper motions. There is a clever way of comparing the motions of RR Lyrae stars (or of any other type within which we know relative absolute magnitudes) without knowing the distances of the individual stars. To simplify the present explanation, we shall assume that all the stars with which we are dealing have the same absolute magnitude; the correction for differences in absolute magnitude is straightforward but would complicate the explanation.

In his work on the stellar luminosity function, Luyten introduced a quantity that he called "reduced proper motion"; it is defined as

$$H = m + 5 \log \mu, \tag{4.6}$$

where m is the apparent magnitude. If we combine this with the well-known relations for transverse velocity and absolute magnitude,

$$V_t = 4.74 r \mu \tag{4.7}$$

and

$$M = m + 5 - 5 \log r, \tag{4.8}$$

it follows easily that

$$H = 5 \log V_t + (M - 8.38). \tag{4.9}$$

Thus H is a logarithmic measure of the star's transverse velocity, offset by a quantity that depends only on the absolute magnitude. The distance of the star has magically disappeared, and transverse velocities can be compared directly. (In taking averages, however, I would expect better results to come from the use of $10^{0.2H}$, but the use of reduced proper motions is so rare that I have never encountered such a treatment.)

Trigonometric parallaxes. Since there are no RR Lyrae stars close to the Sun, the method of trigonometric parallax had not been available for this stellar type until the advent of the high-accuracy measurements by the Hipparcos satellite (Perryman et al. 1997). Even so, the accuracies of Hipparcos parallaxes are too low to make them individually useful (even for the eponymous RR Lyrae itself, at a distance of 250 parsecs). The problem is the Lutz–Kelker correction (Lutz & Kelker 1973, a paper with whose content every astronomer should be familiar)—the most-frequently ignored statistical correction in modern astronomy. The rationale of the L–K correction is simple: since, in the conical volume that a given solid angle covers, the number of stars goes up as the cube of the distance, it is much more likely that a given measured value of the parallax is the result of our having measured a more distant star as too close than the opposite case of having taken a closer star to be more distant. Thus all measured parallaxes are biased, in the sense that a star is probably farther away than we measured it to be. Lutz & Kelker showed that the size of the correction that is needed for this bias depends only on the ratio of the measuring error to the parallax itself. The reason that individual Hipparcos parallaxes of RR Lyrae stars cannot be used is that the Lutz–Kelker correction grows rapidly with the size of the relative error, and diverges completely when the error is 18% as large as the parallax. The Hipparcos parallaxes of nearly all RR Lyraes have errors that fall beyond this limit.

How then can we use these parallaxes? The answer is to average them; the average will then have a small enough error that the Lutz–Kelker correction can be made successfully.

The averaging is not quite trivial, because the stars are at different distances; suffice it to say that the differences can be allowed for (similarly to the magnitude-dependent correction already referred to in the statistical-parallax discussion), effectively reducing all stars to the same distance but giving them appropriate weights.

The Baade–Wesselink method. Another method of estimating RR Lyrae distances depends on the fact that their pulsation involves physical expansions and contractions, which are manifested not only in the characteristic brightness changes but also in a periodic variation in the radial velocity of the star. A time integration of the star's radial-velocity curve gives the total linear amount by which the star's radius changes. Correspondingly, the periodic changes in the brightness and the spectrum of the star allow a calculation of the *fraction* by which its radius changes. Combination of these two results leads to a value for the radius of the star. The temperature of the star's surface can be calculated from its spectrum, and its luminosity then follows as $L = 4\pi R^2 \cdot \sigma T^4$, leading to a value for its absolute magnitude.

There, in principle, is the so-called Baade–Wesselink method. The preceding paragraph aimed at simplicity, but the actual use of the B–W method is totally lacking in that delectable characteristic. The radial velocity is only an effective value over a surface that is moving in a whole range of directions with respect to the line of sight, in a non-static atmosphere whose layers are moving in different ways at a given time. The relative changes in radius are calculated from simultaneous changes of magnitude and surface brightness, the latter derived from a supposedly intimate knowledge of the relation between the latter and the details of the star's spectrum. Furthermore, the magnitude changes are measured in some photometric passband whose bolometric correction must be known at each stage of the pulsation. For the details of these problems I refer you to the result papers referenced below; suffice it to say that the practitioners of the Baade–Wesselink method claim to have gotten all of these details quantitatively right. This has always been a wonder to me, but their results seem to compare favorably with those of other methods of finding out the absolute magnitude of those elusive RR Lyrae stars.

The Large Magellanic Cloud. As a further method of calibrating RR Lyrae absolute magnitudes, we can compare them with other distance indicators, in a stellar system that contains a wide variety of stellar types. The best such milieu is the Large Magellanic Cloud, whose distance can be estimated by means of a wide variety of methods. Since the apparent magnitudes of RR Lyraes in the LMC are known, their absolute magnitude follows immediately from the distance modulus of the LMC—or would, if the distance modulus of the LMC were reliably known. The distance of the LMC can be estimated in ways that are nearly countless: cepheids, Miras, supergiant stars, Supernova 1987A, etc., but the answers have had the unpleasant habit of disagreeing with each other. Values commonly quoted range from 18.2 to 18.7 or more. If the distance modulus of the LMC were to be revealed to reliable perfection, we would know the absolute magnitudes of its RR Lyraes; but in the present state of astronomy that seems an unlikely event.

Results. Papers on the absolute magnitudes of RR Lyrae stars are far too numerous to cite here. I will merely mention two reviews, by Feast (1997) and by McNamara (1997; B–W determinations only), and two recent papers giving results (Fernley 1998a,b). The first of these deals with the metallicity slope coefficient, largely from Baade–Wesselink results; the second paper is a comprehensive discussion of Hipparcos parallaxes, statistical parallax from Hipparcos proper motions, and Baade–Wesselink results. It reaches the result

$$M_V = (0.18 \pm 0.03)([\text{Fe}/\text{H}] + 1.53) + (0.77 \pm 0.15). \qquad (4.10)$$

The properties of RR Lyraes are, of course, discussed in much greater detail by Feast, in Chapter 6.

4.1.2. *Theories of the horizontal branch*

The theory of stellar structure and evolution has reached a point where theoreticians can calculate the absolute magnitude that the horizontal branch ought to have. This number has uncertainties, however, that depend on age, details of chemical composition, and on the theory itself, so that I would not hold this up as one of the primary methods of calibrating globular-cluster distances. (Astronomy is full of opinions rather than certainty; for a view contrary to mine, see Castellani in Chapter 3, Section 3.2.)

4.2. *Main-sequence calibration*

Potentially the best, and certainly the most direct way of establishing a distance scale for globular clusters is to calibrate the absolute magnitudes of main-sequence stars, by comparison with field stars of similar type whose distances have been measured accurately by trigonometric parallax. The problem here is "similar type"; too few nearby stars of low metal abundance are available, and in some cases their characteristics are not well enough known. The stars in question are, of course, the subdwarfs, stars of the halo population that are passing through the Solar neighborhood.

This method has only recently become available, through the advent of the Hipparcos parallaxes, which include values for a number of subdwarfs whose distances had hitherto been only poorly known. There have been several attempts to use these subdwarfs to fix the absolute magnitudes of stars on the main sequences of globular clusters.

Fittings of nine globular-cluster main sequences to local subdwarfs with Hipparcos parallaxes have been made by Gratton et al. (1997) and by Reid (1997, 1998). Their results disagree by small amounts—typically $0\overset{m}{.}1$—but in all cases both investigations find distance moduli that are larger by $0\overset{m}{.}1$ to $0\overset{m}{.}5$ than the values that had previously been accepted. The change in distance modulus has an important impact on the fitting of MS turnoffs to the isochrones from which cluster ages are determined; the ages become smaller by 1–2 Gyr.

In principle this should be a totally reliable method, but in practice there are difficulties. One of these is that many of the best-observed globulars have a quite low metal abundance, whereas most of the subdwarf calibrators are much less metal-poor. This difference necessitates application of a color–metallicity relation. Unfortunately it is in the nature of the process of fitting a sloping but nearly linear sequence that an erroneous color is compensated by a shift in absolute magnitude, resulting in a correspondingly erroneous calibration. Examination of the detailed lists of calibrator stars shows that this criticism applies somewhat the list of Gratton et al., but only where the clusters of lowest metallicities are concerned. (Gratton et al. explain that they rejected the most metal-poor calibrator stars in Reid's list because their parallax errors seemed unacceptably large.)

As already noted, in the use of Hipparcos parallaxes one must be careful about the Lutz–Kelker correction. However, this correction is smaller for subdwarfs than for run-of-the-mill stars, because the known subdwarfs in the solar neighborhood have nearly all been selected by their large proper motions, and this gives them an *a priori* distribution of distances quite different from the r^3 one that Lutz and Kelker had used in their paper. (For a discussion of how to handle this situation, see Hanson 1979.) The L–K corrections needed for these subdwarfs turn out to be rather small, and both investigations seem to have applied them correctly.

The fact of the Hipparcos subdwarfs giving a systematically different result for the globular-cluster distance scale than do other methods is quite disturbing. It is not likely that the difference can be explained away by systematic errors in the Hipparcos catalog, as has been suggested for the discrepant distance that Hipparcos parallaxes gave for the Pleiades (Pinsonneault et al. 1998, Soderblom et al. 1998); the Pleiades argument relies on their being confined to a small region of the sky, whereas the subdwarfs are all around us. This different distance scale for globular clusters—right or wrong—has to be understood in some other way.

There is an insidious suspicion, which is growing in my mind if not in the general astronomical conscience, that this whole game may have rules that are more complicated than the ones we are playing by. The problem that worries me is chemical abundances—I emphasize the use of that last word in the plural. It has been customary for decades to speak of the "metallicity" of a cluster, in terms of a single quantity [Fe/H]. Yet in so doing we have been sweeping under the rug the possibility that the second-parameter problem, recognized for almost as long as we have been using [Fe/H] as the first parameter, might be a manifestation of another independent chemical-abundance ratio. Indeed, in recent years we have seen the appearance of [α/Fe] as a new parameter in the abundances that are used in calculating stellar evolution models. The natural tendency has been to assume that it goes in lock step with [Fe/H]; the correlation is after all explained by a good scenario about what kind of supernovae make what kind of elements. But recent evidence (Barbuy et al. 1999) seems to demolish that scenario, and it may be that the absolute magnitude of the main sequence of a globular cluster depends on several quantities, for some of which we lack measurements for most clusters.

Nevertheless there is a dogged tendency among astronomers to make do with what we have, and for the remainder of this discussion I will banish the suggestion that we are looking for a simplified answer that does not exist. (There is, after all, a saying that God looks after little children, and idiots, and astronomers.)

4.3. *White dwarfs*

One stellar type whose theory can be potentially easy and uncomplicated is the white dwarfs (WDs). They have the advantage that the theory of their structure is relatively simple, and does not involve knowing the metallicity of their progenitors. Now that HST has been able to observe WD sequences in several clusters, it occurred to Renzini and his collaborators to use them as a distance criterion, in a cluster of relatively small distance modulus and small, well-determined reddening, NGC 6752 (Renzini et al. 1996). The idea is to compare their apparent magnitudes with the absolute magnitudes of local white dwarfs, as determined by trigonometric parallaxes. The largest source of potential uncertainty is the masses of the WDs in globular clusters, since the mass affects the absolute magnitude, and local calibrators must be chosen to match the masses of the globular-cluster WDs. The cluster white dwarfs are recent descendants of horizontal-branch stars, however, and theoreticians claim to know the masses of HB stars (probably with considerably more security than their absolute magnitudes). There is, to be sure, some mass loss when a post-HB star goes through the planetary nebula stage, but this is probably not of serious consequence.

White dwarfs come in two main varieties, DA and DB (atmospheres of hydrogen or helium, respectively). Renzini et al. claim to have distinguished the two varieties successfully both in the cluster and in the field, and fit five field DA white dwarfs to more than a dozen in the cluster, and get a distance for which they claim an accuracy of better than 10%. If so, this is the most accurate distance estimate that has yet been given for a globular cluster.

4.4. *Other features of the CMD*

Nearly every imaginable feature of the CMD has been used at one time or another to estimate distances of clusters. For example, the tip of the red giant branch has an absolute magnitude that at near-infrared wavelengths is nearly independent of metallicity. Thus its absolute magnitude has been used as a distance criterion in globular clusters in nearby galaxies (to estimate the distance of the galaxy itself). For clusters in the Milky Way, however, it is inferior to most of the criteria discussed here.

4.5. *Bends in the main sequence*

Although the main sequence is usually portrayed schematically as a diagonal straight line in the CMD, it actually has a conspicuous bend at $M_V \sim 9$, and another one at $M_V \sim 12$. Although the latter feature is probably too faint to be put to practical use, the bend near $M_V \sim 9$ is easily observable in dozens of clusters. Theory predicts that its absolute magnitude should have a weak metal dependence if any. This feature offers two possibilities, which I will call the bold one and the cautious one. The bold approach would be to believe the theoretical value of the absolute magnitude of the bend, and simply to use it to find the apparent distance modulus of each cluster. A more cautious approach would be to say that although we cannot be sure what the absolute magnitude of the bend is, we can at least fit one cluster to another and find the difference of their distance moduli.

As far as I know, no one has yet used this method, although the research group with which I am associated will certainly be investigating it in the near future.

4.6. *Proper-motion dispersions*

Finally, I would like to discuss another method, on whose development my postdoc, Jay Anderson, and I are working personally at the present time. Its idea is to compare the dispersion of the internal proper motions in a cluster with the dispersion of the radial velocities. Since the proper motions are in angular measure and the radial velocities in linear measure, comparison of the two gives the distance of the cluster just as surely and just as fundamentally as the comparison of the annual parallactic motion of a nearby star with the Earth's orbital velocity gives the trigonometric-parallax distance of the star.

This method has been suggested before (Rees 1993), but the proper-motion measurements have never been accurate enough to give distances that are good enough to be useful. Now, however, measurements with HST are carrying astrometry to a new level of accuracy, and proper motions measured over a baseline of only a few years are outstripping ground-based observations whose first epoch goes back to the early part of this century. Furthermore, the ground-based measurements are restricted to the clusters for which good series of plates were taken many decades ago, whereas the shorter time baseline needed by HST insures the accessibility of a large number of clusters, for more than a dozen of which good first-epoch material already exists in the HST Archive.

With the stability and resolution of HST's WFPC2 camera, measurements of star positions can now be made with milliarcsecond (mas) accuracy. The position of a well-exposed star image can be measured with an accuracy of 0.03 pixel, so that the mean of 9 images can produce an accuracy of 0.01 pixel, which in the three Wide Field Camera chips is 1.0 mas; at the superior scale of the Planetary Camera chip this pixel fraction corresponds to 0.45 mas.

Achieving a $\sqrt{N}$ gain in accuracy from averaging N images depends, however, on avoiding systematic errors of measurement. The most serious problems, on which we are concentrating our present efforts, arise from the severe undersampling of star images by the WFPC2 pixels. These are larger than the PSF, whose size is determined almost completely by the diffraction limit of HST's 2.4-m aperture.

The effect of undersampling is to introduce systematic errors that depend on the location of the center of the star image with respect to the center of the pixel that it is in—what I shall refer to as the pixel fraction. If a star is exactly centered in a pixel of a WFC image, only 20% of its light spills over into any of the surrounding pixels, and it is on these small fractions of the total that the positional measurement depends. Even with the most careful determinations of the PSF that needs to be fitted to each image, it is impossible to avoid errors that depend systematically on the pixel fraction. For high-accuracy astrometry it is extremely important to calibrate and remove these errors; this calibration is the key to accuracy. Since the amplitude of pixel-fraction errors tends to be comparable to the accidental error of a single measurement, the averaging of measurements will not properly increase the accuracy unless these systematic errors are removed.

I have dwelt on the errors of measurement, because the correct analysis of these is absolutely essential to the reliable determination of the distance of globular clusters by comparing the dispersion of proper motions with that of radial velocities. The reason is that the observed dispersion of proper motions is the quadrature sum of the true dispersion of proper motions and the dispersion of measuring errors; i.e.,

$$\sigma_{\text{true}}^2 = \sigma_{\text{obs}}^2 - \sigma_{\text{err}}^2 .$$

Let us consider a concrete but hypothetical numerical example: In 47 Tucanae, suppose that we observe a proper-motion dispersion of 0.682 mas/yr, while our measuring error for a star is 0.187 mas/yr—an unprecedented accuracy for such a measurement. When we subtract the errors in quadrature, the result is that the true dispersion is 0.656 mas/yr. Suppose, however, that we have underestimated our errors by 30%, so that the dispersion of errors was really 0.267 mas/yr. In that case the correct dispersion of proper motions would be 0.628 mas/yr, so that we have overestimated our result by 4.5%. At first this would seem to be a good result in any case, *but* our dispersions were determined from a sample of 4000 stars, and the theory of statistical sampling says that a dispersion determined from a sample of size N should have an error of one part in $\sqrt{2N}$, or 1.1%. Even in this case of quite good measurements, a poor error estimate would lead to a tremendous loss of accuracy. (Note: Through the efforts of our collaborators about 4000 radial velocities will be available, also leading to a dispersion that will also have an accuracy of 1.1%; so a failure to get the proper-motion errors right would be a true disaster.)

I emphasize that the figures just given are hypothetical ones, for purposes of illustration. The work is still in progress, and we do not yet know what distance we are going to derive for 47 Tuc.

The example just given illustrates the case of what may well be the smallest relative errors that we are likely to encounter in such proper-motion measurements; 47 Tuc has large motions, and we have an unusually large data set. There are globular clusters for which the error of measuring the proper motion of a star will be comparable to, or even larger than, the dispersion of proper motions. The $\sqrt{2N}$ rule nevertheless predicts that from several thousand stars we should be able to measure a proper-motion dispersion of quite good accuracy—*if* we can make a truly reliable determination of the sizes of the errors.

The estimation of the errors is plagued not only by the existence of systematic errors, but also by the fact that their existence tends to create a correlation between some measurements that should have been independent, because they were made at similar pixel fractions and therefore have similar systematic errors. Errors can be estimated only from residuals, and if the residuals are correlated, the statistics of estimating errors from them acquire a completely new set of rules.

I will not conduct the reader further into this chamber of statistical horrors, except to indicate that it exists and that arriving at a reliable answer depends critically on emerging from it correctly. The only further point that I want to mention is that there are ways of planning a set of observations so as to minimize the effect of pixel-fraction errors. These involve taking the set of images at an evenly spaced set of pixel-fraction offsets, the strategy already referred to as dithering. It is also noteworthy, as mentioned previously, that for more than one reason good dithering also leads to a dramatic increase in photometric accuracy.

4.7. *Corrections for interstellar absorption and reddening*

Since most of the methods of estimating the distances of globular clusters depend on comparing absolute magnitudes with apparent magnitudes, it is absolutely essential to have good estimates of the amount of interstellar absorption by which each cluster is dimmed. These absorption estimates usually come, in turn, from estimates of the reddening, via the relationship

$$A_V = c \cdot E(B - V),$$

where the constant of proportionality c is, for reasons that go beyond the scope of the present discussion, generally taken to be about 3.1 or 3.2. Reddenings may be estimated from the colors of stars in the same field, or of features in the CMD of the cluster itself. Since the intrinsic colors of CMD features depend on metallicity, the metallicity of the cluster must first be estimated by some independent means. An exception to this restriction, however, is the extreme blue horizontal branch (if the cluster has one); for reasons explained in Section 4.1, its color is nearly independent of metallicity.

Other methods come from study of the integrated colors of the clusters, in many wavelength bands, and their integrated spectra. (In this way a cluster as a whole is treated as one would treat an individual star, but the calibrations are of course different.) Two notable compilations of reddening values are by Zinn (1980) and by Reed, Hesser, & Shawl (1988).

A completely different method of estimating absorption comes from radio observations of interstellar hydrogen. Burstein & Heiles (1982), making use of the close kinship of interstellar gas and dust, showed that the total Galactic absorption in a given direction is closely correlated with the total column density of hydrogen in that direction, as measured with the 21-cm line. They provided detailed maps from which reddening estimates can be made. Such estimates are useful, of course, only for clusters that lie outside the layer of gas and dust of the Galactic disk, but most clusters satisfy that criterion.

It should be noted that the comparison of proper-motion dispersions with the dispersion in radial velocities, being purely geometrical, offers a route to absorption-free distances. But the absorption problem then turns itself around, when we come to use such cluster distances to find the absolute magnitudes of stars in them.

(This discussion of distance measures has necessarily been brief. A more extended discussion can be found in Chaboyer 1998.)

5. Luminosity Functions and Mass Functions

5.1. *The two functions*

One of the fundamental questions about a globular cluster is how many stars of each type it contains. This question can be answered in either of two ways, each of which is fundamental for a different set of stellar types and for a different reason. We can specify the *luminosity function*, the number of stars at each luminosity; or we can specify the *mass function*, the number of stars at each mass.

Since there is no way of measuring the masses of individual stars, however, what we actually observe is the luminosity function (LF); we can find the mass function (MF) only by calculating it from the LF by use of the mass–luminosity relation (MLR). Specifically, we define $L(M)\,dM$ to be the number of stars between absolute magnitude M and $M + dM$. (Note that the M is usually marked as pertaining to a particular photometric band, e.g., M_V.) We define the mass function $f(m)$ in an analogous way. We can then write (where for the sake of being specific I will choose the V band)

$$L(M_V)\,dM_V = f(m)\,dm.$$

The expressions on the two sides of the equation are equal, because they are merely different ways of expressing the same number of stars. The equation can then be rearranged to read

$$f(m) = L(M_V) \cdot \frac{dM_V}{dm},$$

which shows how an MF is calculated from an observed LF. Notice that the MLR enters implicitly in the correspondence between the two arguments M_V and m, and also explicitly in the multiplication by the *slope* of the MLR. It is for this latter reason that any MF that we derive is so sensitive to the details of the mass–luminosity relation that was used.

It is well to emphasize once again here the differences between the theoretical quantities and the observational quantities. An important part of making an MLR available is to convert it from the theoretician's M_{bol} and $\log T_e$ to M_V and $(B - V)_0$, where B and V refer to the actual observational passbands that were used for the observations.

There is no empirical mass–luminosity relation for the range of metallicities that globular clusters occupy. Only two visual binaries are known among the local stars of lower-than-solar metallicity—85 Peg and μ Cas—and they are only mildly metal-deficient. Even for these two stars accurate masses have only recently become known. (Let me remind the reader that the mass that we derive for a visual binary depends on the cube of its distance; only with Hipparcos parallaxes have the distances of these two stars become reasonably well known.) The fact is that for the MLR that applies to a given globular cluster we are completely reliant on theories of stellar structure, for the chemical composition that applies to that cluster.

There is another interesting way of looking at the relation between LF and MF. In a very meaningful sense it is the MF that is fundamental: it expresses the mass spectrum of fragmentation of the pre-stellar blobs of gas that were to become the stars of the cluster. As such, it is likely to vary rather smoothly with mass. The LF, on the other hand, arises from the agency of the MLR operating on the MF; thus any erratic features of the MLR will be reflected in the LF. This is known to happen in the solar-metal-abundance population, which has a kink at $M_V \sim 8$ that produces an even stronger kink in the LF, because of the presence of the MLR slope in the transformation equation. (There is an excellent discussion of this phenomenon in D'Antona 1998.) And of course the LF must encounter a catastrophe at the lower limit of hydrogen burning; we will discuss that at greater length in Section 6.

For one part of the magnitude range in a globular cluster, however, it is the luminosity function that carries the fundamental physical information. At and above the main-sequence turnoff, the masses of stars hardly differ at all. To put this a little differently, a tiny difference of mass corresponds to a large difference in evolutionary state, at this fixed moment in time. Thus above the MSTO the LF expresses the rate of evolution rather than the mass. Comparison of theory with observation is usually made by merely fitting the CMD to isochrones for different ages, which have been calculated from the theory; in most cases no use is made of the information in the luminosity function, which I believe has been seriously under-utilized.

5.2. *Observational determination of LFs*

The luminosity function of a globular cluster can be derived from the same data that gave rise to its CMD. Although it is possible in principle to derive an LF from counts made in a single passband, in practice it is safer to have a CMD, primarily to correct the counts for the field stars of the foreground and background. There are examples in the astronomical literature where single-band counts have been "corrected" in a disastrously erroneous way, from field-star counts made in a comparison field. Furthermore, now that LFs reach to really faint magnitudes, it is necessary to exclude white dwarfs in order to have a luminosity function that refers purely to the main sequence.

In principle the derivation of a luminosity function from a color–magnitude diagram is trivially easy: just count up the number of cluster members in each interval of luminosity—typically, bins of half a magnitude or a magnitude—and assign error bars on the basis of the Poisson uncertainty in the number of stars in the bin. There are, however, some practical problems. The most serious of these is completeness. Much of the interest in globular-cluster LFs is at the faint end, and it is just there that the incompleteness of the sample becomes most serious. The derivation of the faint end of an LF usually requires that more labor be spent on completeness tests than on the CMD itself.

The type of completeness test that is used almost universally is to add artificial stars at random locations in the image and then to put the image through the same star-finding and photometry procedures that were used before. The fraction of the artificial stars that are successfully found and measured serves as a measure of the completeness. The process also yields photometric data: the measuring error, and also the photometric bias that arises from the fact that a star image that is accidentally too faint is more likely to be lost than is one that is accidentally too bright. The artificial-star tests must be made at all magnitudes, since even a well-exposed star image can be lost if it happens to fall on a much brighter star.

There are two necessary requirements for artificial-star tests: the artificial stars must resemble real ones as closely as possible, and one must not throw in so many of them as to make a significant increase in the degree of crowding. The first of these requirements is easily met by creating the artificial stars out of the point spread function, with a suitable amount of random noise added. DAOPHOT, for example, offers the ability to carry out this process.

It is the requirement that the addition of artificial stars not significantly increase the crowding that makes the completeness tests so burdensome in computing time. Most researchers consider the limit to be the addition of 10% as many artificial stars as there are real stars in the image. Since several hundred artificial stars are needed at each magnitude level, this requirement necessitates the measurement of a large number of images—considerably more labor than was involved in the original CMD study.

In an effort to lessen this labor, Piotto and Anderson (see, e.g., Piotto et al. 1999) have devised a procedure that allows the addition of a large number of artificial stars at once, without affection the crowding in any way that matters. This is done by arranging the artificial stars "in a spatial grid such that the separation of the centers in each star pair [is] two PSF radii plus one pixel." In this way the artificial stars are unable to interfere with each other, and the probability of recovering a given artificial star is unchanged from what it would be if that star were the only one added to the image. The grid can be used in multiple independent experiments just by shifting the entire rigid grid by a random amount each time.

An even more economical artificial-star test can be used if the basic photometric method used is one that finds and measures one star at a time, unlike DAOPHOT, for example, which measures stars in groups of images that overlap enough to affect each other. In that case, artificial stars can be thrown in one at a time, searched for and measured, without the remeasurement of all the stars in the image each time a set of artificial stars is added, which is the expensive necessity in the more conventional method.

5.3. *From the LF to the MF*

Although it is the luminosity function that is observed, what matters in the understanding of star clusters is their mass functions—the true manifestation of stellar nature, as distinct from the impression that their radiation gives us. (Let me repeat here, however, a distinction that I made earlier: the part of the LF around and above the main-sequence turnoff is indeed of direct interest, as an observational indicator of the rate of post-MS evolution. But what I will be referring to in the remainder of this section is the part of the main sequence below the turnoff, where the role of the LF is only as an intermediary toward the determination of the mass function.)

As already indicated, the transformation of the LF into an MF depends crucially on the mass–luminosity relation. I say crucially, because not only does the MLR map luminosities into masses; more important, the transformation from one function to the other depends on the *slope* of the MLR.

It is also worthy of note that the LF→MF transformation depends on one other observational quantity, the distance modulus of the cluster. The LF is observed with respect to apparent magnitude, whereas the MLR is stated in absolute magnitude. Where the MLR is laid down on the LF obviously makes a difference to the resulting MF; in fact, the result of this process may actually carry some weight in the choice of a distance modulus, since some values of the latter may lead to wiggles in the MF that seem unnatural. (Note also the extreme importance of the distance modulus in age determination; not only must the magnitude of the turnoff fit, but the LF in that region must closely fit the predicted evolution rate.)

One more consideration affects the derivation of a correct mass function for a globular cluster: radial mass segregation. Because stellar encounters in a cluster tend to produce equipartition of kinetic energy between stars of different mass, the stars of lower mass acquire higher velocities and therefore tend, on the average, to be farther from the center of the cluster. Since LF determinations usually involve only a small part of the radial extent of the cluster, the local mass function of the region observed must be transformed to a global mass function of the cluster. This step can usually be carried out in a straightforward way by dynamical modeling of the cluster, but it must not be ignored, since the corrections can be quite sizable. Figure 10 illustrates this for a modeling of the highly concentrated cluster NGC 6397.

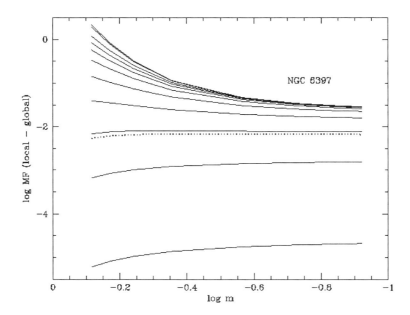

FIGURE 10. Corrections from local to global MF, calculated from a multi-mass King model of NGC 6397. Each curve shows for some radial distance in the cluster the quantity that must be subtracted from log MF at each mass in order to correct the local MF to a global one. Although radii are not given explicitly, the upward progression of the curves corresponds to the density increase from the edge of the cluster to the center. The dotted line refers to the radius at which the observations discussed in the following section were taken.

5.4. *Observed mass functions*

The best mass functions that exist today are those that come from the Hubble Space Telescope, largely on account of the faintness that its observations reach.

Figure 11 shows nine of these mass functions, as converted from LFs by an MLR of Alexander et al. (1997). (Note that although the are plotted as log N vs. log m, they represent dN/dm, i.e., number per unit interval of m rather than log m.)

Since on this logarithmic plot MFs are often fitted approximately by straight lines (equivalent to a power law between the quantities whose logarithms are plotted), it is well at this point to digress on the notation that is commonly used for such power laws. When the mass function is written as $f(m)dm$, it is customary to write the power law as

$$f(m) \propto m^{-\alpha},$$

while a mass function written as $\phi(m)d\log m$ is commonly written as

$$\phi(m) \propto m^{-x}.$$

Since these are merely different ways of expressing the same information, it is clear that

$$f(m)dm = \phi(m)d(logm).$$

But a little manipulation shows that

$$m^{-x}d(logm) = m^{-x}\left(\log e\right)d\ln m$$

$$= \log e \cdot m^{-(1+x)}dm.$$

Comparison of this with the power-law form of $f(m)$ shows immediately that $\alpha = 1 + x$.

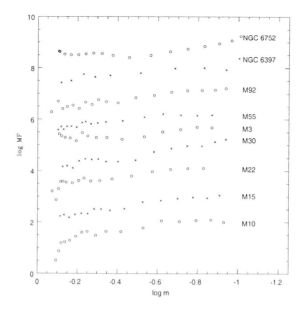

FIGURE 11. Mass functions of 9 globular clusters. In each case, the LF was transformed to an MF using an MLR from Alexander et al. (1997), and a multi-mass King model was then used to transform the local MF to a global one. The sharp bend at the left end of the M10 MF is probably due to a small mismatch of distance modulus or of the age corresponding to the MLR that was used.

(The Salpeter law that fits the upper end of young main sequences has $x = 1.35$, $\alpha = 2.35$.)

The lower ends of the MFs of globular clusters are similar to that of the stars of the Galactic disk, whose mass-function slope is found by Gould, Bahcall, & Flynn (1997) to change from $x = 1.21$ above about half a solar mass to $x = -0.44$ below that point. The clusters shown in Fig. 11 have MF slopes at their low-mass ends that range from $\alpha = 0.0$ to 1.0, straddling the Gould et al. value of $\alpha = 0.56$. (Reader beware! Gould et al. use the symbol α for what we have been referring to as x.)

If a mass function slopes upward steeply enough at its low-mass end, it can imply that the low-mass end dominates the numbers, or even the total mass. If $\alpha \geq 1$, then the total number of stars diverges as $m \rightarrow 0$; if $\alpha \geq 2$, the mass also diverges. The globular clusters whose MFs are exhibited here thus have large numbers of low-mass stars, but their MF slopes are far from implying dynamical dominance by the low-mass stars. Another conclusion is that these mass functions imply (if we extrapolate them) that globular clusters contain a large number of brown dwarfs, but that those stars do not contain a large fraction of the total mass of the cluster.

One should take careful note of the MF slopes that are given above. It is a common error, and sometimes a disastrous one, to assume a Salpeter mass function for the entire range of stellar masses. The Salpeter function is a good approximation for the stars of the upper main sequence of a young population, but it decidedly *not* a good approximation for stars of lower mass. Those who fall into this error soon recognize the mass divergence and try to compensate for it by using a cut-off at some low value of the mass, but the approximation to the low-mass end of the MF is so poor that the results of such a procedure can be ridiculously bad.

Interestingly, the mass functions in Fig. 11 differ considerably from one cluster to another. The relative fewness of low-mass stars in NGC 6397 was commented on by Piotto, Cool, & King (1997); they attributed this fact to loss of low-mass stars in that cluster as a result of ejection by internal relaxation and detachment by tidal shocks of the Galactic plane. More recently, De Marchi et al. (1999) have found an even more striking deficiency of low-mass stars in NGC 6712, another cluster whose Galactic orbit makes it particularly vulnerable to tidal shocks. I will leave detailed discussion of these dynamical processes, however, to Elson, in Chapter 5.

6. Low-Mass Stars

Finally, I want to discuss the observational study of the faintest stars of the main sequence. Such studies can make two important contributions:

(1) The main importance of faint MS studies is their ability to provide data that can be used to check theories of the structure of such stars. Such checks have two different types. (a) The theory should correctly predict the relation between luminosity and radius, whose equivalent in observational coordinates is the color–magnitude diagram, and (b) it should agree with the observed mass–luminosity relation. (The ability of globular-cluster observations to check the latter is a new idea, which will be explained below.)

(2) The behavior of mass functions (MFs) at the low-mass end suggests that globular clusters contain a considerable number of brown dwarfs (King 1998), as already noted. Although there is hope of soon observing a few of these directly in one cluster, M22, by their microlensing of stars in the Galactic bulge (an HST project that is likely to be carried out in 1999), in nearly all other clusters the only way of estimating the number of brown dwarfs is to extrapolate the lower end of the mass function.

The lower end of the main sequence is an intriguing region. I say "lower end" because the MS actually does end (not by a true termination, but by a hopeless dearth of stars). Below a mass limit of 0.075–0.095 $m_\odot$ (the location of the limit depends on metallicity), a pre-stellar configuration is unable to develop a high enough central temperature to ignite and maintain the hydrogen-burning reactions that provide the luminosity of main-sequence stars. Below the H-burning limit, a star reaches degeneracy before it is able successfully to ignite hydrogen, and it is thereafter unable to continue converting the gravitational energy of its contraction into heat. It shines only from its internal heat, and fades rapidly, whereas the stars above the limit maintain nuclear reactions that allow them to continue to shine undimmed. After the passage of a Hubble time, a large gap in luminosity has opened between the main-sequence stars and the brown dwarfs.

The boundary in mass between luminous stars and brown dwarfs is thus very sharp. To repeat, pre-stellar configurations just above the H-burning limit have been able to sustain H-burning and maintain a luminosity in the stellar range, while those that were just below the limit have by now faded deep into the darkness of the brown-dwarf condition. Theory predicts that the boundary between these two extreme states will span only a miniscule range of mass. In any reasonable MF, very few of the stars in a cluster will fall in this tiny range of mass, which after more than 10 Gyr must span a luminosity range of many magnitudes. Thus just above the critical mass, the luminosity function (LF) must begin a precipitous drop.

In the globular cluster of smallest distance modulus, NGC 6397 $[(m - M)_I \simeq 12.05]$, the precipitous drop in the luminosity function takes place around $I \sim 24$, and in M4 $[(m - M)_I \simeq 12.4]$ it is predicted to be seen only a few tenths of a magnitude fainter. We have already observed this transition region in NGC 6397 (King et al. 1998, discussed below), and we have an HST program to observe three fields in M4 in a similar way.

One of the problems of observing faint stars in these two clusters is that they both lie

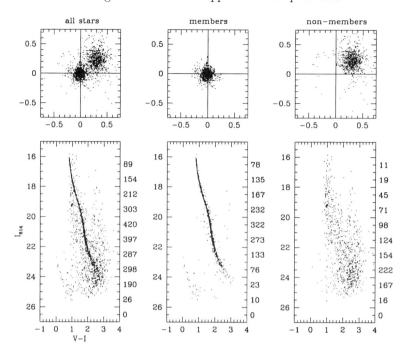

FIGURE 12. Proper-motion distributions, above, and color–magnitude diagrams, below. The scale of the proper motions is displacement in WFC pixels over the 32-month time baseline; a full WFC pixel of displacement would correspond to 37.5 mas/yr. Since all reference stars were cluster members, the zero point of motion is the mean motion of cluster stars. Left: the entire sample; center: stars within the proper-motion region described in the text; right: stars outside this region. Numbers at right are stars per unit-magnitude bin.

in rich Galactic fields, where at the faintest magnitudes the clusters stars are quite lost among the more numerous field stars. (Note the lowest part of Fig. 2.) The cluster stars can be separated out quite effectively by proper motions, however; this mere separation does not require nearly as much accuracy and care as the measurement of the dispersion of proper motions, which has already been described. An example is given in Figure 12; it was actually this step which allowed us (King et al. 1998) to derive the last magnitude and a half of the LF of NGC 6397, the region where the steep drop occurs.

The abruptness of the plunge in the LF allows us to use it in a new application, which actually constructs a part of the MLR from the shape of the LF. What allows us to do this is precisely the fact that the region spans such a small mass range. This means that the MF in this mass range can be estimated quite reliably by extrapolation of the values of the MF at slightly higher masses; then from this extended MF we can infer the masses of the individual stars in this range, as follows: The mass function, by definition, tells us how many stars there are per unit interval of mass, centered on a given mass. If we order the stars by decreasing mass (which we can do by ordering them in decreasing luminosity), the MF will tell us explicitly the mass interval Δm between successive stars. To restate this in different words, the MF can be thought of as dN/dm, where N is the cumulative number of stars, and the Δm between successive stars is then just the dm that corresponds to $dN = 1$.

What we do then is to (1) arrange the stars in order of faintness, (2) use the MLR to calculate the mass of some star that is near the onset of the plunge, and (3) use the Δm's described above to find the mass of each star that is fainter than that starting point. Note

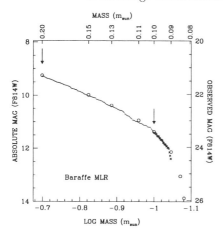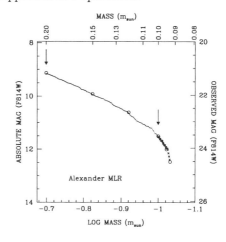

FIGURE 13. Empirical extension of the MLR, for two different MLRs (circles), from Baraffe et al. (1997) and from Alexander et al. (1997). In each case the line shows the power-law fit to the MLR, while the crosses are the empirical extension. The arrows identify the interval that was used for the power-law fit.

that the masses were inferred from the MF, without using the luminosities, except to set the order in which the stars were arranged. We can therefore plot luminosity against mass for these stars, and thus get an empirical MLR for them. The result depends on the MLR that we used in getting the MF from the LF; thus it tells how *that* MLR *ought* to behave just above the H-burning limit. The location and character of the turndown in the LF therefore provide an important constraint for theories of the structure of low-mass stars.

In the above procedure, it is important to allow for the increasing incompleteness of the data as they approach our magnitude limit. This is easily done, by counting each observed star as $1/c$ stars, where c is the completeness fraction. We simply take Δm to be the dm that corresponds to $dN = 1/c$.

One aspect of this process deserves special note: in using an MLR to convert an LF to an MF, we must know the distance modulus, because the MLR is specified in *absolute* magnitude. Thus the procedure, and its results, depend on the value we have chosen for the distance modulus of the cluster. We intend to make a thorough study of the distance modulus of each cluster and its uncertainty—but we note that once our LFs become available, the procedure can be carried out by anyone who prefers a different distance modulus (or MLR).

The faint LF recently derived for NGC 6397 by King et al. (1998), which features a dramatic turndown at $I \sim 24$, makes it possible to illustrate this procedure of extracting the MLR from the LF. In Figure 13 we demonstrate it for two different theoretical MLRs, using the procedure that was described above.

We thus use the faintward extension of the CMD as a check not only on the mass–radius relation but also on the mass–luminosity relation.

Though in the above data we clearly observe the dropoff of the LF related to the H-burning limit, the derivation of the implied shape of the MLR is statistically weak, in that the extension that was made relied on only 22 to 29 stars (the actual number of crosses in the two panels of Fig. 13), and on only 3 stars in the region where the MLR slope is steepest. This is the reason why we want to increase our observational base next year. Observation of M4 will also extend our study to a different metallicity ([Fe/H] = −1.3, compared with −1.9 in NGC 6397).

7. Conclusion

This chapter has tried to set forth some of the salient facts about globular clusters, and has concentrated in particular on the methods by which those facts are assembled. (Some areas have been omitted, however: The observations of density distributions and velocity distributions, which relate directly to the dynamics of the clusters, will be presented in Chapter 5, in conjunction with the discussion of dynamics there. Also, the motions of the clusters in the Milky Way will be presented in Chapter 2.)

The author wishes to thank Jay Anderson in particular for his contribution to some of the research that is described here. This work was supported by Grant AR-7993 from the Space Telescope Science Institute.

Appendix A

A.1. *Introduction*

Since no one else was covering binaries in globular clusters, I agreed to add a section that discusses them, attempting to answer three questions: How many binaries are ther in clusters? How do the binaries evolve? What dynamical effect do binaries have on the cluster?

Finally, at the end is a definition of a set of units that is especially useful for Galactic calculations.

A.2. *The Binary Fraction*

A.2.1. *Radial-velocity searches*

A good general discussion of this is in the middle section (by Pryor) of MacMillan, Pryor, & Phinney (1998).

One way of finding binaries is by means of their variable radial velocity. This needs observations at three separate times, at least; if you have only two observations of a star's velocity that differ, you cannot be sure that there is not simply an erroneous measurement. Naturally, even more than three observations (of stars whose radial velocities are suspected of variations) are an even better idea.

There are strong selection effects in such observations; they obviously discriminate against the wider binaries, whose orbits have smaller velocities that may also take years to change appreciably. However, binaries in globular clusters should *a priori* have a restricted range of separations. For dynamical reasons that will be explained below, there should not be any binaries with separations of more than a few a.u., i.e., with periods longer than just a few years. Even so, even among the allowable separations there is some bias against the larger ones.

Another range of separations against which radial-velocity observations discriminate is the close binaries. Most of the radial-velocity observations have been confined to red-giant stars, whose radii are so large that any close binaries among them would long since have merged.

Finally, at any separation and velocity we will lose some pairs because their orbital inclination is unfavorable.

When all of these considerations are put together, even the small percentage of stars that are observed to have variable radial velocities implies a binary fraction of at least 10 to 20%, or perhaps even more.

A.2.2. *Photometric searches*

Width of the main sequence. Another way of finding binaries is to note that the point in the CMD that represents the combined light of an unresolved pair of stars usually does not fall on any of the usual sequences. Such a pair is unlikely to involve an evolved star as a primary, because (1) the rates of evolution are such that it is improbable that both members of a pair will be in the brief stages of post-MS evolution at the same time, and (2) a main-sequence companion to a (much brighter) evolved star will not change its magnitude enough to be noticeable. It is on—or rather just *off*—the main sequence that we should look for photometric indications of binarity.

The simplest kind of binary to find is an equal pair. Since a factor of 2 is equivalent to $0.^m76$, equal pairs should form a parallel main sequence 3/4 of a magnitude above the MS. (See the beautiful example given by Elson in her Fig. 16, in Chapter 5.)

When a binary consists of two unequal MS stars, the result is more complicated. If the secondary is much fainter than the primary, the combined color and magnitude of the pair will differ only a little from those of the primary. The displacement always leaves the star on the red side of the MS, however. If we consider a set of binaries in which the relative magnitude of the secondary increases, the equivalent CMD point follows a curving track away from the main sequence and increasingly upward. (For a general discussion of the behavior of such pairs, see Romani & Weinberg 1991.)

Thus the way to investigate this whole class of binaries is to look for a broadening of the MS on the red side that is not matched by a spread to the blue. To detect such a spread, two precautions are needed: (1) The photometric accuracy must be high. (2) Careful tests must be made, by adding artificial stars to the image, to see how much of such a spread to the red is caused by the mere accidental superposition of star images; this is an important test that has not always been made. Also, because their enhanced mass tends to cause binaries to collect at the center of a cluster, the best place to look for them is at the cluster center, where the superior resolving power of HST is a virtual necessity.

The best HST study was made at the center of NGC 6752, by Rubenstein & Bailyn (1997). Their superb photometric accuracy depends mainly on their careful dithering (Bailyn, private communication). Rubenstein & Bailyn estimated that between 15 and 38% of their main-sequence stars were binaries. Noting, however, the extent to which dynamical segregation tends to concentrate binaries to the cluster center, Rubenstein & Bailyn conclude that 16% is a reasonable upper limit to the fraction of binaries in the cluster overall.

Variable stars. A number of types of variable stars that are binary have been found in globular clusters (see Hut et al. 1992 for a general discussion). Several main-sequence eclipsing stars are known, two novae (in M80 in 1860 and in M14 in 1938), and U-Geminorum-type cataclysmic variables in several clusters.

Neutron-star binaries. I will merely mention here the low-mass X-ray binaries, and the millisecond pulsars (most of which are binary); both these types have a much higher frequency in globular clusters than in the field. They are discussed in detail in Chapter 7, by Canal.

A.3. *The Dynamical Evolution of Binaries*

Each individual binary in a globular cluster tends to evolve dynamically, as a result of encounters with single stars. Unlike the case of encounters between two single stars (see

Chapter 5), for binaries the important encounters are the close ones, in which the impact parameter is no larger than the separation between the components of the binary. The evolution of binaries is a watershed effect: the wide binaries get wider and the close binaries get closer. The standard terminology for these two types is "hard" and "soft" (Heggie 1975).

The dividing line between hard and soft binaries is at the gravitational binding energy that is equal in magnitude to the average kinetic energy of a single star. For a soft binary the encounter is essentially between one component of the binary and a third star; its effect is to give the binary component a higher velocity with respect to its companion and thus to loosen the binary. The time scale for break-up of such a binary by successive encounters was calculated long ago by Ambartsumian (1937). His formula, when applied to the conditions in a globular cluster, predict that all soft binaries will have been broken up long ago.

A hard binary acts quite differently. Since its components have an orbital motion that is faster than the velocity of the average passing star, the tendency is for the component of the binary to give energy to the intruder. But this slows down the component of the binary; the orbit then becomes smaller, and the binary harder.

The results of encounters are stochastic, and not every encounter goes in the direction just indicated. But the tendencies just described are the expected ones, and once a binary is far on either side of the dividing line in binding energy, the probability is very strong that the evolution will go in the expected direction.

The eventual fate of a hard binary in a cluster is perhaps surprising. The harder the binary gets, the larger are the energy kicks that it gives to passing stars. But in an encounter between a binary and a single star, momentum is conserved, and the binary experiences a recoil. Just as it can eject another star from the cluster, it can eject itself by this recoil, and this is the ultimate fate of a hard binary.

It is interesting to note another dynamical behavior of binaries that favors the formation of pairs that contain a neutron star. In an encounter of a more massive star with a binary, there is a probability, which increases as a high power of the mass, that the more massive star will eject one component of the binary and take its place. Since neutron stars have a higher mass than average, they tend to inhabit the center of the cluster and participate readily in such "exchange encounters." It is this circumstance that undoubtedly accounts for the large number of low-mass X-ray binaries in globular clusters—and perhaps for the millisecond pulsars too.

A.4. *The Effect of Binaries on the Dynamical Evolution of a Cluster*

Binaries play an important dynamical role in clusters after core collapse occurs. Interactions with hard binaries then pump kinetic energy into the other stars, slowing the collapse, and even causing the core to re-expand if there are enough binaries. The near-magical aspect of the behavior of binaries is that they will necessarily appear in sufficient numbers; as the star density in the core increases, with it rises the ease with which new binaries are formed.

The formation of new binaries takes place by either of two processes. One formation mechanism is a very close encounter between two stars, so close that tides are created that dissipate enough of their orbital energy that this pair that was previously unbound becomes bound. Since the process requires two stars, its probability obviously goes as the square of the density. The other process requires three stars and therefore goes as the cube of the density in the core. It consists of an encounter between three stars at once, in which the trajectories are such that the stars emerge from their encounter as a single star plus a bound pair. (If it is difficult to imagine such an encounter, simply picture an encounter in which a single star has just broken up a binary, and reverse the direction of motion of each star.)

What makes the formation of sufficient binaries inevitable is that a collapsing core keeps raising its density until this heightened density causes the formation of enough binaries.

The story has not ended with the stabilization of the core, however. As described above, each hard binary eventually ejects itself by a recoil. When there are too few binaries to continue to stabilize the core, contraction sets in until more binaries form ... and so on.

A.5. *Galactic Units*

Finally, one more item that is quite separate from binaries:

There is a system of units that is particularly useful in calculations that involve the dynamics of the Galaxy or of globular clusters. The unit of length is 1 parsec, the unit of mass is 1 solar mass, and the unit of time is 0.978×10^6 years. (In any calculation that does not require high precision, this last can simply be taken to be a million years.)

The advanatage of this system is that 1 length unit per 1 time unit is exactly 1 km/s. (Cf. the well-known rule of thumb that 1 km/s is a parsec in a million years.) The only additional piece of information that is needed is that in these units the value of the constant of gravitation is 1/233. With this information one can do all calculations in the familiar, convenient units, without laborious conversions to and from cgs units. This should be a useful tool for employing many of the dynamical formulas given in Chapter 5.

REFERENCES

Ajhar, E. A., Grillmair, C. J., Lauer, T. R., Baum, W. A., Faber, S. M., Holtzman, J. A., Lynds, C. R., & O'Neil, E. J. 1996, AJ, 111, 1110

Ambartsumian, V. 1937, Astron. Zh., 14, 207

Alexander, D. R., Brocato, E., Cassisi, S., Castellani, V., Ciacio, F., & Degl'Innocenti, S. 1997, A&A, 317, 90

Anderson, A. J. 1997, Ph.D. thesis, U.C. Berkeley

Anthony-Twarog, B. J., Twarog, B. A., & Craig, J. 1995, PASP, 107, 32

Baade, W. 1944, ApJ, 100, 137

Bailyn, C. D., & Pinsonneault, M. H. 1995, ApJ, 439, 705

Baraffe, I., Chabrier, G., Allard, F., & Hauschildt, P. H. 1997, A&A, 327, 1054

Barbuy, B., Renzini, A., Ortolani, S., Bica, E., & Guarnieri, M. D. 1999, A&A, in press

Binney, J., & Merrifield, M. 1998, Galactic Astronomy (Princeton: Princeton Univ. Press)

Burstein, D., & Heiles, C. 1982, AJ, 87, 1165

Chaboyer, B. 1998, in Post-Hipparcos Cosmic Candles, eds. A. Heck & F. Caputo, (Dordrecht: Kluwer), in press

Chamberlain, J. W., & Aller, L. H. 1951, ApJ, 114, 52

Chiappini, C., Matteucci, F., Beers, T. C., & Nomoto, K. 1998, preprint astro-ph/9810422

Cool, A. M., Piotto, G., & King, I. R. 1996, ApJ, 468, 655

D'Antona, F. 1998, in The Stellar Initial Mass Function, 38th Herstmonceux Conference (ASPCS 142), eds. G. Gilmore & D. Howell (San Francisco: ASP), p. 157

De Marchi, G., Leibundgut, B., Paresce, F. & Pulone, L. 1999, A&A, in press

Feast, M. W. 1997, MNRAS, 284,761

Fernley, J., Carney, B. W., Skillen, I., Cacciari, C., & Janes, K. 1998a, MNRAS, 293, L61

Fernley, J., Barnes, T. G., Hawley, S. L., Hanley, C. J., Evans, D. W., Solano, E., & Garrido, R. 1998b, A&A, 330, 515

Fusi Pecci, F., Buonanno, R., Cacciari, C., Corsi, C. E., Djorgovski, S. G., Federici, L., Ferraro, F. R., Parmeggiani, G., & Rich, R. M. 1996, AJ, 112, 1461

Gould, A., Bahcall, J. N., & Flynn, C. 1997, ApJ, 482, 913

Gratton, R. G., Fusi Pecci, F., Carretta, E., Clementini, G., Corsi, C. E., & Lattanzi, M. 1997, ApJ, 491, 749

Greenstein, J. L. 1939, ApJ 90, 387

Hachenberg, O. 1939, Zs.f.Ap, 18, 49

Hanson, R. B. 1979, MNRAS, 186, 875

Heggie, D. 1975, MNRAS, 173, 729

Helfer, H. L., Wallerstein, G., & Greenstein, J. L. 1959, ApJ, 129, 700

Hoyle, F., & Schwarzschild, M. 1955, ApJS, 2, 1

Hut, P., McMillan, S., Goodman, J., Mateo, M., Phinney, E. S., Pryor, C., Richer, H. B., Verbunt, F., and Weinberg, M. 1992, PASP, 104, 981

Johnson, H. L., & Morgan, W. W. 1953, ApJ, 117, 313

King, I. R. 1998, in Workshop on Substellar-Mass Objects, ed. J. Lunine (Washington: National Academy Press), in press

King, I. R., Anderson, J., Cool, A. M., & Piotto, G. 1998, ApJ, 492, L37

Landolt, A. U. 1992, AJ, 104, 340

Lindblad, B. 1922, ApJ, 55, 85

Lutz, T. E., & Kelker, D. H. 1973, PASP, 85, 573

McMillan, S. L. W., Pryor, C., & Phinney, E. S. 1998, in Highlights of Astronomy, Vol. 11 (from a Joint Discussion at 1997 Kyoto IAU), in press

McNamara, D. H. 1997, PASP, 109, 857

Morgan, W. W. 1956, PASP 68, 509

Morgan, W. W. 1959, AJ, 64, 432

O'Connell, D. J. K. (ed.). 1958, Stellar Populations (Amsterdam: North Holland)

Ortolani, S., Barbuy, B., & Bica, E. 1990, A&A, 236, 362

Perryman, M. A. C. et al. (20 authors). 1997, A&A, 323, L49

Pinsonneault, M. H., Stauffer, J., Soderblom, D. R., King, J. R., & Hanson, R. B. 1998, ApJ, 504, 170

Piotto, G., Cool, A. M., & King, I. R. 1997, AJ, 113, 1345

Piotto, G., Zoccali, M., King, I. R., Djorgovski, S. G., Sosin, C., Dorman, B., Rich, R. M., & Meylan, G. 1999, AJ, in press

Popowski, P., & Gould, A. 1998, ApJ, 506, 259

Reed, B. C., Hesser, J. E., & Shawl, S. J. 1988, PASP, 100, 545

Rees, R. 1993, in The Globular Cluster Galaxy Connection (ASPCS 48), eds. G. H. Smith & J. P. Brodie (San Francisco: ASP), p. 104

Renzini, A., Bragaglia, A., Ferraro, F. R., Gilmozzi, R., Ortolani, S., Holberg, J. B., Liebert, J., Wesemael, F., & Bohlin, R. C. 1996, ApJ, 465, L23

Reid, I. N. 1997, AJ, 114, 161

Reid, I. N. 1998, AJ, 115, 204

Rich, R. M. 1996, in The Galactic Halo .... Inside and Out (ASPCS 92), eds. H. Morrison & A. Sarajedini (San Francisco: ASP), p. 24.

Rich R. M., Mighell, K. J., & Neill, J. D. 1996, in The Galactic Halo .... Inside and Out (ASPCS 92), eds. H. Morrison & A. Sarajedini (San Francisco: ASP), p. 544

Rich R. M., Sosin, C., Djorgovski, S. G., Piotto, G., King, I. R., Renzini, A., Phinney, E. S., Dorman B., Liebert, J., & Meylan, G. 1997, ApJ, 484, L25

Rich, R. M., Ortolani, S., Bica, E., & Barbuy, B. 1998, AJ, 116, 1295

Roman, N. 1950, ApJ, 112, 554

Romani, R. W., and Weinberg, M. D. 1991, ApJ, 372, 487

Rubenstein, E. P., & Bailyn, C. D. 1997, ApJ, 474, 701

Sandage A. 1954, AJ, 59, 162

Soderblom, D. R., King, J. R., Hanson, R. B., Jones, B. F., Fischer, D., Stauffer, J. R., & Pinsonneault, M. H. 1998, ApJ, 504, 192

Sosin, C., Dorman B., Djorgovski, S. G., Piotto, G., Rich R. M., King, I. R., Liebert, J., Phinney, E. S., & Renzini, A. 1997, ApJ, 480, L35

Stetson, P. B. 1987, PASP, 99, 191

Yanny, B., Guhathakurta, P., Schneider, D. P., & Bahcall, J. N. 1994, ApJ, 435, L59

Wheeler, J. C., Sneden, C., & Truran, J. W. 1989, ARAA, 27, 279

Zinn, R. 1980, ApJS, 42, 19

STEVEN R. MAJEWSKI, a native of Chicago, received a BA with honors from Northwestern University in 1983, with majors in physics, mathematics, and integrated science. Since his graduate work at the University of Chicago's Yerkes Observatory, from which he received his Ph.D. in 1991, his research has concentrated on the evolution of galaxies and stellar populations, both from the perspective of studying extragalactic systems to high redshifts as well as through detailed study of the spatial, kinematical and abundance distributions of populations in the Milky Way and its satellite system.

In 1990 Majewski began postgraduate work, first as a Carnegie Fellow and then as a Hubble Fellow, at the Carnegie Observatories in Pasadena, CA. He is grateful to maintain connection with the Observatories as a Visiting Associate.

Since 1995, he has been on the astronomy faculty at the University of Virginia where his research group focuses on astrometry, photometry and spectroscopy of stellar systems in the Milky Way. One goal of the research is to understand the history of satellite mergers with the Milky Way. Soon the Virginia group hopes to turn some attention to preparatory science for the Space Interferometry Mission, which will be launched into orbit in 2005. In 1997 Majewski was awarded a David and Lucile Packard Foundation Fellowship and a National Science Foundation Career Award.

Stellar Populations and the Formation of the Milky Way

By STEVEN R. MAJEWSKI[1]

Department of Astronomy, University of Virginia, Charlottesville, VA 22903-0818, USA

[1]David and Lucile Packard Foundation Fellow; Cottrell Scholar of The Research Corporation

I review topics in the area of Galactic stellar populations and the formation and evolution of the Milky Way, with particular emphasis on the role of globular clusters in tracing stellar populations and unraveling the Galactic history. While clusters provide a means of determining some global properties of stellar populations, our understanding of stellar populations serves also to guide us to scenarios for the origin and evolution of the family of Milky Way globular clusters.

1. Introduction

1.1. Stellar Populations – General Concepts

The goal of stellar population studies is to understand the formation and evolution of galaxies through investigation of the detailed distributions of properties – including, but not limited to, stellar types, kinematics, chemical abundances, ages, and spatial distributions – of its constituent luminous parts. Inherent to this endeavor is the notion that stars in galaxies can be categorized into groups – *populations* with shared, or well-defined distributions of, properties. Since stars generally form in associations and clusters†, these families of sibling stars may represent the smallest viable population "unit" – the "simple stellar population" ("SSP"; Renzini & Buzzoni 1986) – of coeval, initially chemically homogeneous stars on similar orbits through a galaxy. In the context of this discussion, then, a galaxy is made up of populations, each which is assembled from particular combinations of SSPs:

$$GALAXY = \sum_{i=1}^{m} a_i(POPULATION_i)$$

$$POPULATION_i = \sum_{j=1}^{n_m} b_j(SSP_j).$$

The larger the relative sizes of homogenized SSPs, and the fewer the number of them that constitute a galaxy (i.e., the smaller are m and n_m), the easier it should be to unravel its history. In the other less tractable extreme, the SSPs might be extraordinarily small and multitudinous, at the level of individually forming stars.

In principle, a galaxy can be broken down into an expansion of smaller population units, and, by retrieving an age-dated collection of components of this expansion, one can put together a timeline of formation. More often than not, identification of *individual* SSPs is difficult in complex galaxies. In this circumstance, the hope of "stellar population" studies is that the SSPs have been strung together in somewhat simple patterns constituting what I will call *principal component* stellar populations ‡ (Figure 1). In this

---

† Although this may not be true for the first stars that may have formed in the early universe, the so-called "Population III".

‡ Use of this terminology here is intended as an allusion to principal component analysis.

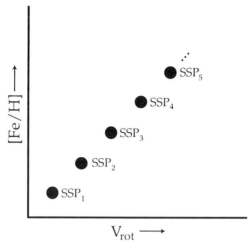

FIGURE 1. Construction of a fictional, star forming "disk stellar population" out of a series of steadily metal-enriching SSP "bursts" of increasing rotational velocity about the galactic center.

way, the problem is reduced to the (hopefully simpler) task of identifying these *patterns* characteristic of each principal population. Examples of principal component populations in the Milky Way might be "the disk", "the halo", or "the bulge", each constructed of smaller, simple stellar populations characteristic of specific sites and epochs of star formation (perhaps star clusters or associations).

Thus, when confronted with a galaxy having a complex mixture of stellar populations we seek correlations between various *observable* attributes, such as

- **spatial distributions**   e.g., density laws
- **kinematics**   velocities, velocity dispersions
- **chemistry**   e.g., mean [Fe/H], chemical abundance *patterns* ([O/Fe], [Ca/Fe], [Zn/Fe], ...)
- **ages**   reflected, e.g., in the types of stars seen (see Figure 2),

in order to find and characterize the principal component populations that will allow us to reconstruct a complete, physical, *galactic chemodynamical evolutionary model* of the entire system. This ultimate model of the system must necessarily incorporate all baryonic components, and for each principal stellar population might include the *model* attributes: $\mathbf{f} = \mathbf{f}(\vec{x}, \vec{v}, t)$, the evolution of the phase space distribution of stars; $\mathbf{g} = \mathbf{g}(\vec{x}, t)$, the evolution of the gas density; $\mathbf{X} = \mathbf{X}(\vec{x}, t, X_1, X_2, X_3, ...)$, the evolution of the detailed abundances of atomic species X_i in the interstellar gas out of which stars form; $\mathbf{SFR} = \mathbf{SFR}(\vec{x}, t)$, the *star formation rate*, the number of stars formed per unit time interval; and $\mathbf{IMF} = \mathbf{IMF}(\vec{x}, t)$, the instantaneous *initial mass function* which denotes how the new stars are distributed by mass. In general, these model functions are not observable, at least not completely. The physics of the system drives their evolution, and, if our model is an accurate representation of the Galaxy, the model functions, in turn, will be able to predict the proportions of observed stellar types, i.e., the a_i and b_j in the schematic equations above.

Unfortunately, determining this array of model descriptors is impossible for any galaxy. In the first place there are limitations of physics, in that some aspects of the evolution of galaxies may well be hopelessly irrecoverable, erased by "the operation of contingent processes that cannot, even in principle, be inferred from observations of its present

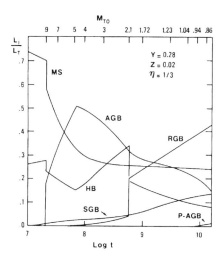

FIGURE 2. The luminosity-weighted relative contribution of various evolutionary stages to the integrated bolometric luminosity of a simple stellar population as a function of age (bottom scale) and mass at the turnoff of the main sequence (top scale). This example model is for a specific composition of helium, Y, and metals, Z. "MS" = main sequence, "SGB" = subgiant branch, "RGB" = red giant branch, "HB" = horizontal branch, "AGB" = asymptotic giant branch, and "P-AGB" = post-asymptotic branch stars. From Renzini & Buzzoni (1986).

state" (Searle 1993). For example, the processes leading to dynamical relaxation alter stochastically the kinematical attributes of stars in dense regions. Fortunately, as we shall see (section 3.4.3), at least some regions of galaxies avoid this regime.

Secondly, technological limitations in the observer's domain prevent us from measuring detailed distributive properties of stellar systems in external galaxies, though great strides are being made in this area for Local Group galaxies. Whereas previously our knowledge of stellar populations in galaxies was limited to that which could be ascertained from their integrated light, now color-magnitude diagrams ("CMDs") of resolved stars on images from the Hubble Space Telescope yield detailed mappings of the SFR histories of our neighbors (see below). This remarkable advance is a product of the unique time stamps afforded by the orderly progression of stellar evolution phases in SSPs. These various phases separate into branches of the CMD (see the discussion of stellar evolution and the branches of the CMD by Castellani and by King, this volume), and the presence or lack of representative stellar evolution phases, or the relative numbers of stars between phases, may be exploited to age date simple population systems (Figure 2). Because the age of a stellar population is tied directly to the mass of stars evolving off the main sequence, particularly useful are stars in stellar evolutionary phases specific to a narrow stellar mass range. An example of a logic chart for time stamping a stellar population that utilizes these well-defined age indicators is shown in Figure 3. Intrinsically bright age indicators are, of course, especially useful for the study of more distant galaxies.

By virtue of the Vogt-Russell theorem (Vogt 1926, Russell et al. 1927), which states that the mass and chemical composition of a star are sufficient to define uniquely its structure (which in turn determines, albeit not necessarily uniquely, its position in the CMD) at any given age, the shape of the CMD for an ensemble of stars in an SSP is a function of both its age and chemical composition (with relative densities modulated by the initial mass function). The metallicity/age dependence of the CMD for the SSPs represented by globular clusters is discussed by Castellani and by King (this volume).

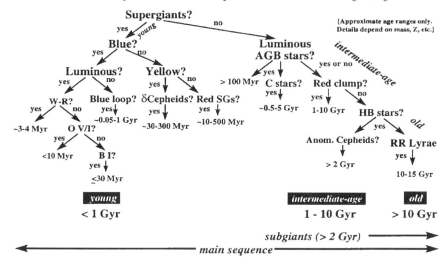

FIGURE 3. Age indicator logic chart for stellar populations. The age resolution is greatest for stellar evolutionary phases corresponding to the more quickly evolving high mass stars. From Grebel (1998).

While some combinations of age/abundance do yield similar morphologies in the CMD†, the CMD is a powerful tool for ascertaining information on *both* the age *and* mean abundance of an SSP.

Clearly, the CMDs of more complex systems consisting of multiple stellar populations will be commensurately more complicated. CMDs for simple agglomerations of SSPs, as are now being found in some of the dwarf galaxies of the Local Group, can often be well represented by a superposition of a small number of single age/abundance CMDs. A good example is the Carina dwarf galaxy, a satellite of the Milky Way that appears to have formed stars in three bursts, each which leaves a characteristic hallmark in the galaxy's CMD (Figure 4). From deconstruction of the CMDs of complex systems, we may begin to connect some of the chemical and star formation history of these systems.

A convenient way to visualize these connections is by way of the "Hodge population box" (Hodge 1989), a three dimensional representation with the axes of age, mean metal abundance ([Fe/H])‡, and star formation rate. The Hodge population box for a rather simple system like Carina, would look something like Figure 5. Typical galaxies, especially more massive ones, have more complicated star formation histories that include more extended periods of star formation and enrichment. An example of a Hodge population box for a galaxy with a more complicated star formation history is shown in Figure 6. A great achievement of the last decade is that with large aperture telescopes, the precision afforded by CCD detectors, and the ability to resolve stars in nearby galaxies, approximate star formation and abundance distributions are being worked out for a

† These degeneracies can often be broken by constructing CMDs with different filter systems.
‡ In principle, equivalent representations that chart the enrichment in other chemical species could be formulated. However, it is traditional to evaluate the overall level of enrichment via the ratio of iron to hydrogen. Moreover, we are not presently able to discern much more than the *mean* level of enrichment of stars in other galaxies, although the measurement of chemical abundance *patterns* for stars in the Milky Way, as well as in intergalactic *gas clouds* (via absorption lines by the gas in the spectra of background quasars) have become cottage industries.

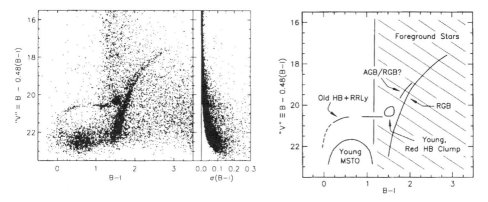

FIGURE 4. A color-magnitude diagram of the Carina dwarf galaxy, a satellite of the Milky Way that appears to have experienced three major bursts of star formation. The individual burst populations are discerned (see Figure 2) by a distinctive horizontal branch with blue/red extensions and an RR Lyrae gap, most characteristic of old, $10-15$ Gyr populations; a distinctive red clump, redward of the red HB extension for the older population, and characteristic of intermediate-aged populations, in this case approximately 7 Gyr old; and a main sequence with a rather blue turn-off indicating a rather recent (a few Gyr old) burst of star formation that still has rather massive stars in the core hydrogen burning phase. From Smecker-Hane et al. (1994).

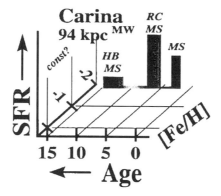

FIGURE 5. Hodge population box for the dwarf galaxy Carina. The three bursts of star formation for this galaxy seem to be of nearly the same metallicity. The types of stars used to time stamp the bursts are indicated above their peaks. From Grebel (1998).

majority of the Local Group (see summary of Hodge boxes for nearby galaxies by Grebel 1998).

Clearly, completing our chemodynamical models requires tying together the chemical enrichment history of galaxies with the dynamical evolution of the stars and gas within. One might imagine visualizing the dynamical history of a galaxy with a *dynamical population box*, such as that shown for a fictitious galaxy in Figure 7. In this case two axes of the box are given by the star formation rate and age, as before. The third axis is a measure of the relative distribution of the stars formed at an epoch in the disordered motion of velocity dispersions (σ) versus the ordered motion in rotation (V_{rot}): $\log(V_{rot}/\sigma)$†.

† It is common in Milky Way studies to refer to rotational velocities as measured with respect to a single axis of rotational symmetry in the galaxy. With a single defined axis it is possible to have populations with negative rotational velocities, when they are in retrograde motion compared to a defined axis. Because there is little evidence suggesting differing axes of symmetry for different populations in the Milky Way, and to preserve the commonly used sense of V_{rot},

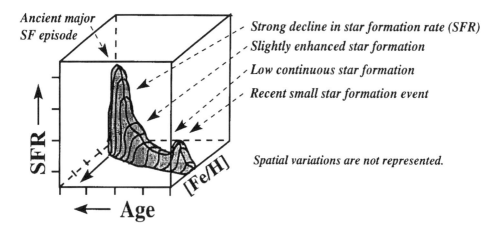

FIGURE 6. Hodge population box for a fictitious galaxy with an early, strong star forming episode, followed by an overall declining star formation rate until a recent smaller star formation event. Modified from Grebel (1998).

The quantity V_{rot}/σ can be thought of as a measure of the relative importance of rotational support to pressure support for the stellar population. As an example, the stars found in a galaxy that formed similarly to the model described by Eggen, Lynden-Bell & Sandage (1962) – with a rapid collapse of more or less randomly moving gas followed by the formation of a disk with spin-up (see Section 3.1 below) – might have a dynamical population box like that shown in Figure 7.

Unfortunately *radial velocities* for stars in external galaxies are difficult enough to obtain; the hope of measuring complete internal kinematics with *proper motions* for stars in external galaxies must await the promise of new technologies. Projects as technologically advanced as the planned the Space Interferometry Mission, which will have microarc-second positioning capability, will deliver astrometric precision to measure transverse motions of bright stars in only the very nearest galaxies.

In the Milky Way at least we have the unique opportunity to measure a full array of properties for individual stars and clusters. Thus, the Milky Way is presently the only "laboratory" galaxy where we can get detailed information on the kinematics of stellar populations to tie together with the chemical, age, and spatial properties toward our goal of synthesizing a complete chemodynamical model of formation. However, it is common to assume (and there is little reason to think otherwise) that the Milky Way is representative of galaxies of similar Hubble type. The status of our knowledge of our "laboratory" in this context is the emphasis of the remainder of this discussion. In Section 4 we conclude by attempting to assemble "population box" representations for principal component populations of the Milky Way.

Before proceeding, it is worth noting one possible ambiguity of the "population box".

as a quantity measured with respect to the rotational axis of the young Milky Way disk, I will occasionally employ the quantity $\log(|V_{rot}|/\sigma)$ here. However, in the general case when populations may not share common axial symmetries, it may be more physical to think of V_{rot} as the velocity pertaining to the angular momentum axis *for each population*. An illustration of the difference is had by imagining $\log(V_{rot}/\sigma)$ for a stellar population concentrated to an annulus on a polar orbit around the Milky Way, like what one might get from the tidal disruption of a satellite galaxy (Section 3.4.3): V_{rot} is clearly different when measured with respect to the Galactic disk and with respect to the angular momentum axis of the population itself!

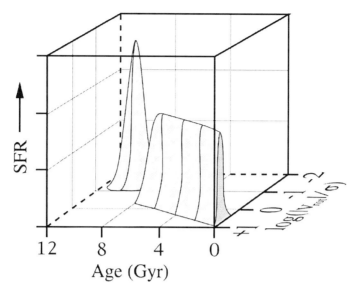

FIGURE 7. Dynamical population box for a fictitious galaxy with an early rapid (i.e., $V_{radial} \gg V_{rot}$) collapse of material (originally moving almost completely randomly) followed by the formation of a disk with increasingly rapid rotation ($V_{rot} \gg V_{radial}$) due to the conservation of angular momentum (i.e., "spin-up").

The intention is to demonstrate the properties of the star forming gas as presumably "stored" in the properties of the extant stars that we actually observe to derive these data. Unfortunately, several problems conspire to thwart a simple mapping from observed to original properties. For example, in their evolution past core hydrogen burning, stars can alter chemically their stellar atmospheres through convective mixing and dredge up of products of the nucleosynthetic yield from their fusion cores. To a lesser degree, the process of astration, the accumulation of enriched interstellar gas as a star passes through the disk (Clayton 1989) will also alter the observed abundance from the birth abundance. Similarly, a number of processes (some discussed in the next section) can alter the dynamics of stars. Generally these processes, like the scattering of disk stars off of giant molecular cloud complexes via the Spitzer-Schwarzschild (1953) mechanism, serve to increase the random motions of stars. Thus, one must always bear in mind the evolution from the *birth* values to the *observed* values when discussing a property of a stellar population, or visualizing it in a population box as we do here. In the present paper, we assume that the observed [Fe/H] distributions are relatively unchanged from the observed ones, but we do concern ourselves with possible changes between the birth dynamics and observed dynamics of stars.

1.2. *Globular Clusters as Stellar Population Tracers*

Since globular clusters are the main topic of this Winter School, it is worth emphasizing their importance to stellar population studies. Clusters, open and globular, provide us with unique tracers of stellar populations. Obviously, because of their luminosity, they are easy to find and easy to see to great distances. More importantly, perhaps, clusters seem to have been formed in relatively brief star formation bursts with a resultant yield of stars having great chemical homogeneity that testifies to efficient mixing of cluster precursor gas. This is evidenced, for example, by the remarkably small spread in the mean metal abundance of stars within each cluster, which is typically given by $\sigma([Fe/H]) < 0.10$ (see summary by Suntzeff 1993, and Gratton, this volume). This is to be contrasted with a large spread in abundances from cluster to cluster. Clusters may well be the closest thing we have to cohesive, "simple stellar populations" and may be paradigms for the building blocks out of which at least some of the stellar populations in galaxies are constructed (the possibility of specific populations formed directly from *globular* clusters is addressed below). In the Milky Way, globular clusters represent the oldest sites of star formation still cohesively bound; globular clusters therefore hold clues to the early chemodynamics of the Milky Way.

Through the color-magnitude diagram, the distances of clusters can be estimated fairly readily, while the age and mean metallicity are, in principle, ascertainable, at least in a relative sense. Moreover, through high resolution studies of the coeval stars in clusters, we are making some progress toward understanding sources of observed stellar variation in chemical abundance patterns, though it is still not clear to what degree these variations are a result of primordial inhomogeneities in the pre-cluster gas and how much they are the result of the mixing of stellar interiors during the evolution of the cluster stars (Kraft 1994, Sneden et al. 1997, Gratton, this volume). These variations are often seen among the elements of the CNO cycle, as well as other light elements like Na and Al.

Finally, compared to individual field stars at the same distance, the orbits of globular clusters (and dwarf satellite galaxies) can be determined with much better precision, since cluster mean proper motions can be determined as the *average* of a large number of member stars.

While globular clusters have played, and continue to play, a central role in stellar population studies, it is important to be cognizant of their limitations. Foremost among these shortcomings for Milky Way studies is the limited number of globular clusters. It is unlikely that the some 150 known Galactic globular clusters (Harris 1996) represent a complete census (e.g., some clusters probably live in highly dust obscured regions of the Galaxy), but it may well be close to complete. Nor is it likely that the *present* family of globulars represents the *initial* retinue of the Milky Way, as a number of processes act to destroy globular clusters (see Weinberg 1993; Elson covers these topics extensively in her contribution to this volume). The usual means is by expansion of the cluster "halo" to the point where stars on the outermost orbits are liberated from the gravitational hold of the cluster. Several processes will act in clusters regardless of outside forces:

• *Evaporation* operates as stars in the cluster interact through two body encounters. Gradually, through two body relaxation, the cluster approaches equipartition of kinetic energy and this produces a tendency toward higher velocities for the lower mass stars, which can escape if they overcome the cluster escape velocity (Ambartsumian 1938, Spitzer 1940, 1958). As low mass stars escape, the binding energy of the cluster decreases, allowing even more stars to reach the lowering escape velocity. While models of this process based on the properties of Galactic globular clusters indicate timescales for total evaporation that exceed the age of the Galaxy by a factor of $2 - 20$, the change in the cluster mass function is noticeable in a Hubble time (Johnstone 1993).

• *Core collapse* in clusters occurs because two-body relaxation preferentially removes the "hotter" stars near the core. Because the lost stars are preferentially the less massive, *mass segregation* hastens the collapse (Spitzer 1969). The core reacts to create dynamical energy to stave off further collapse, and two-body relaxation transports this energy outward, increasing again the number of stars that can achieve escape velocities. Evaporation rates in post collapse globulars may be several times greater than normal clusters (Lee & Goodman 1995). Rotation of the cluster apparently acts to enlarge the regimes of both normal and post-collapse evaporation (Longaretti & Lagoute 1997).

• *Mass loss by stellar evolution* also decreases the binding energy of the cluster, assisting with evaporation. Mass loss occurs through supernovae explosions, ejection of planetary nebulae, and stellar winds. These processes, because they are mainly associated with high mass stars, are most important for the early evolution of clusters. Energetic disruption of the cluster by supernova explosions is also possible at early times.

Each of these mechanisms for shedding stars is enhanced in the presence of the Galactic potential in that the cluster becomes "tidally truncated" at the approximate point where the Galactic force exceeds the cluster force and effectively reduces the required velocity of escape. For clusters of mass M on nearly circular orbits around the Galaxy, the tidal radius of the cluster is given by $R_t = R_G(M/3M_G)^{1/3}$, where R_G and M_G are the Galactocentric radius of the orbit and the mass of the Galaxy interior to this orbit.

The Galactic potential also works actively to destroy clusters in several ways:

• *Disk shocking* of the cluster occurs as it plunges through the disk. The transient perturbation acts to compress the cluster in the direction perpendicular to the disk, resulting in the acceleration of cluster stars. The kinematical "heating" of stars in the outer regions of the clusters from this "compressive gravitational shock" enhances the ability for some to escape (Chernoff et al. 1986, Spitzer 1987).

• *Bulge shocking* occurs when a cluster passes near the Galactic center. Both disk and bulge shocking were probably more important for the primordial cluster population because these processes selectively act on clusters with highly eccentric orbits (Aguilar et al. 1988). This also suggests the likelihood of differences in the distribution of orbital types between destroyed and extant clusters.

• *Bar shocking* may increase cluster destruction rates above those from bulge shocking. The effect, however, appears to be small and active only on clusters that pass within a few kiloparsecs of the Galactic center (Long et al. 1992).

• *Tidal gravitational shocking* is similar to the above processes, and occurs when the cluster passes another spherical mass (Spitzer 1987). The predominant affect is by passage of giant molecular cloud complexes in the Galactic disk, which, for less dense clusters (not globulars) can destroy them in one encounter. Globular clusters can be destroyed by repeated shocks from molecular cloud complexes, but this is limited to particular orbits, specifically, prograde orbits of low eccentricity and low inclination to the disk (Surdin 1997). The effect of globulars on each other is minimal (with timescales 3 − 4 orders of magnitude longer than their lifetimes), though cluster-cluster shocking may have been important for large clusters at early times (see Elson chapter).

• *Galactic tides* will strip stars that reach the cluster tidal radius with sufficient energy. The stars that come off with negative relative orbital energy with respect to the cluster should form a leading débris stream at a slightly smaller orbital radius, and those that come off with positive relative orbital energy will form a trailing débris stream at a slightly larger orbital radius. This process is seen to occur in at least some of the dwarf satellites in the Milky Way (see Section 3.4.3 below). More recently, Grillmair (1998), collaborators, and others (see summary in Grillmair 1998) have enumerated 20 Galactic and M31 globular clusters showing evidence of tidal tails.

• *Dynamical friction* occurs as a cluster continuously loses orbital energy to stars (and dark matter) in the field through which it passes (Chandrasekhar 1943a,b; see also Saslaw 1985). Gradually, the cluster spirals towards the center of the potential. Both tidal stripping and evaporation of stars from a cluster will increase as the cluster orbit decays by dynamical friction. This process would act preferentially to deplete the interior parts of galaxies of its most massive clusters (Tremaine et al. 1975).

• *Field star diffusion* through globular clusters results in the star's deceleration

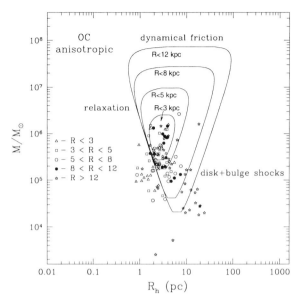

FIGURE 8. Representative "vital diagram" for the present Milky Way cluster system, from Gnedin & Ostriker (1997). The distribution of clusters is shown in the mass-radius (R_h) plane, where R_h is the half-mass radius of the cluster. The regime where various destructive processes dominate are shown outside the triangular-shaped regions, which define the region of survivability over the next Hubble time for clusters of different Galactocentric radius. A significant fraction of "lucky survivor" clusters presently reside outside their respective vitality zone; these will be destroyed in the next Hubble time.

by dynamical friction, and the loss of kinetic energy to the heating of the cluster. This process is a small effect, of a similar low order of magnitude as disk and bulge shocking. However, it would preferentially act on disk clusters, where the field star population is higher. Interestingly, this process also allows for a tangible number of *captures* of field star "immigrants" by globular clusters, which might explain the appearance of some anomalous stars observed in clusters (Peng & Weisheit 1992). Thus, even globular clusters may not be "pristine" SSPs when observed later.

Despite the numerous destructive mechanisms at work on globular clusters, the net effect has traditionally been thought to be relatively small. Aguilar et al. (1988) concluded that among the *present* set of globular clusters, evaporation dominates the destructive processes, and they derived a destruction rate of only about 4 clusters per Hubble time. But Lee & Goodman (1995) suggest that a majority of the present 30 or so postcollapse Galactic globular clusters (Djorgovski 1993) will be destroyed by enhanced evaporation in the next Hubble time, and they claim this as a conservative evaluation of the cluster destruction rate among the *present* sample. More recently, Gnedin & Ostriker (1997) reevaluated the combined destructive processes using improved codes and accounting for the fact that tidal shocking also induces an additional source of relaxation within the cluster that increases the energy dispersion among the stars (Kundić & Ostriker 1995). They determined a much larger destruction rate than Aguilar et al. for the present clusters, with more than half and as many as 90% expected to succumb to destructive forces in the next Hubble time. A summary of their results is presented in Figure 8.

Several conclusions follow from an analysis like that resulting in Figure 8. The first is that the initial globular cluster population of the Milky Way may have been quite larger than it is today. Those clusters no longer with us have dispersed their stars into the field populations of the Galaxy. A second conclusion relates to the Galactocentric radial dependency of the survivability of a cluster, whence there are particularly strong effects on the innermost clusters. The influence on inner clusters, combined with the effects of bulge and disk shocking on objects of larger mean orbital radius but highly eccentric orbits, raises important questions regarding the relationship of the presently observed clusters both to the *primordial* cluster population as well as to the field star population that the observed clusters may, or may *not*, trace. I return to the question of a contribution of destructed globular clusters to the formation of the halo in Section 3.4.3 and to the formation of the bulge in Section 3.5.

2. The Size and Shape of the Milky Way and its Stellar Populations

In my view, interpretation of the chemodynamical properties of stars and clusters in the various populations of the Milky Way is highly dependent on an understanding of the *spatial structure* of these populations. Assigning stars and clusters to one or another of the Galactic populations is made an especially troubling task because of the significant overlap in chemistry, kinematics and spatial distributions between populations. Untangling them would be aided if we could at least determine where in the Galaxy certain populations dominate.

2.1. *Assessing Spatial Distributions of Populations With Stars*

The earliest advances in understanding the "sidereal universe" were based in the science of star counting. For example, the idea that the Milky Way was a disk-like structure (as proposed by Thomas Wright and Immanual Kant in the 1700s) became evident with the basic telescopic observation (first done by Galileo) of resolving the "via lactea" into a dense sea of *stars*, much more populated than other parts of the sky. However, many ideas concerning the structure of the Galaxy were based more on mere speculation rather than hard-core data until William Herschel began a systematic approach to determining the distribution of stars. Herschel may be said to have begun the science of statistical astronomy†. By counting stars in defined areas of the sky ("star-gaging"), and assuming that stars had the same intrinsic luminosity and were distributed equally throughout the Galaxy, and moreover, assuming that his telescope could see to the edge of the Galaxy, Herschel (1785) produced a three dimensional model of the universe. Naturally, the Sun was found to be near the center of this distribution, which was shaped like a flattened (axis ratio 5:1) American football. By 1817, Herschel recognized various problems with the assumptions used to construct his model (for example, by noting himself that stars in binary systems were often of different luminosities) and concluded that (1) the number of stars in a field related not only to the size of the Galaxy in that direction but to the stellar density, and (2) even his great telescopes might not be able to "fathom the Profundity of the milky way." It became evident that constructing a model of the stellar system without knowledge of the intrinsic brightnesses (distances) of stars beset any direct approach.

As if lack of distances were not enough of a problem for gauging the Milky Way, the possibility of the absorption of starlight in space (first postulated by H.W.M. Olbers in 1823

---

† Students inclined to the history of science will find fascinating reading in E. R. Paul's book *The Milky Way Galaxy and Statistical Cosmology 1890–1924*, from which some of the following (highly selective) discussion is based. Parts of Paul's book are also presented in articles in the *Journal of the History of Astronomy*. Reid (1993) provides additional historical context.

and confirmed by Trumpler in this century) complicated matters more, although postulation of this process did provide apparent help with certain other problems. Friedrich Struve (1847), building on Herschel's work and concerned with the inconsistencies in his assumptions, applied the first statistical approaches to understanding the *density laws* of stars. In so doing, and as a means also to avoid Olbers' paradox‡, Struve became convinced that interstellar absorption did indeed occur¶.

Even with the development of extensive catalogues of hundreds of thousands of stars, like the *Bonner Durchmusterung*, and the promise of even more systematic work with *photographic* starcounts to fainter magnitudes, e.g. the *Carte du Ciel* and the *Cape Photographic Durchmusterung* of David Gill (see Reid 1993), most workers in "Galactic structure studies" at the end of last century were still relying on the traditional "direct" analyses of the data, a method introduced and developed by Struve, William Herschel and Herschel's son, John. No doubt frustrated by his own attempts over a decade and a half with this same approach, Hugo von Seeliger (steeped in a mathematical physics training at Leipzig, and heavily influenced by Neumann and Gauss) introduced a powerful new mathematical approach to starcounts around 1900. Von Seeliger's "Fundamental Equation of Stellar Statistics" got around the "vicious circle"† of the interdependence of stellar density laws and the stellar *luminosity function* – i.e. the fact that full knowledge of one requires full knowledge of the other – that had plagued all previous work. A quarter century earlier, the Swedish astronomer Gyldèn suggested that the brightnesses of stars might be distributed according to an analytical frequency function (he suggested a Gaussian form), but this idea had lain dormant (and apparently hidden from the non-Swedish community) until von Seeliger because the theory of integral equations was still so new that Gyldèn himself could not solve his integral expression! Now von Seeliger adopted the idea of a luminosity function (of unknown shape, of course, but constrained by $\int \Phi(M)dM = 1$) and combined it with the density law (also unknown, of course) $D(r)$, for an expression in terms of the observable starcounts per magnitude, $A(m)$, in a given direction of the sky:

$$A(m) = \int \Phi(M)dM \int D(r)dr$$

Several points are worth noting. First, the simple elegance of the Fundamental Equation belies the fact that it is not uniquely invertible. At some level, we remain in the "vicious circle" of before. However, von Seeliger provides us with a statistical means by which we may compare combinations of functional form for the luminosity function and density law against the observations. So, for example, if one assumes a standard form for the luminosity function, then a global form of the density law may be derived (more accurately, *tested*) by fitting counts in various directions of the sky. On the other hand, if one has a sense of the density law, then we may arrive at a functional form of the luminosity function. These examples suggest the idea of an iterative approach to the solution of von Seeliger's equation. I address this and other approaches to the problem in the next section.

Von Seeliger understood well that application of his formula was limited by the quality of the data, and he devoted the next decades to refining his solutions as new catalogues

‡ Olbers pointed out that if space were isotropically filled with stars and were also infinite, then the sky should be uniformly as bright as the surface of a star, since any line of sight should intercept a star.

¶ Alas, while absorption *is* a key factor in Galactic studies, Struve's suggestion that it was the solution to Olber's paradox was incorrect.

† Paul (1993), p.76

of data were produced. This problem was not lost on the other industrious statistical astronomer of the time, Jacobus Kapteyn, in Groningen. The director of an astronomical institution without a telescope, Kapteyn first volunteered to measure and reduce Gill's plates from the southern hemisphere *Cape Durchmusterung‡*. With a voracious appetite for data, and a knack for convincing directors of major observatories to contribute large amounts of telescope time¶ for his (aptly named) "attack" to address systematically the problem of understanding "the sidereal world", Kapteyn (1906) devised the *Plan of Selected Areas* ("SAs"). The SA's are a regularly spaced grid of survey regions around the sky. Apart from starcounts, Kapteyn was extremely motivated to create this *Plan* by his extensive efforts to understand Galactic kinematics. There ensued a period of great activity, whereby substantial amounts of effort the world over were devoted to contributing photometry, astrometry, and spectroscopy of stars in Kapteyn's 206 SAs. The grand scope and initially perceived importance of the *Plan* was such that coordination was essential, and this prompted the eventual creation of *IAU Commission 32: Selected Areas* as well as the *Subcommittee on Selected Areas of IAU Commission 33: Structure and Dynamics of the Galactic System*. Coordination of the *Plan* was also the subject of two of the earliest IAU Symposia (Nos. 1 and 7).

Since the mid-part of this century – when it was discovered that the spiral nebulae were extragalactic systems, and coincident with the rise in emphasis on star *clusters* as a tool for Galactic astronomy (see Paul 1981) – activity on the SAs has unfortunately strongly declined (IAU Commission 32 no longer exists). However, the wisdom and value of a systematic and coordinated astrometric, photometric and spectroscopic approach to studying the Milky Way is now obvious. There is growing evidence that various subsystems of the Galaxy (e.g., the bulge with its bar, the disk with its warp, and the apparently dynamically unrelaxed halo with its gaseous and stellar, tidal streams; see below) are highly asymmetric, and therefore not described adequately by global models derived from only a few lines of sight (as is common practice). Kapteyn's original vision of a fully integrated photometric, astrometric and spectroscopic survey has never been fully realized. Ironically, the decline in SA activity has overlapped with the development of modern instrumentation that might be brought to bear on the program with far more efficiency, precision and depth than Kapteyn could have imagined. Fortunately, with the development of high speed photographic plate scanners able to produce huge imaging databases like the *Digitized Sky Survey* and the APM catalogue, and astrometric databases soon to be produced by the US Naval Observatory and the Minnesota groups, as well as the imminent production of the Sloan Digital Sky Survey, aspects of the Kapteyn vision may soon be revisited with a vengeance.

2.2. *Modern Starcount Analyses and Galactic Structure*

In modern usage, the von Seeliger equation is given by

$$A(m, S) = \sum_i A_i(m, S) = \Omega \sum_i \int \Phi_i(M, S) D_i(\vec{r}) r^2 dr,$$

where $A(m, S)$ refers to the differential counts of stars of type S, Ω is the area of the sky surveyed in steradians, $D_i(\vec{r})$ is the density law as a function of position in the Galaxy, and $\Phi_i(M, S)$ is the luminosity function, in this case presented as the relative number of stars per magnitude of type S. As usual, the absolute magnitude of the star, M, is

---

‡ Today an astronomer (especially one so highly placed in the astronomical community) making such an offer would make a prized collaborator!

¶ "Kapteyn presented the unique figure of an astronomer without a telescope. More accurately, all the telescopes of the world were his." Frederick Seares (1922).

related to the apparent magnitude of the star via

$$m = 5\log(r) + M - 5 + a(\vec{r}),$$

which makes the integral over r calculable. The function $a(\vec{r})$ accounts for the three dimensional distribution of absorbing material along the line of sight†. Because we recognize that the Galaxy consists of separate principal populations (disk, halo, etc.), which may (or may not) be discrete and follow their own density laws, the starcounts are given as the sum of i individual populations.

The introduction of the stellar type, S, is meant to address the problem of degeneracy of contributors to a given apparent magnitude bin (e.g., a nearby, faint star and a distant, but bright star). That is, it is a means by which to simplify the convolution of luminosity function and density law that makes the equation uninvertible. As demonstrated by the work last century, the starcounts problem becomes intractable if you do not have some assessment of the absolute magnitude, and therefore distance, of each star. Alternatively, if one knew in advance that stars of a certain *type* (however defined) had the same magnitude, and one could identify stars of that type *a priori*, then the problem simplifies to the star-gauging of the past. For example, if one only sought stars of spectral type G2V (stars like the Sun) then one could obtain the density law for that stellar type readily through measuring distances by apparent magnitudes relative to the Sun (*sans* the effects of absorption). Clearly, the narrower the definition of S, the more accurate the solution becomes. In practice, spectral types would not be easy to obtain for lots of stars. But fortunately, unlike our predecessors of last century, we can gain ready access to *some* information about stars to practical magnitude limits that allow us to improve our approach to the luminosity function portion of the von Seeliger integral. For example, with a few photographic exposures, one might identify certain classes of variable stars, like RR Lyrae, that have more narrowly limited mean magnitudes.

More commonly, starcounts are measured in pairs of filters, so that S may be restricted to stars of certain colors. An example of such a data set is the field star CMD shown in Figure 9 (left panel). Note that restricting starcounts to stars of specific colors still yields degeneracies in absolute magnitudes, but in general, these degeneracies are bimodal at most colors. For example, at $B - V = 1.0$, the color of K type stars, we will see contribution from K giants ($M_V \sim 0$) and K dwarfs ($M_V \sim 6$). Depending on the density law and its interplay with the growing volume element along the line of sight, it is often the case that the contribution from one or the other of these luminosity classes will dominate at certain apparent magnitudes.

Thus, a common approach to the starcount problem is the following:

a) Measure starcounts $A(m, color)$.

b) Assume a given luminosity class for stars of this color, which then narrows considerably the range of absolute magnitudes for the stars. For example, when looking at high Galactic latitudes and faint ($V \gtrsim 17$) magnitudes, it is often safe to assume that one is dealing entirely with dwarf stars. This particular assumption works best for red stars, where the absolute magnitude separation of dwarfs and giants is greatest‡, but it becomes unreliable for colors near the main sequence turn off.

c) From an established form of the color-magnitude relation for the appropriate lumi-

---

† An often used mapping of reddening is that given by Burstein & Heiles (1978, 1982), but improved, higher resolution maps based on COBE/DIRBE and IRAS/SISSA far infrared data have been derived by Schlegel et al. (1998).

‡ If the density law for the population from which the giants are seen falls faster than the volume element grows (i.e., as r^3), as might be the case for the halo, then the counts for the giants falls off. At faint enough magnitudes, presumably you run out of Galaxy.

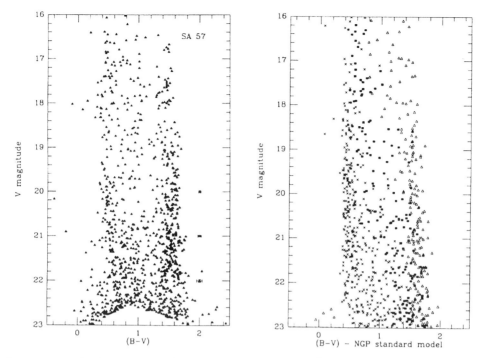

FIGURE 9. *Left*: Counts in the $(V, B - V)$ plane for 0.3 deg^2 at the North Galactic Pole (SA57) from the photographic survey of Majewski (1992). The field star CMD shows a characteristic two-ridge feature (first discussed by Kron [1980]) in this particular color plane. The left ridge is caused by the build up of stars at the color of the main sequence turnoff for populations of the age of the upper disk and halo. The right ridge is cause by the "saturation" of $B - V$ colors for late type stellar spectra due to the growth of TiO absorption in both the B and V bands. *Right*: Starcount model fit to the data on the left. Crosses are for model halo stars, squares are for IPII stars, and triangles are for thin disk stars. From Reid & Majewski (1993).

nosity class, assign an absolute magnitude and, from the apparent magnitude for each star, derive a photometric parallax, $r(m, M)$.

d) Fit model functions to the count density, $A(r)$.

This approach is particularly well suited for study of the disk because of its expected primary dependence on height from the Galactic plane (other populations of the Milky Way show a stronger dependence on the radius from the Galactic center). An application of the technique is shown in Figure 10.

As noted in the first careful studies with this approach in the 1980's (Brooks 1981, Yoshii 1982, Gilmore & Reid 1983), fitting the disk density law with an analytical function requires at least two components, to account for an apparent break around 1 kpc from the Galactic plane. The expected "old disk" or "thin disk" population dominates the first kiloparsec or so. Beyond this dominates the Intermediate Population II, a population first formerly characterized at the historic 1957 Vatican Conference on Stellar Populations (O' Connell 1958). It is common to refer to this component as the "thick disk" after the parlance of Gilmore & Reid, who noted similarities of this Galactic population to thick disks seen in some edge-on spiral galaxies a few years earlier (Burstein 1979, Tsikoudi 1979).

It is convenient to fit these two apparent populations with exponentials, such that

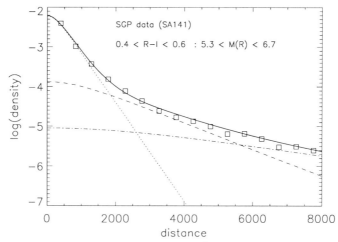

FIGURE 10. Density distribution for stars in 1.85 deg^2 towards the south Galactic Pole field SA141 from the author's CCD starcount collaboration with I.N. Reid, M. Siegel & I. Thompson. Distances are derived from $(R - I)$ colors, which are chosen because of the minimal metallicity effects in the $M_I(R - I)$ relation. These colors are also ideal for sensitivity to late type stars (although bluer stars are shown in the example here as they probe to greater distances). The fits shown are for an old, thin disk of scaleheight 350 pc (dotted line), an Intermediate Population II/thick disk with scaleheight 1300 pc and 2.2% relative normalization to the thin disk (dashed line), and a halo, in this case modeled with an exponential with scaleheight 3500 pc and 0.15% relative normalization (dot-dash line). The solid line is the sum of the models.

$$A(Z)/A(0) = \rho_{\mathrm{IPII}}e^{-Z/h_{\mathrm{IPII}}} + (1 - \rho_{\mathrm{IPII}})e^{-Z/h_{\mathrm{thin}}}$$

where $A(0)$ is the number of stars seen locally, h_{IPII} and h_{thin} are the exponential scale-height of the IPII and thin disks respectively, and ρ_{IPII} is the local fractional contribution of IPII stars to the old thin disk stars. Note that while exponentials are convenient functions, they are not necessarily physical. Increasingly, researchers are turning to $sech^2Z$ functional forms, which (1) are not singular at $Z = 0$, (2) have some physical motivation from dynamical analyses of isothermal disks, and (3) approach the functional form of exponentials at large Z.

Several words of caution are in order. The first relates to the assumption that the counts are not affected by contamination from stars of other luminosity classes. Indeed, the question of contamination by giants was one of the early main criticisms of the photometric parallax count studies that required an extra component, when good fits to starcount data using another approach (discussed below) were found to be adequate without this extra component (Bahcall & Soneira 1980; Figure 11). While the veracity of the two component fit has now been borne out (cf. Casertano et al. 1990), ultimately, contamination by giants *does* have some effect that, while small in many cases, ought to be properly assessed in all starcount attempts. Second, metallicity variations certainly affect color-magnitude relations, but, it is not typical to have abundance information on a star by star basis. In this case, metallicity corrections may only be applied in a statistical way. Typically, this means assuming a $d[\mathrm{Fe/H}]/dZ$ dependence, which then implies a dM/dZ dependence. However, this is not a worry-free procedure when attempting to find $A(Z)$, because the scale of the abundance gradient determines the degree of Z-compression in stellar densities (as stars are made increasingly subluminous, their derived photometric parallaxes decrease). Alternatively, one may attempt to do starcounts in a bandpass system that minimizes metallicity effects in the color-magnitude relation (as is done in Figure 10).

Third, the discussion to this point has completely ignored density (and other) variations as a function of other spatial dimensions of the Galaxy. In more general studies, it is common to assume that the disk components have exponential dependences in the Galactocentric radial direction as well. In addition, we have completely left out the important consideration one must give to the Malmquist bias, an effect of the intrinsic absolute magnitude spreads we will encounter in our starcount sample, however restricted in stellar type we attempt to make our sample. A thorough description of the bias is beyond the scope of this lecture, but is discussed in the lectures by Sandage (1995). In brief, if $A(m)$ is increasing, then the Malmquist bias says that the mean absolute magnitude of stars in our sample will be systematically higher than one would expect in a volume limited sample. This occurs because at a given apparent magnitude the volume element occupied by intrinsically brighter stars contributing at that apparent magnitude is larger than the volume element for intrinsically fainter stars contributing at that apparent magnitude. Thus, more intrinsically bright stars can find their way into a specific $A(m)$ bin than can intrinsically fainter stars.

Finally, the overall problem of contamination of star counts by *extragalactic* objects (QSOs and compact galaxies) becomes a problem at faint magnitudes (see Reid & Majewski 1993).

A second approach to the starcounts problem is to use computers to integrate the von Seeliger equation directly (after assuming functional forms for the luminosity function, density law, and color-magnitude relation for each expected stellar population) and then numerically generate model $A(m, S)$ to compare to data (Bahcall & Soneira 1980, Reid & Majewski 1993). While it is possible to construct luminosity functions of disk stars from analysis of the solar neighborhood, the lower local density of individual stars from other Galactic components make generation of an observed luminosity function much more problematical. However, globular clusters are especially convenient to determining metal-poor luminosity function templates since, with all stars at the same distance, one need only count the numbers of stars as a function of apparent magnitude to get $\Phi(M)$, once the cluster distance is known (see the discussion of cluster luminosity functions by King in this volume). Because of the problem of the degeneracy of the $M(color)$ relation across luminosity classes, it is more physical to replace a single $M(color)$ relation used in combination with $\Phi(M)$, and instead employ a more sophisticated $\Phi(M, color)$ array, called a *Hess diagram* (see models by Robin & Crézé 1986, Méndez 1995, for example). In general, the Hess diagram varies greatly with age and abundance (and possibly the $\vec{r}$ distribution) of the stellar populations. The results of a computer modeling approach to the starcount problem are shown in Figure 9 (right panel).

The lack of a systematic starcount attack along many lines of sight, as envisioned and attempted by Kapteyn and von Seeliger, has not prevented a number of researchers (including the author!) from declaring "best" model parameters for the various Galactic components. Table 1 gives the parameters from a sampling of many published starcount analyses, to give the flavor of the general form and ranges of parameters adopted for density laws, scaleheights, etc. The density law typically adopted for the Galactic halo is based on observations of the spheroids of external galaxies, which seem to show surface brightnesses, $I(R)$ (in solar luminosities pc^{-2}), that fall off as a "de Vaucouleurs (1948) $R^{1/4}$ law":

$$I(R) = I_e 10^{-3.33[(r/r_e)^{1/4}-1]} = I_e \exp\left\{-7.669[(r/r_e)^{1/4}-1]\right\}$$

where r_e is the *effective radius*, or the radius that encloses one half of the spheroidal

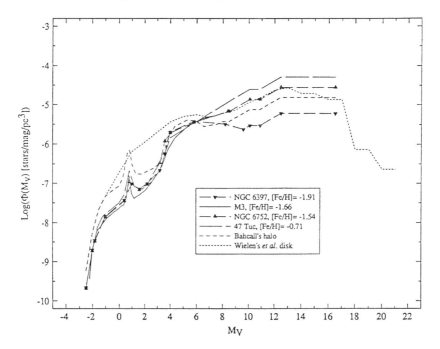

FIGURE 11. Luminosity functions for clusters of various metallicities. M3 is often adopted as representative of the Galactic halo, while the luminosity function for the intermediate metallicity cluster 47 Tucanae is often adopted for the Intermediate Population II thick disk. The local peak near $M_V = 1$ is due to the horizontal branch. Also shown is the Wielen et al. (1983) luminosity function for disk stars, and the halo luminosity function adopted by Bahcall (1986) and collaborators. Note that the larger number of giants in the latter luminosity function compared to, say, that of the M3 function shown, will imply a larger contribution of halo starcounts at magnitudes and colors normally associated with the Intermediate Population II; thus Bahcall's model with this halo luminosity function gave reasonable fits to the data without inclusion of an IPII component. From Méndez (1995). See the model of Méndez & van Altena (1996).

light, and I_e is the brightness at r_e. This equation is derived from noting that for most spheroids, a plot of surface brightness (in mag arcsec^{-2}) versus $R^{1/4}$ is fairly linear. Note that with this profile, the peak surface brightness is given by

$$I(0) = 10^{3.33} I_e \sim 2000 I_e.$$

Applying this two-dimensional *brightness law* of external spheroids as projected on the sky to a three-dimensional *density distribution* useful for studies in our own galaxy requires the Young (1976) deprojection:

$$\rho(R) = \rho_o \exp\left[-7.669(R/R_e)^{1/4}\right]/(R/R_e)^{0.875}.$$

All studies presented in Table 1 use the latter density law for the halo, and universally accept an effective radius $R_e = 2700$ pc. The *local normalization* of the halo with respect to the thin disk, ρ_o, is typically found to be around 1 star out of 700 in the solar neighborhood.

An alternative density law often used for spheroidal components is the *power law* relation, given by

MODEL	HALO $\rho_0(\%)$	HALO c/a	IPII DISK $\rho_0(\%)$	IPII DISK h_Z	IPII DISK CMD	THIN DISK $h_Z(1)$	THIN DISK $h_Z(2)$	COMMENTS
a......	0.15	0.85	2	1200	47 Tuc	325	325	"Standard"
b......	0.025	0.9	2	1200	47 Tuc	325	325	Hartwick (1987)
	0.27	0.5						
c......	0.15	0.8	4	1000	47 Tuc	249	249	Kuijken & Gilmore (1989)
d......	0.15	0.8	11	940	OD	270	270	Sandage (1987)
e......	0.15	0.8	2.5	1400	Mid	325	400	Reid & Majewski (1993)
f.......	0.15	0.8	2.5	1400	Mid	325	400	Sommer-Larsen & Zhen
	0.09	0.4						(1990)

TABLE 1. A sampling of starcount models from the literature, showing the relative density normalization locally and axial ratio of the halo; the local normalization, scaleheight and color-magnitude diagram (where "47 Tuc" is from the cluster, "Mid" is an intermediate aged CMD, and "OD" is an old disk CMD) used for the IPII thick disk; and the scaleheight of the thin disk broken into bins by luminosity, as $M_V < +9$ and $M_V \geq +9$, respectively, to account for likely differences in mean age. The "standard" model is based on a review of literature by Reid & Majewski (1993).

$$\rho(R) = \rho_o \frac{(a_o^n + R_o^n)}{(a_o^n + R^n)}$$

where R_o is the Galactocentric radius of the Sun, a_o is a *core radius* (typically about 1 kpc for the halo), and ρ_o is as before. The power, n, of the power law is usually found to fit well the distribution of halo tracers, like RR Lyrae stars and blue horizontal branch stars, when $3 < n < 4$. For the inner bulge (< 1 kpc), a power of something like $n \sim 1.8$ seems to apply.

Whether a power law or de Vaucouleurs law is adopted, it is common to account for the fact that the halo of the Galaxy (and other spheroids) is not perfectly spherical, but show some flattening with minor to major axis ratio (c/a). Thus, a correction to the density law is applied in the Z-direction as

$$Z \longrightarrow (c/a)Z.$$

Typical flattenings for the halo are found to be something like $(c/a) \sim 0.8$.

The examples of starcount models shown in Table 1 show relatively good agreement, but this is partly a result of some data sets being in common. It is important to note that several studies (two of them are shown in Table 1) indicate the need for *two* separate halo components. These *dual halo* models are discussed at greater length in Section 3.4.2.

Unfortunately, the general agreement of the starcount models breaks down with consideration of the disk. A contentious issue is the density law of the IPII, with a variety of relative normalizations and exponential scaleheights derived using a large range of possible tracers and starcount analyses (Figure 12). Part of the discrepancy derives from the problem that the IPII size is intermediate between that of the thin disk and the halo: While the thin disk and halo dominate the starcounts at bright and faint magnitudes, respectively, the IPII-dominant regime overlaps considerably with each of these two other populations, and this makes it difficult to "extract" from the mix (see model example in Figure 9). Moreover, fitting the exponential density law described above to stars predominantly in the Z-distance regime that the IPII dominates (wherever that may be) apparently yields some amount of degeneracy, in that the values of ρ_{IPII} and

$h_{\rm IPII}$ tend to play off of each other. (This seems especially a problem when starcounts have a rather bright magnitude limit that restricts the range of accessible extent of IPII dominance.) Figure 12 demonstrates the range of scaleheights and normalizations derived for the IPII from (top panel) starcount models and photometric starcount analyses, and (bottom panel) various objects thought to trace the IPII population. Also shown are lines indicating the surface mass density (mass per pc^2) of the IPII (as projected on the Galactic plane) relative to that of the thin disk, as a function of the density law parameters. As can be seen, even though the derived normalization and scaleheight ranges are large (an order of magnitude for the normalization), the derived mass density of the IPII is rather more constrained to be within $5 - 20\%$ that of the thin disk, and generally clustered around 10%. In summary, while there is agreement that by surface mass the IPII clearly represents a significant component of the Galaxy (by comparison, the mass of the halo is another factor of 10 less), it is not clear how this mass is distributed (how "thick" *is* the thick disk?).

In my opinion, the importance of solving the problem of how the IPII mass is distributed cannot be understated, especially when the appreciable kinematical and chemical overlap of the IPII with other populations is considered. Figure 13, which shows the relative fraction and actual count contribution of the IPII as a function of height above the Galactic plane for a variety of IPII density laws against constant halo and thin disk density distributions, demonstrates the essence of the problem. If one wants to select stars to characterize the IPII for, say, an assessment of its chemical and kinematical properties, from where does one take the representative sample? Without knowledge of the density law, it is not clear at what heights the IPII dominates. Just as troubling is the problem that for practically none of the examples shown in Figure 13 does the IPII suffer minimal contamination of stars from either the thin disk or halo (in contrast to the situation for either of the latter two components). Unless one is able to find some tracer object that is assuredly only found in the IPII, one is forced to make certain assumptions about the properties of the IPII (and the other populations) in order to sort its stars out. A certain amount of circularity then follows: E.g., if one sorts stars on the basis of abundance, then one may not address abundance questions but might be able to address kinematics, as long as there are no correlations between abundance and kinematics. However, it is often the case that such correlations *do* exist. Moreover, the distributions of abundance and kinematics of the halo and disk are known to have large dispersions that likely overlap the properties of the IPII significantly.

Figure 13 also illustrates how easy it is to generate artificial gradients in properties when they may not exist. Imagine, for example, single-valued parameters (e.g., [Fe/H], rotational velocity) attributed to each the thin disk, IPII and halo. Without knowing *a priori* the population assignments of any particular star, a tally of the mean *observed* value of this parameter as a function of height above the Galactic plane will yield relatively smooth gradients. A more complex, and even more difficult to deconvolve, example would be if some or all of the populations *themselves* showed gradients as a function of Z. Finally, with the ability to generate parameter trends and gradients with the combinations of homogeneous populations (see Section 1.1), how can one be assured of even the *number* of discrete components that might be required (cf. Lindblad 1927)? The issue of a discrete versus continuous disk/IPII is addressed again in Section 3.3.3.

Thus we find ourselves in a highly disagreeable situation. Separation of populations in the range $1 < Z < 10$ kpc is generally an extremely difficult, and risky, proposition. It is common practice to make certain assumptions about the stars in populations that overlap in order to ascertain certain other properties; but it is appropriate always to bear in mind that in science it is often the case that initial predilections drive an experiment

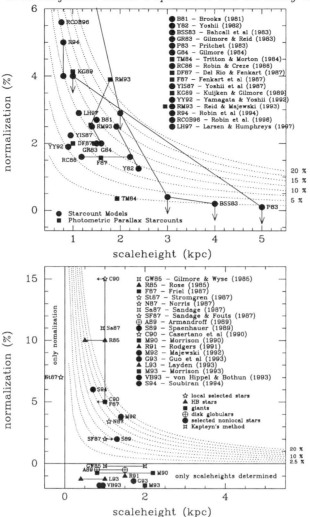

FIGURE 12. Intermediate Population II/thick disk scaleheights and local normalizations as a fraction of the thin disk population, as derived from a number of starcount analyses (*top panel*) and for various types of tracer objects (*bottom panel*). Dotted lines show loci of IPII mass density relative to the thin disk, when a scaleheight of 325 pc is adopted for the latter.

to reinforce those predilections, whether they are correct or not. This conundrum has a certain resonance with the state of the starcounting endeavor before von Seeliger's great insight and statistical prowess were brought to bear on that problem. Overcoming the present impasse may require a similar advance in the level of statistical sophistication. Progress in this direction may well be in the direction of multidimensional, univariate mixture models discussed by Nemec & Nemec (1991, 1993) and others.

2.3. *The Role of Globular Clusters in Assessing Spatial Structure*

One of the first great successes in the use of *globular clusters* for stellar population studies was as a probe of the size and shape of the Milky Way by Harlow Shapley. Shapley's work altered the Kapteyn and von Seeliger starcount view of the "universe" which, in spite of its great successes, still found the Sun to be near the center of the stars in the Milky Way system. Shapley, like John Herschel before him, noted an excess of clusters toward

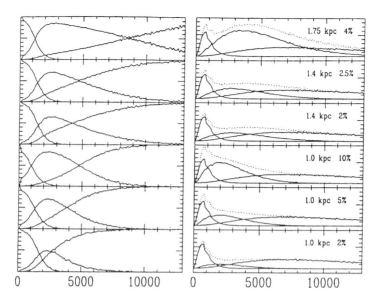

FIGURE 13. Relative fraction (*left*) and relative number (*right*) of stars as a function of Z (in kpc) contributed by the Intermediate Population II/thick disk for various scaleheights and local normalization combinations. In all cases the density laws of the halo (right-most peaking solid line) and thin disk (left-most peaking solid line) were kept the same. In the right panels, the dotted line shows the sum of the starcounts for all three populations.

the direction of the sky near Sagittarius. With the assumption that the globular clusters traced the true shape and extent of the Galaxy, Shapley suggested that the center of the Galaxy was in the direction of the apparent center of the globular cluster system (Figure 14).

Globular clusters were also central to what has now become known as the "Great Debate" between Heber Curtis and Harlow Shapley before the National Academy of Sciences (Washington, D.C.) in 1920 (Curtis 1921, Shapley 1921). The center of the debate, "The Scale of the Universe", turned in large measure on the question of the distances to globular clusters, as each protagonist identified the globular cluster system as a part of the Milky Way with remote members that defined its extent.

The outcome of the debate is commonly synopsized as "Curtis was right for the wrong reasons and Shapley was wrong for the right reasons". Shapley's position was that the Milky Way's diameter was likely to be at least 300,000 light years (90 kpcs), and that the Sun was some 50,000 light years (15 kpc) from the center (modern studies are obtaining a solar Galactocentric distance half of this; cf. Olling & Merrifield 1998). At the time, this was considered an immensely vast scale, and it seemed likely that the resultant volume defined by such a Galaxy contained all objects in the observed universe. It should be noted that the debate predated the revolution in understanding of "spiral nebulae" brought on in the next decade by groundbreaking work in the decade before. This included Slipher's work (beginning in 1912, cf. Slipher 1913) at Lowell Observatory to measure the redshifts of the spiral nebulae and Hubble's discovery of Cepheid variables in them at Mt. Wilson. Indeed, van Maanen's (1916, *et seq.*) claims for astrometrically measurable *rotations* in spiral nebulae (later shown by Hubble [1935] to be artifacts of systematic error) greatly influenced many astronomers, including Shapley, in their belief that the spiral nebulae were much more local, star forming clouds, rather than "island

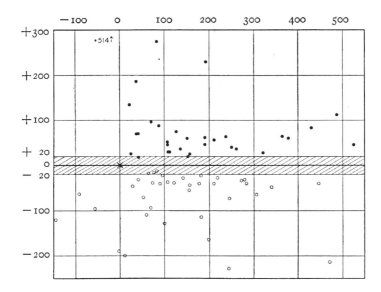

FIGURE 14. Shapley, working at the Mt. Wilson Observatory, was involved in a survey to estimate distances to globular clusters. These results, and those of others, he compiled into this representation of the cluster system projected on the XZ-plane, where X is the line containing the solar neighborhood and the Galactic center, and Z is perpendicular to the Galactic plane. From this distribution, he estimated both the size of the Milky Way and the distance of the Sun from the Galactic center. The units shown are 100 pc, as measured from the Sun (location of the "×"). From Shapley (1918).

universes" like the Milky Way, as postulated (with little physical foundation) long before by Thomas Wright, Heinrich Lambert and Immanual Kant. As one line of support that his measurements of rotation in the spiral nebulae were real, van Maanen pointed out the contrast of the magnitude of his spiral arm proper motions to his finding (van Maanen 1925, 1927) of minute internal motions in globular clusters!†

In hindsight, Shapley's major "failing" in the debate was in assuming (quite understandably, even as admitted by Curtis, who was generally more cautious) that the Cepheids and blue stars in clusters were similar to those found near the Sun; thus Shapley overestimated distances in his cluster system. The problem, of course, is that he confused much more luminous, Population I, main sequence B stars with fainter horizontal branch stars. Moreover, in the case of the Cepheids, Shapley could hardly have known (although he acknowledged the possibility and Curtis certainly highlighted it) that Population I Cepheids (the *classical Cepheids*, which are high mass supergiant stars) near the Sun, and Population II Cepheids (low mass, metal-poor *W Virginis stars*), the type found in globular clusters, follow different period-luminosity relations. The W Virginis stars are

† To be fair, van Maanen's claims for rotational motion in spiral nebulae were not inconsistent with a number of related findings (including Slipher's [1914] and Wolf's [1914] findings of internal motions via *radial velocities*) and theories by predecessors, nor even similar measurements by other astronomers in what was a rather active field of endeavor (see Berendzen & Hart 1973, Heatherington 1975). However, van Maanen, by his reputation as a meticulous observer, may well have been the most trusted of the astrometrists working on the problem. He was at least the most vocal and most published in this research field. Surprisingly, Curtis' (1915) own work in this area on some of the *same* plate material yielded no detectable motions in the spiral nebulae, but Curtis' results seem largely to have been ignored.

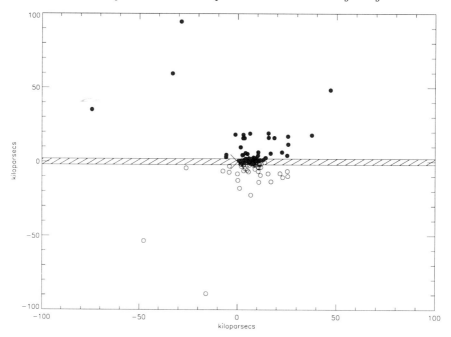

FIGURE 15. Distribution of the presently known globular cluster system in the XZ-plane, centered on the Sun. Compare to Shapley's distribution in the previous Figure. Note the number of clusters in the Galactic plane not included by Shapley and the rounder shape of the inner cluster group. Data from Harris (1996).

fainter than the classical Cepheid prototypes adopted as standard candles by Shapley. Thus, while Shapley's ultimate estimate for the size of the Milky Way as defined by clusters, i.e. probably greater than about 100 kpc, is approximately correct, his distance scale for individual clusters was exaggerated. The new outer cluster limit of about 100 kpc comes mainly from globular clusters found after Shapley's work (a number of them, the "Pal" clusters, were found during the first Palomar Observatory Sky Survey). A modern view of the distribution of the Milky Way globulars is shown in Figure 15.

In contrast to Shapley, Curtis maintained allegiance to the island universe theory. This was motivated, in part, by his own observations of globular clusters and analysis of their distance, which severely *underestimated* the size of the Galaxy as being at most 30,000 light years (9 kpc) in diameter, a size that could easily exclude spiral nebulae by almost any accounting of their distance. The basis of Curtis's underestimate of distance was that he had assumed, on the basis that the mean spectral type of globular clusters looks like that of the Sun, that the average stars seen in clusters were dwarfs of luminosity comparable to the Sun. As well argued by Shapley, Curtis ignored a serious systematic effect: "in a distant external system we naturally first observe its giant stars ... the comparison of averages means practically nothing because of the obvious and vital selection of brighter stars in the cluster". In the end, the globular cluster distance scale is now established between the scales of Shapley and Curtis (see discussions by Feast and by King in this volume).

2.4. *Modern Descriptions of the Galactic Globular Cluster System*

In his contribution to this Winter School, Harris discusses the presently known spatial, chemical and kinematical distributions of the Milky Way globular cluster system,

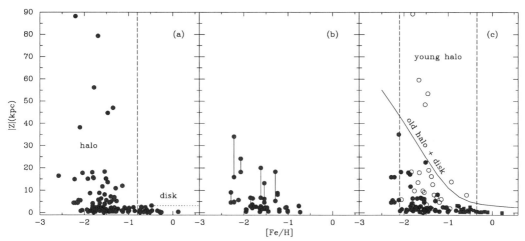

FIGURE 16. (a) The familiar division of globular clusters into disk and halo systems, adapted from Figure 1 of Zinn (1985), and including his metallicities and Z distances. Panels (b) and (c) are discussed in Section 3.4.1. (b) Calculated orbital Z_{max} from Luis Aguilar (UNAM) for those old halo and disk globulars having proper motions (see Table 2 of Majewski 1994a, Dinescu 1998) and connected by vertical lines to the present Z distances for these systems. In many cases, $Z \sim Z_{max}$ so that the pairs of points overlay one another. (c) Data points show the RHB "young halo" (*open circles*), BHB "old halo" (*filled circles*) and disk (*solid squares*) globular clusters, with updated Z and [Fe/H] taken from Harris (1996). *Dashed lines* show the range of metallicity spanned by the RHB "young halo" globulars, while the *solid, curved line* shows, schematically, a cluster paradigm (Zinn 1993a) in which the BHB "old halo" and disk globulars represent one system, perhaps connected through a dissipational collapse of the disk.

based on his very useful compilation of cluster data (Harris 1996, also available on-line at http://physun.physics.mcmaster.ca/Globular.html). Therefore, I briefly concentrate here (and in later sections) on only those aspects of the distributions that are germane to this discussion, and some for which I present a different interpretation than Harris.

Studies of the globular cluster system in the past few decades have made evident the existence of at least two subsystems. A *disk system of globulars* (Zinn 1985, Armandroff 1989) is found concentrated towards the plane of the Galaxy (the darkened patch of points in Figure 15), while the *halo system of globulars* is represented by the more extended distribution of points in Figure 15. A conventional description of the two systems is given by the [Fe/H]-$|Z|$ distribution, as presented by Zinn (1985), reproduced here in Figure 16(a). Zinn recommended a division between the disk and halo systems at [Fe/H]$=-0.8$ (and a division near -1.0 is commonly used today), so that, by definition, the disk globulars are more metal rich than those in the halo. The range of distances from the Galactic plane, $|Z|$, is much wider for the metal-poor, "halo" globulars than the metal-rich, "disk" clusters. More recently, Zinn (1993a) has proposed a different division of the globular cluster system into *three* components after taking into account not only the metallicities of the clusters, but the character of their horizontal branches. This new description is addressed in Section 3.4.1.

2.5. *Pushing the Envelope*

The presently known halo globular clusters show a rather spherical density distribution, with a majority of clusters within 30 kpc of the Galactic center (Figure 15). After this radius, the density apparently drops, so much so that the region between about $40 < R_{GC} < 80$ kpc has been characterized as a globular cluster "gap" between the inner and outer halo systems (there being one known cluster in the gap, according to Harris' 1996 compilation). The five outer (past the gap) globulars inhabit $80 < R_{GC} < 120$ kpc, with the most distant known outlier, AM-1, at $R_{GC} = 120$ kpc.

By the earlier definition of the extent of the Galaxy as given by globular clusters, this would define the limit of the Milky Way. However, beginning in the 1930s with discoveries of the Fornax and Sculptor systems by Shapley, a number of dwarf satellite galaxies of the Milky Way have been found. The most recently found dwarf satellite is, ironically, also the closest one – Sagittarius, which is at $R_{GC} = 16$ kpc (Ibata et al. 1995). The present count of satellite galaxies of the Milky Way is eleven, with Leo I the most distant at 280 kpc from the Galactic center (Lee et al. 1993). The recently discovered (van de Rydt et al. 1991) Phoenix system is at $R_{GC} = 417$ kpc, but it is not yet certain whether this system is bound to the Milky Way. Among the satellite galaxies, the most massive members, the Large and Small Magellanic Clouds, Fornax, and Sagittarius all have globular cluster systems – an important clue to the formation of some of the Milky Way globulars (see Section 3.4.1 below). Note that the globulars of Fornax, located at $R_{GC} = 143$ kpc, are technically the most distant globular clusters associated with the Milky Way system. Depending on whether or not Phoenix is included, the known extent of the Milky Way system is therefore established to be at least 280 kpc, but perhaps more than 400 kpc, in radius. It is not unreasonable to expect that more dwarf satellites of the Milky Way may remain to be found. Because of their very low surface brightnesses, dwarf galaxies are hard to spot, but several groups are now using sophisticated algorithms to search for faint brightness enhancements in scans of POSS-II plates and ESO-SERC plates and other wide field data. This work has activated a new round of discoveries of Local Group dwarfs (e.g., the Antlia dwarf, Cas dSph, And V and And VI (=Peg dSph); Whiting et al. 1997, Armandroff et al. 1998, Jacoby et al. 1998, Karachentsev & Karachentseva 1999).

Note that the R_{GC} of these outer Galactic satellites are in the same regime as the extent of Lyman α absorbers found in QSO absorption line studies (Lanzetta et al. 1995). If the Milky Way extends to $> 300 - 400$ kpc in distance, and M31 and M33 have similarly extended outer parts, then we begin to reach the regime where the outer parts of these individual systems begin to overlap. Considering that the *dark matter halos* of galaxies are likely to be more extended than the luminous tracers, it may be that the primary galaxies of the Local Group are not really isolated, but, at the least, floating in a shared dark matter soup.

What about the extent of the Milky Way as gauged by individual field stars? This is not well studied and may be difficult to establish without having full velocity information to check whether outliers are bound to the Milky Way. The possibility that some dwarf satellites are tidally stripped of stellar and cluster débris (see Section 3.4.3 below) begs the question of whether there may be extremely distant, now isolated stars bound not to the Milky Way or other galaxies, but to the Local Group. Stars released on hyperbolic orbits are a characteristic of galaxy interactions (Toomre & Toomre 1972) and could lead, perhaps, to "intergalactic tramps". Intergalactic planetary nebulae have been identified in the Fornax (Theuns & Warren 1997) and Virgo (Méndez et al. 1997) clusters, but very little work has been done on finding very distant stars *bound* to the Milky Way, let alone "extra-Galactic" stars in the Local Group. The latter would be especially useful for gauging the extent of the Local Group dark matter. Gould et al. (1992) report finding dwarf stars at a distance of about 100 kpc, near the Sextans dwarf galaxy, while the author and collaborators have recently found giant stars more distant than 140 kpc. Richstone et al. (1992) sought distant main sequence turn off (MSTO) stars in small area (0.002 deg^2), but very deep ($B \approx 27$) CCD frames that reach potentially to 500 kpc for stars at main sequence luminosities. While Richstone et al. found no excess signal above the expectations for contamination of their sample from compact galaxies, from statistical arguments these authors place a lower limit on $(M/L)_{\text{LocalGroup}}$ of 400.

3. Survey of Age, Kinematical and Chemical Distributions in Stellar Populations

Our goal is to recognize chemodynamical patterns in stellar populations that will give some idea of the progression of events during the evolution of the Galaxy, so that we may synthesize models to explain the origin and distribution of stellar populations in the Milky Way, and, by analogy, similar spiral type galaxies. In the interests of the limited time in these lectures, and influenced by the predilections and expertise of the author, here the focus will be on the observer's role of seeking patterns among age, chemistry and kinematics, rather than the host of chemodynamical modeling efforts being conducted at a number of centers around the world.

A great deal of work has been done in the last decade or so to establish the chemodynamical and age characteristics of the Galactic stellar populations. My approach to covering this wealth of material is to survey various approaches utilized to establish population properties and summarize results of these approaches in an attempt to assemble a first-order picture (via chemical and dynamical "population box" representations) of the evolution of the Galaxy. A number of the observational techniques, while discussed specifically in certain contexts, lend themselves to broader use in population studies.

3.1. *Brief History of Kinematical and Chemical Studies of the Milky Way*

The modern era of Galactic structure studies may be identified with a shift in preoccupation from the simple clarification of the order and shape of stellar systems within the Galaxy to the application of the properties of these systems toward ascertaining the evolutionary history of the Milky Way. This change in mindset was technically feasible only after great theoretical and observational advances in the 1950s that made possible the joining of Galactic structure, kinematics, chemistry, and age into a unified evolutionary context. The development of a theory of stellar evolution not only made possible a means for dating the ages of stellar clusters through use of the color-magnitude diagram (Sandage & Schwarzschild 1952) but also led to an understanding of the process of nucleosynthesis (Burbidge et al. 1957) and the idea of the progressive chemical enrichment of the Galaxy. Once it was appreciated that stellar spectra held key information on the level of stellar enrichment via the strength of absorption lines (Chamberlain & Aller 1951), classification of individual field stars into *relative* age groups became possible. These absorption lines are not evenly distributed across the visible spectrum, and it was found that some broad band filters are particularly sensitive to variations in abundance. For example, weak-lined stars show an *ultraviolet excess* of flux (normally measured as a particularly blue $U - B$ color for stars of a given $B - V$ color, Wildey et al. 1962) due to suppression of line absorption normally concentrated in the near ultraviolet. The relative age scale of field stars could be calibrated to that of star clusters, once it had been shown (Sandage & Walker 1955) via ultraviolet excess measurements that the peculiar, weak-lined spectra exhibited by some stars were shared by globular cluster stars.

Decades earlier, through the work of Kapteyn, Stromberg, Oort, Lindblad, and others, great advances had been made in the understanding of stellar kinematics and in the identification of kinematic subsystems within the Galaxy. Through Baade's (1944) insight, the association of these kinematic groups with specific structural components in galaxies was made: *Population I* were the metal-rich stars in disks moving with rotational velocities like the Sun, while *Population II* objects Baade associated with the spheroidal parts of galaxies (bulges and halos), which he thought to be primarily metal-poor. The final link between the structural-kinematic stellar populations and the age-metallicity groups came with the discovery by Roman (1954) that metal deficiency in stars is typically correlated with high velocities with respect to the Sun, while higher metallicity stars move with velocities similar to the Sun and other disk stars.

With loose connections established between kinematical, chemical, age, and structural groupings of stars within the Galaxy, the stage was set for the development of the modern evolutionary pictures that accounted for the existence and properties of the various stellar populations. The first major breakthrough along these lines, the landmark paper by Eggen, Lynden-Bell & Sandage (1962, ELS hereafter), remains the foundation for modern discussions of Galaxy formation. It also set the precedent for the construction of modern Galactic structure surveys. ELS compiled ultraviolet excesses, radial velocities, and proper motions for nearby stars, which ELS selected from a catalogue of large proper motion stars (i.e., typically stars with large velocities with respect to the solar neighborhood) combined with a catalogue of well studied, bright stars without kinematical biases (the latter sample would be dominated by the overwhelming number of stars in the local disk population). From their combined sample, ELS discovered smooth correlations between ultraviolet excess and (a) the orbital eccentricity of the stars, (b) their angular momentum, and (c) the $|W|$ velocity (the velocity perpendicular to the Galactic disk). Based on these correlations ELS constructed a formation model that incorporated the new age-dating techniques from stellar evolutionary theory as well as a dynamical analysis of stellar orbits. In this picture, the Galaxy formed from the rapid collapse out of the general Hubble expansion of a metal-poor, roughly spherical, primordial density fluctuation. During this collapse, condensations of gas were created and star formation turned on. The orbits of these stars today were presumed to reflect the kinematical state of the gas from which they formed, while the stellar abundances were a function of the level of chemical enrichment of that gas (via the fusion of "metals" in the cores of massive stars which belched forth these processed elements into the interstellar medium in supernova explosions). Thus, the earliest stars formed had the weakest metal abundances, and were born into highly radial orbits with the momentum of the initial collapse. Radial collapse was unable to continue as far as that in the direction perpendicular to the Galactic plane because of the increase in rotational velocity and centrifugal acceleration due to angular momentum conservation. Dissipation of energy via cloud-cloud collisions enabled the gas eventually to settle into a flattened, rapidly-rotating disk. Continued star formation during the contraction allowed progressive enrichment of the interstellar medium. It is from the flattened, metal-rick disk of gas that new, circularly-orbiting stars are forming today (Figure 10 shows the relative flattenings of various Galactic components as viewed near the Sun).

Based on the apparent age of the globular clusters, which were attributed as members of the first formed stellar population, ELS assumed the earliest star formation to have started 10 Gyr ago†. From the $|W|$ velocities, the maximum distance from the Galactic plane, Z_{max} can be calculated under an assumed Galactic potential. From the range in Z_{max}, ELS estimated the vertical collapse to have been a factor of 25. From the

---

† The true age of the globular clusters is still much debated, and derived values have encompassed a range up to a factor of two higher than ELS assumed (Chaboyer et al. 1996). However, some recent age determinations have come almost back down to the ELS value, near 10-12 Gyr (e.g., Salaris & Weiss 1997, Reid & Gizas 1998). Throughout the present contribution I've used an age scale for the Galaxy and globular clusters about in the middle of the range typically discussed.

ratio of apogalactica of high eccentricity to circular orbits for stars with similar angular momenta, the radial collapse was estimated to have been a factor of 10. Finally, the rate of the collapse was determined from the dynamical constraint that in order to form the old stars (i.e. the low metallicity stars) on highly radial (i.e. high eccentricity) orbits, the radial velocity of the initially collapsing gas from which they formed must have been less than the rotational velocity. In other words, the rate of the collapse must have been more rapid than the Galactic rotation period, or ~ 0.2 Gyr. Because this is approximately the timescale for freefall collapse from the estimated initial gas distances, ELS concluded that the halo of the Galaxy formed during a collapse without pressure support. As support for this idea, ELS noted: (a) if the gas had been hot enough to provide pressure support to slow the collapse, it would have also prevented the small-scale collapses which form stars, and (b) the five halo globular clusters with accurate photometry at that time all had nearly the same age, which suggests a brief formation epoch for the halo globular system.

While the following decades brought criticisms and refinements of the ELS model, and the rise of a competing picture for the formation of at least some part of the halo (see Section 3.4), the ELS picture incorporated the basic elements that are the foundation of modern chemodynamical models, and the general ideas they presented are likely relevant for at least some part of the formation of the Galaxy (see Section 3.3.3, for example). It has been a primary endeavor of Galactic astronomy since ELS to refine and elaborate on their themes with ever improving data to constrain Galactic formation models.

3.2. *The Thin Disk*

3.2.1. *Age and Star Formation History of the Disk*

A number of means have been used to determine the distribution of ages of stars in the disk of the Milky Way. In this subsection I concentrate on age dating the Baade Population I *thin* disk, but remind the reader that the problem of truly separating tracers of the thin disk from those of the IPII – *indeed, whether this is even possible if they form a single, contiguous population* – remains a nagging concern for understanding the elder part of the thin disk. Because of remaining uncertainties in the age scale, I will try whenever possible, to discuss ages in relative terms; whenever absolute values for ages are given, the reader should bear in mind that the absolute age scale still varies among researchers by up to a factor of two.

Traditionally, a maximum age of the thin disk of about 10 Gyr was adopted from an estimated age (now questioned) of the open cluster NGC 188 (VandenBerg 1985). In the last decade, a number of new techniques have been used to age date the disk, with results from slightly less than 10 Gyr to several Gyr more:

• *Nucleocosmochronology* utilizes measures of ratios of long-lived isotopes to estimate ages. To first order, when a star is formed, its outer atmosphere provides an unchanging sample of the composition of the Galactic gas at that time, modified only by the decay of radioactive species. For example, one may use as a chronometer variations in the line strengths of the radioactive thorium isotope ^{232}Th, which has a half-life of 20 Gyr (i.e., close to a Hubble time), to that of a stable element, like Nd (which has an absorption line very near the Th line at 4019 Å, a useful coincidence for high resolution spectroscopic studies). Butcher (1987) attempted the experiment with nearby G dwarfs of very different assumed ages (postulated using a variety of means) and found *no* variation in ^{232}Th/Nd over the entire range of ages. This means that either (1) all of the ^{232}Th was formed in a single event before all of the stars were formed, (2) all of the stars were in actuality formed only a short while ago, so that no significant decay has occurred, or (3) the synthesis

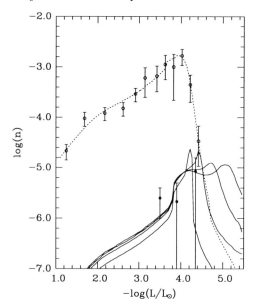

FIGURE 17. Observed white dwarf luminosity function for the disk (*upper dots*), and the halo (*lower dots with large error bars*). Also shown are models for the disk (*dotted line*) at 12 Gyr and for halo white dwarfs of ages 10, 12, 14, and 16 Gyr (*solid lines*). The halo white dwarf luminosity function is not yet constrained by the data. From Isern et al. (1997).

of ^{232}Th/Nd is increasing fractionally with time. From measures of other transuranic elements in meteorites, we know that the duration of nucleosynthesis *before* formation of the solar system (itself at least 4.5 Gyr old) is about 5.4±1.5 Gyr (Fowler 1987), so (2) is excluded, but so too is (1) if the solar system measure gives a *maximum* age. Analysis of option (3) yields a maximum age for the disk of about 10 Gyr. Unfortunately, this type of analysis is subject to a number of uncertainties, especially over the amount of infall of gas to the disk from outside – something sure to affect isotope ratios.

• The *white dwarf luminosity function* provides a useful limit on the age of the Galactic disk, because, for all but the youngest populations, stars evolving off the main sequence have masses ($< 8M_\odot$) that result in the formation of white dwarfs as their final evolutionary stage. White dwarfs are essentially cooling embers, for which the thermodynamical physics, in principle, can be estimated. Since the age of the disk is finite, the allowed cooling time for the oldest stars is constrained, and this should lead to a sudden cutoff at the faint end of the luminosity function corresponding to the limit of white dwarf cooling in this time (D'Antona & Mazzitelli 1978). Indeed, such a cutoff was found by Liebert et al. (1979); a recent compilation of the data is presented by Isern et al. (1997; Figure 17). A significant amount of theoretical work has been devoted to understanding what the cutoff means in terms of age limits.

Modeling the luminosity function requires an understanding of the various stages of white dwarf cooling, and the details of each type of cooling dependent on the white dwarf mass and chemical composition (see Isern et al. 1997 for a more complete description): (1) *neutrino cooling* from p–p chains in the hydrogen layer; (2) *fluid cooling* by gravothermal shrinking; (3) *crystallization*, which takes advantage of two new sources of energy, the latent heat which makes a small (5%) contribution to the luminosity of the star, and the sedimentation of heavy elements towards the center of the star, which results in the release of gravitational energy; and (4) when the star is almost completely solidified,

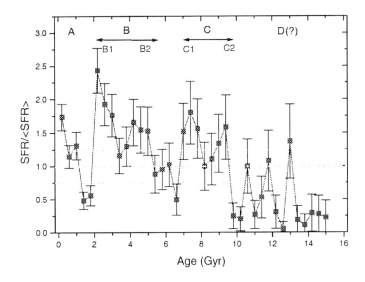

FIGURE 18. Disk star formation rate given by stellar chromospheric ages, as corrected for sampling effects by Rocha-Pinto et al. (1998). The various star formation maxima, labeled as "bursts A", "B", and "C", correspond roughly to those found by Barry (1988) with the same technique, and roughly correspond to the age distribution by Twarog (1980) from Strömgren photometry of disk stars. The data are incomplete at the oldest ages, where there may have been an additional burst of star formation.

Debye cooling takes over. With a model for cooling time that incorporates these various stages, one may estimate the expected luminosity function of white dwarfs from

$$n(\log{(L/L_o)}) = \int_{M_{min}}^{M_{max}} \Phi(M)\Psi[T - t_{cool} - t_{MS}]\tau_{cool}dM$$

where M is the main sequence mass of the white dwarf progenitor stars which have a luminosity function $\Phi(M)$, $\tau_{cool} = dt/dM_{bol}$ is the characteristic cooling time at each white dwarf bolometric luminosity, M_{min} and M_{max} give the range of masses of main sequence stars able to produce white dwarfs of luminosity $\log{(L/L_o)}$, t_{cool} is the time required to cool to this luminosity, t_{MS} is the main sequence lifetime of a star of mass M, and T is the age of the white dwarf population.

Note that because the observed luminosity function is also dependent on the star formation rate, $\Psi(t)$, which is creating stars according to the luminosity function $\Phi(M)$ at any given time t, the *shape* of the white dwarf luminosity function can also give information about the *history* of star formation. According to Isern et al. (1997), the best fitting star formation rate to the disk white dwarf luminosity function shows an age of 12 Gyr for the first star formation (but at a relatively low rate) followed by a rise in activity after $1 - 3$ Gyr and reaching a peak $4 - 6$ Gyr after that. The star formation rate has remained constant or slowly declined since this time. Compare this disk star formation history with that presented next (Figure 18). Note that the limit to the disk age as set by this technique is still controversial at the level of about 20%, partly due to the data (which are heavily dependent on a only a small number of known white dwarfs at the turnover point of the luminosity function; e.g., Wood & Oswalt 1998), and partly to differences in the adopted cooling rates and white dwarf core compositions (yielding some age estimates as low as 8 ± 1.5 Gyr; Leggett et al. 1998).

• *Stellar chromospheric activity* declines with stellar age, and the amount of activity can be measured via the strength of the CaII H and K lines in F and G dwarf stars (Soderblom 1985, Barry 1988, Henry et al. 1996†). Results for a sample of 730 stars selected from the literature and recalibrated with a new calibration of the chromospheric line strength-age relation are given by Rocha-Pinto et al. (1998, Figure 18). Interestingly, the results of this work imply a rather variable star formation history for the Galactic disk, with a variation of perhaps more than an order of magnitude in the star formation rate. For example, a rather strong SFR is indicated at intermediate ages, and a rather low, but non-negligible, SFR is indicated some $10-12$ Gyr ago, consistent with the white dwarf luminosity function results. A major decline in the SFR several Gyr ago can also be seen and has been long established (Vaughan & Preston 1980, Henry et al. 1996). A somewhat "bursty" star formation history for the disk begs the question of possible *triggers* to set off the cycles. One trigger that has been studied several times relates to the passage of the Magellanic Clouds through the disk; however, recent work to check the timing of the Magellanic orbit (still somewhat uncertain, of course) against the timing of the disk star formation rate maxima yields no satisfactory connection (Rocha-Pinto et al. 1998).

• *Strömgren photometry* is a very useful technique for age-dating stars, as the intermediate band filters of the Strömgren system are extremely sensitive to changes due to the evolution of stars away from the zero-age main sequence. A $uvby\beta$ photometric study by Twarog (1980) shows the same increase in the number of stars of age $\sim$4 Gyr old (Meusinger 1991) indicated by the chromospheric age data (those shown as "burst B" in Figure 18). The highest age found among the stars in the Edvardsson et al. (1993) Strömgren photometry and spectroscopic sample is 12 Gyr.

• *The red edge of the red HB clump* is well defined for MSTO stars with $0.8 M_\odot < M < 1.3 M_\odot$ (i.e., ages between 2 and 16 Gyr) and is only a function of metallicity. Using a CMD of disk stars from HIPPARCOS, Jimenez et al. (1998) were able to identify the abundance of the most metal-rich red clump stars as no more than [Fe/H]=+0.3, and from this assessment an appropriate isochrone can be selected to match the red envelope of HIPPARCOS subgiants. Under the conservative assumption that the most metal-rich subgiants have [Fe/H]=+0.3, Jimenez et al. derive a minimum age for the disk of 8 Gyr. However, it is unlikely that the oldest disk stars are this metal rich, and therefore the age of the red envelope subgiants is likely to be higher than this limit.

• *Open clusters* are commonly regarded as a purely thin disk population, especially as the lifetime of the typical open cluster was thought to be relatively brief, $\sim$200 Myr or less (Janes et al. 1988). These notions have no doubt been influenced by the concentration of work on relatively nearby examples. Indeed, of the some 1200 open clusters known (Lyngå 1987), less than one quarter have received more than superficial attention, and fewer still have quality CMDs. However, new surveys (e.g., Phelps et al. 1994) of open clusters reveal a substantial number of very old open clusters (age $\geq$ 5 Gyr), and the age of the oldest known open cluster, Be 17, suggests that the age of the disk is at least $\sim$12 Gyr old (Kałużny 1997, Phelps 1997). That long-lived population open clusters exist is unexpected given ideas of cluster dynamics (e.g., Wielen 1977), which predict that clusters moving near the Galactic plane will have short lifetimes. What seems to permit the long-lived open cluster population to exist is that they are located mostly in the outer disk, where disruptive encounters with molecular clouds (e.g., Spitzer 1958) are less frequent. Moreover, from their observed distances from the plane, many old open

---

† Henry et al. (1996) note the possibility of some age ambiguity depending on what part of the "solar minimum-maximum" cycle a star is in when observed.

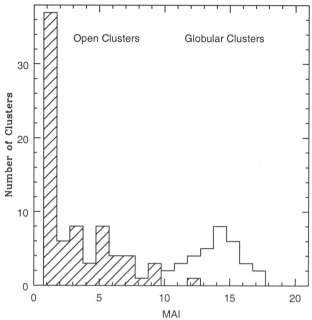

FIGURE 19. Open cluster and globular cluster age distributions. Open cluster ages were determined using the morphological age index (MAI) of Janes & Phelps (1994) for clusters listed in Phelps et al. (1994), as well as photometry of several new clusters discussed in Majewski et al. (1997). Following Janes & Phelps (1994), globular cluster ages are from Chaboyer et al. (1992). Several young globular clusters (Terzan 7, Arp 2 and IC 4499) were added to the list, with ages (10.9, 10.9 and 12.6 Gyr, respectively) determined from published photometry and the MAI. The clusters Lyngå 7 and BH176, both thought to be young globulars or old open clusters, were omitted because of difficulties in establishing ages from the published photometry. From Majewski et al. (1997).

clusters must have Z-motions that take them away from the plane for much of their orbit, so that their survivability is greatly enhanced relative to the rest of the open clusters. Curiously, some of these old open clusters are *several kpc away* from the plane – too far to have been "heated" to this distance by secular dynamical processes. These clusters had to have been *formed* far from the Galactic midplane – which would seem to be an important clue to the history of the old disk.

Before the recent work to uncover new examples of ancient open clusters, it was commonly believed that there was a fairly long pause between the epochs of globular cluster and open cluster formation. However, incorporating the recent discoveries into a comparison of the relative ages of the open clusters and globular clusters as determined in a homogeneous way (using a "morphological age index" in the color-magnitude diagram of the cluster stars; Phelps et al. 1994) reveals that a gap no longer exists (Figure 19). Moreover, the ages of the *oldest* known open clusters are now beginning to overlap the ages of the *youngest* globular clusters. Thus, the disk open cluster system appears to have begun formation even while the halo globular cluster system (which contains the young globular clusters shown in the Figure) was still forming. While the age distribution shown for the open clusters in Figure 19 is influenced heavily by selection effects (research emphasis has been predominantly on the youngest and oldest open clusters), it is interesting to note that present population of Galactic open clusters appears to have formed over the lifetime of the disk.

Curiously, while the oldest open clusters seem to be located outside the solar circle, with no age gradient found for radii larger than the solar circle, this is in direct contradiction to the findings for solitary disk *stars*. Edvardsson et al. (1993) conclude from their well studied sample of F and G dwarfs that the disk formed "inside out", and question whether stars more than $10 - 12$ Gyr formed in the disk at the radius of the solar circle. From the radial dependence of the distribution of $[\alpha/\text{Fe}]$ ratios to $[\text{Fe/H}]$ (see below), which is a function of the enrichment history of the gas, and, in turn, the star formation history of the disk, Edvardsson et al. claim that star formation in the disk proceeded quickly at early times near the center of the Galaxy, while star formation was increasingly drawn out as a function of radius. This "inside-out" formation is a characteristic of recent chemodynamical models of the disk (e.g., Burkert et al. 1992, Chiappini et al. 1997).

Perhaps the resolution of this apparent contradiction lies in the discussion of cluster survivability in Section 1.2 above. In the inner parts of the disk, clusters are subject to far more disruption; thus we would expect a larger contribution of old cluster stars to the field at smaller radii than we would farther out in the disk.

3.2.2. *The Age-Metallicity Relation*

As discussed in Section 3.1, historically it was believed that the metallicity of a star correlates to its age according to some *age-metallicity relation* (AMR). However, the observed age-metallicity distribution for open clusters (Geisler et al. 1992, Friel 1995) *and* disk stars (Marsakov et al. 1990, Edvardsson et al. 1993) almost completely dispels this notion. As can be seen in Figure 20, at any given age stars of a wide range of $[\text{Fe/H}]$ can form, and, moreover, this is also the case in the abundance *patterns*, such as the $[\alpha/\text{H}]$ and neutron capture elements, like barium. The mean increase in these abundances over the life of the disk has been only a few 0.1 dex. This great inhomogeneity in element abundances at any age suggests that chemical mixing in the disk is not very efficient.

Closer inspection of Figure 20 shows, however, that there do appear to be *radial* abundance gradients. For $[\text{Fe/H}]$ in the open clusters, the gradient is found to be -0.091 ± 0.014 dex kpc^{-1} (Friel 1995). This is similar to the gradient found in the Edvardsson et al. stellar sample. Based on the observed radial abundance gradient, it appears that over the history of the disk it was *position* in the disk that played a dominant role in determining a star/cluster metallicity, so that, remarkably, at almost any given disk abundance we can find both old and young disk stars.

It should be pointed out that the sample used to produce Figure 20 is a highly selected one, and not representative of the relative metallicity distribution in the disk. Several groups have updated the thin disk metallicity distribution via surveys of G dwarfs in the solar neighborhood. The results of these studies are shown in Figure 21. Note that the lack of a significant number of very metal poor stars in the solar neighborhood (the "G-dwarf Problem") has long been a problem for simple chemical evolution models of the Galaxy that follow the enrichment of gas in "closed-box scenarios". That simple model scenario is unlikely to be a good one, given the likely possibilities for pre-enrichment of disk gas or infall of enriched gas onto the disk, both explanations for the paucity of extremely metal poor disk stars (see Gratton's discussion in this volume).

If we now simplistically adopt the large scatter in the metallicity distribution of disk stars as shown in Figure 21, the age-metallicity relation as shown in Figure 20, and the chromospheric age distribution shown in Figure 18, we may attempt construction of an approximate Hodge population box for the Galactic disk. The result is shown in Figure 22. In this Figure, we also anticipate the addition of the IPII, discussed next.

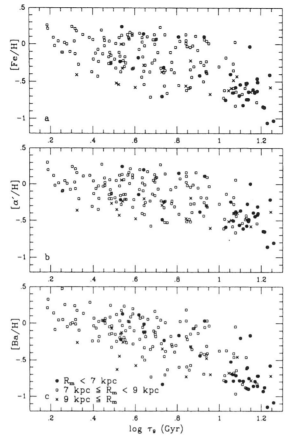

FIGURE 20. The age-metallicity distribution for a sample of F and G dwarfs in [Fe/H], α'-elements (where α' indicates that a mean of the logarithmic silicon and calcium abundances is used), and the neutron capture element barium as a function of the logarithm of the age in gigayears. The symbols indicate stars in different zones of mean Galactocentric radius, R_m, as derived from the stars' orbits. Note that no appreciable age-[Fe/H] relation is found, but radial abundance gradients do seem to exist. From Edvardsson et al. (1993).

3.3. *Intermediate Population II, Thick Disk*

3.3.1. *Age*

The general consensus is that the IPII component is old–, approximately the age of the halo. The precise relative timing of the first star formation in the halo and IPII bears critically on some formation scenarios, but is still somewhat uncertain. As usual, complications in age dating the IPII extend from problems in separating pure IPII samples reliably from those of the halo (particularly the "low" halo) and the old, thin disk, both populations that have significant overlap with the IPII spatially, kinematically, and chemically.

• *Disk globular clusters* are commonly attributed as tracers of the IPII on the basis of their vertical scaleheight (see Figure 16), velocity dispersion, and the known orbit of the prototype disk globular 47 Tuc. Unfortunately, few disk globulars have been age-dated relative to other globulars or in an absolute sense, but the numbers are growing. 47 Tuc appears to be some $1-2$ Gyr younger than the oldest halo globulars, when measured on the same age scale (Chaboyer et al. 1992), but perhaps several Gyr *older* than some

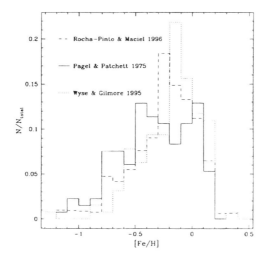

FIGURE 21. The observed G dwarf metallicity distribution for stars in the solar vicinity, as derived by three groups. From Chiappini et al. (1997).

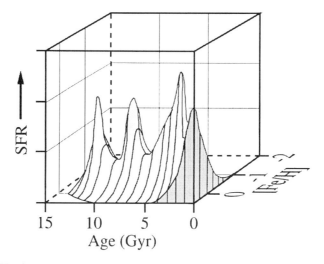

FIGURE 22. The SFR-abundance population box for the Galactic disk, synthesized from our survey of results for disk stars and clusters. The figure shows the three SFR maxima ("bursts" A, B, and C) from Figure 18, and one additional burst (D?) near 13 Gyr that corresponds to the formation of the IPII thick disk (discussed in Section 3.3). The abundance spreads are constrained by the metallicity distributions for thin disk G dwarfs presented in Figure 21, and the metallicity distribution for the IPII presented in Figure 23. The overall spread in abundance with age is adopted from the Edvardsson et al. (1993) data presented in Figure 20. A possible IPII metallicity spread to a "metal-weak thick disk" for the oldest "burst" is hidden from view in this representation.

of the youngest halo globulars (see below). From a variety of studies, it appears that many of the disk globulars studied – 47 Tuc, NGC 6352, NGC 6760 and M71 – are coeval (Hodder et al. 1992, Fullton 1995, Grundahl 1996, Salaris & Weiss 1998), and, in the age dating studies with the shortest cluster timescales, the same age as the oldest disk white dwarfs (Reid 1998, Salaris & Weiss 1998). However, one disk globular, the more metal rich NGC 5927, exhibits a younger age by $2-3$ Gyr than these other disk clusters

(Fullton 1995). Another disk cluster, NGC 6553, has yielded a range of ages less than or equal to the old halo clusters (Demarque & Lee 1992, Ortolani et al. 1995). On the basis of their globular cluster populations, it would appear that the IPII began forming soon after the halo began forming but before the halo stopped making globular clusters. Note that the question of the population assignment of metal-rich clusters within a few kiloparsecs of the Galactic center has become a matter of debate (see Section 3.5): It has been proposed that these metal-rich clusters, including those with younger ages just mentioned, may be members of a *bulge* population of clusters (Minniti 1995, 1996).

• *Tidal circularization of binary stars* has yielded similar ages for halo and IPII binaries (Latham et al. 1992, Carney 1993). The basis of the technique is that short period binaries have their orbits circularized by tidal friction mechanisms, e.g., turbulent viscosity in the convective envelopes of stars acting on the equilibrium tide (Zahn 1977). Thus the orbital period at which the transition from elliptical to circular orbits due to the tidal interaction is a function of time.

• The existence of an apparent disk-like *RR Lyrae* population has been attributed to the IPII. More metal rich members, if IPII, set the age of the IPII to > 11Gyr (see Figure 3), via comparison to the age of onset of an RR Lyrae population in the Magellanic Cloud clusters (Rodgers 1991).

• The apparent *main sequence turn-off* for IPII field stars could be used to establish a formation age. The color of the turnoff for stars of intermediate metallicity is consistent with the IPII stars being as old as 47 Tuc (Carney et al. 1989, Rose & Agostinho 1991, Gilmore et al. 1995). Moreover, if there were a younger component to the IPII, then one would expect to find stars bluer than $B - V = 0.5$ at the intermediate metallicities expected to be most representative for the IPII (Norris & Green 1989). Few such stars are found in the distance range $1 < Z < 5$ kpc in the magnitude complete surveys of Croswell et al. (1991) and Majewski (1992). Thus, it would appear that the bulk of the IPII stars were formed at the same time, *unless* the mean abundance for the IPII is significantly higher than that of 47 Tuc, as has been suggested recently by Reid (1998).

• From *Strömgren photometry* a more specific timeline emerges. For a reasonably large sample of stars from the halo and "high velocity disk stars" (IPII) it has been established that the IPII is 1-2 Gyr younger than the inner halo, but perhaps a little older, or as old as, the outer halo (Marquez & Schuster 1994). One explanation is that the formation of the outer halo and IPII are related, perhaps in a puffing of a primeval thin disk to produce an old thick disk at precisely the time the halo was apparently accreting its outer halo. Perhaps the same "fragments" (Searle & Zinn 1978) that formed the outer halo were responsible for dynamically heating the thin disk at early times. In this case, the IPII, thick disk stars could represent the population of primeval thin disk stars that were dynamically heated, whereas the thin disk we see today represents those disk stars formed after this event, as well as any primeval thin disk stars left relatively unaffected by the event. Presumably, the thick disk could be composed, partly or wholly, from the *remains* of the shredded, merged satellite galaxy or "fragment".

From the present data, the most consistent time sequence that emerges is that the bulk of the IPII stars and globular clusters appear to have formed at about the same time, which was after the inner halo formed but at the same time or before the outer halo formed. Such a timeline has important implications for the formation of the IPII. For example, if the IPII formed from a merger event puffing up a previously formed disk, this event must have occurred quite early in the history of the Galaxy. However, the age sequence with the thin disk is also consistent with a formation scenario wherein the IPII may have been the less dissipated precursor to the present thin disk.

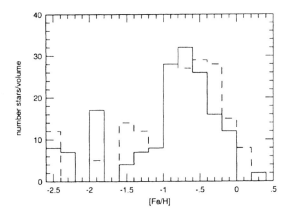

FIGURE 23. Abundance distribution of IPII/thick disk stars for a sample at $Z = 1.0$kpc (*solid histograms*) and $Z = 1.5$ kpc from Gilmore et al. (1995). Certain assumptions have been made to clean the histogram of halo and thin disk stars.

3.3.2. *Abundance*

It has been considered well established that the mean metallicity and *peak* of the metallicity distribution for IPII stars is near [Fe/H]~ -0.6 (see summary in Majewski 1993; Gilmore et al. 1995). However, one of the surprising results of work on presumed IPII stars in the past decade is the possibility of a metal poor tail to the IPII (Norris 1986, Morrison et al. 1990, Majewski 1992, Morrison 1993, Beers & Sommer-Larsen 1995). Clearly there is some controversy over this "metal weak thick disk". For RR Lyrae stars with kinematics like those expected for the IPII, Martin & Morrison (1998) report stars as metal poor as [Fe/H]$=-2.05$, yet, from a very similar data set, Chiba & Yoshii (1998) report essentially *no* thick disk stars with [Fe/H]$<= -1.6$! Moreover, recent work with HIPPARCOS data by Reid (1998, 1999) argues that commonly used abundance scales for disk stars may be substantially underestimated, by as much as 0.4 dex. If so, such a change would affect the distributions shown in both Figures 21 and 23.

Apart from new issues regarding the metallicity scale, there are the longstanding problems associated with difficulties in separating IPII stars from other components. Nevertheless, there have been some attempts to determine a metallicity distribution for IPII stars; the abundance distribution for a sample of IPII stars at $Z = 1.0 - 1.5$ kpc is shown in Figure 23. Though the sample size is relatively small in this case, and whether or not the metallicity scale adopted needs to be shifted or compressed, an apparent tail to low abundances makes the distribution decidedly non-Gaussian.

There seems to be some consensus that there is almost no vertical abundance gradient in the IPII (Yoss et al. 1987, Carney et al. 1989, Majewski 1992, Gilmore et al. 1995). However, Chiba & Yoshii (1998) find some evidence for an abundance gradient in the metal weak portion of the IPII, which they argue implies a dissipative IPII formation (Section 3.3.4) after major parts of the halo formed.

Combining the age timeline from the last section with a smoothed version of Figure 23, we might expect the age–abundance distribution of the IPII to look something like that shown in Figure 31 at the end of this article. Note the asymmetric tail to lower abundance in that representation. A schematic Hodge population box of the combined thin and IPII disk was shown in Figure 22, earlier, with the IPII as part of the possible "burst D" in the star formation rate history as shown in Figure 18.

3.3.3. *Kinematics of the Disk Populations*

For this discussion, we adopt the coordinate system of velocities (U, V, W), which correspond to motion with respect to the Local Standard of Rest in the directions toward the Galactic Center (X), in the direction of the motion of rotation (Y) and the direction towards the North Galactic Pole (Z), respectively. The IAU adopted standard velocity of the Local Standard of Rest, the motion expected for an object at the Galactic radius of the Sun orbiting the Galactic Center in true circular motion (i.e., in the absence of *peculiar motions*), is $\Theta_{LSR} = 220$ km s^{-1}, but recent work suggests that this value may be closer to 184 ± 8 km s^{-1} (Olling & Merrifield 1998). The difference between the V velocity of a star and Θ_{LSR} gives the rotational velocity of the star. Another name for V is the *asymmetric drift* velocity, which is 0 for the LSR.

Perhaps in no other property is the problem of the division (?!) of the thin and thick disk components less certain than in the kinematics, with widely reported differences in the derived kinematics for the IPII among different groups. For example, the rotational velocity of the IPII has been reported to be anywhere from a single value in the range $20 - 100$ km s^{-1} to a gradient between these values as a function of Z-distance (see summary in Majewski 1993). At least three uncertainties confound the problem: (1) uncertainty in the relative density laws (Figure 13) between Galactic components, (2) uncertainty regarding a gradient in the kinematics, as opposed to, say, a single valued rotational velocity, for the IPII (gradients in kinematics are expected for the thin disk due to secular heating of stars as they scatter off of giant molecular clouds complexes and other perturbations), and (3) the question of whether the thin disk and IPII are distinct components, or rather, parts of a single, contiguous population as described by Norris (1987). An illustration of the problem is given by the distribution of kinematical properties of *all* stars in complete samples at the Galactic poles. Figure 24 summarizes the mean asymmetric drift velocities and their dispersion as a function of mean Z of the sample. While there is general agreement in the global trend of kinematics between the surveys, the derived kinematical properties for specific Galactic components range broadly among these same samples. For example, Soubiran (1993) finds a rotational velocity of 179 ± 16 km s^{-1} for her IPII, Spaenhauer (1989) finds a rotational velocity of $140 - 160$ km s^{-1}, while Majewski (1992) claims a gradient in IPII velocities from around 200 km s^{-1} near the Sun to about 100 km s^{-1} some 5.5 kpc above the plane. These differences are clearly related to disagreements in how the stars above the plane are distributed into populations (refer, again, to Figure 13): The IPII parameters (h_{IPII}, ρ_{IPII}) adopted for each of the three surveys just mentioned are (0.7 kpc, 6%), (1.3 kpc, 2%) and (1.4 kpc, 3.8%), respectively. Majewski (1993, see Figure 6 in that paper) demonstrates that similar effects are seen in the derivation of rotational velocities from IPII tracer (i.e., selected star type) surveys when the mean heights of the samples are considered. Ultimately, of course, the decomposition of the disk into components (if indeed this is appropriate as opposed to a single, continuous population) must satisfy self-consistently and simultaneously the kinematical, chemical and starcount constraints, and may require multivariate analyses such as those described by Nemec & Nemec (1991, 1993), Soubiran (1993), and others.

Nevertheless, without such complete modeling some broad statements can still be made about the kinematical properties of the thin disk and IPII. First, given the relative *spatial* dimensions of the thin and thick disks and the results of Figure 24, it should be evident that the kinematics of the IPII/thick disk are more extreme (larger velocity dispersions, slower rotational velocity) than those of the thin disk. Then, by bearing in mind the relative ages of the IPII (which is apparently uniformly old) and thin disk

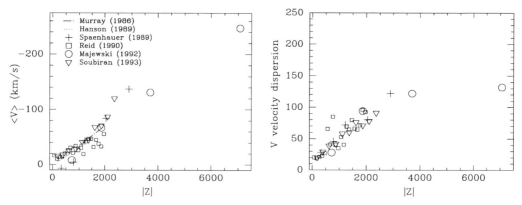

FIGURE 24. The run of asymmetric drift velocity and dispersion as a function of height (in pc) above the Galactic plane for complete surveys of proper motions at the Galactic poles. Error bars have been left out for clarity, but generally range from a few km s^{-1} to $\sim$ 20 km s^{-1} vertically. General agreement is seen between the surveys for the *global* kinematics of all stars, even when the individual surveys derive widely different values for the IPII kinematics after distributing the stars (differently!) into populations. From Majewski (1994a).

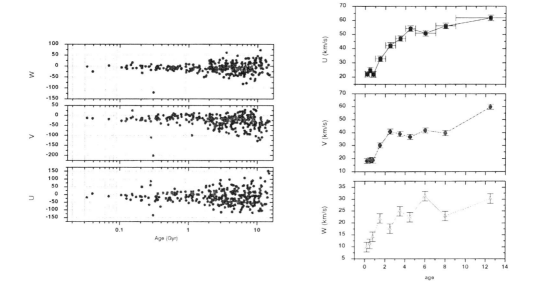

FIGURE 25. The run of the velocity components U, V and W (*left panels*), and their dispersions (*right panels*) with age for a set of disk stars having ages determined from chromospheric HK measures by Rocha-Pinto et al. (1998). The spread in velocity dispersion with age is obvious in the thin disk, though note the apparent saturation of the increase beyond several Gyr. From Rocha-Pinto et al. (1998).

(which apparently spans a range of ages from 0 up to that of the IPII) populations, we may surmise a timeline for the formation of stars now at different velocities: At the extremes, the older stars of the IPII have hotter kinematics while the younger stars of the disk have colder kinematics. From dynamical considerations, this is what one might expect – hotter kinematics are required to support a more extended spatial structure.

Figure 25 shows that this basic timeline is, in fact, a more general *age-velocity relation* (AVR), well established for some time for the thin disk. The growth in the velocity

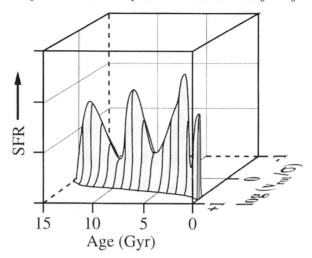

FIGURE 26. *Observed* dynamical population box for the disk populations of the Galaxy.

dispersion of thin disk stars as a function of age is obvious, as is the trend to a slower rotational velocity (increasingly negative V velocity) with age. Note that most of the increase in velocity dispersion occurs early on (see below). It is reasonable to assume that the age-velocity trends shown proceed through to the even older stars of the IPII, since it is known that the kinematics as well as the vertical distribution of the IPII stars are more extreme than those shown in Figure 25. From the approximate AVR we observe (Figure 25), and the disk star formation history observed (Figure 18), we may assemble an *observed* dynamical population box for the disk populations, as shown in Figure 26. Note that without knowledge of the *birth* kinematics of the populations, or how disk stars may have evolved to the *observed* distributions, we cannot construct a reliable dynamical population box for the stars *as they formed*.

3.3.4. *Formation Scenarios for the Disk*

We now have collected several important observational facts about the disk population(s). First, the IPII appears to be an old population, formed at about the same time as the halo, and before much of the thin disk. Second, the almost non-existence of an AMR and large scatter in abundances for the disk populations at any age hints at very poor mixing of the pre-stellar gas. Third, there is an observed AVR across the disk populations.

How did the observed AVR for disk stars arise? Why are the older disk stars dynamically hotter, and therefore more spatially extended, than the younger thin disk stars? Basically there are two possibilities: (1) the older stars were born that way, or (2) the older stars were dynamically heated and diffused from colder, thin disk-like orbits.

In fact, one or the other of these two answers form the bases for most creation scenarios for the IPII thick disk. A detailed accounting of the basic models and their variants is described in Majewski (1993). Here we concentrate mainly on the two families of IPII formation corresponding to the two methods of creating the observed AVR. I call these two families the (1) "top down" models, wherein the AVR is more or less created in place as gas from the forming Galaxy settles into an ever thinner, dynamically colder, plane; and (2) the "bottom-up" models wherein an initially thin, relatively cold population is dynamically heated in some way. As in many situations in astronomy, the true history of the disk may well involve some combination of the extremes.

"Bottom-up" Formation

Among dynamical diffusion processes, there is a firm theoretical foundation for *secular* dynamical heating. We have already seen in Section 1.2 that open and globular clusters that travel through or within the disk see time varying gravitational fields that act to accelerate stars within, and cause eventual destruction of the cluster through evaporation. These same fields act on individual stars once they have escaped from clusters. The predominant secular heating process is the Spitzer-Schwarzschild mechanism (1953), whereby the accumulated effect of repeated collisions of stars with large, $10^{5-6} M_\odot$ molecular cloud complexes is a steady increase in the velocity dispersion of a coeval disk population with time, according to

$$\sigma[v(t)] \sim \sigma[v(0)][1 + (t/\tau)]^{1/3},$$

where $\sigma[v(0)]$ is the initial velocity dispersion of the stellar population, and τ is the characteristic timescale for energy exchange in the collision process. The value of τ is about several 0.1 Gyr. This approximate functional form for the rate of stellar dispersion increase is evident in the data in Figure 25.

The leveling out of the velocity increase for older stars reflects a *saturation effect* with the scattering process: As the velocity dispersion of a population increases, the average time between encounters increases (as stars spend more time away from the thin molecular cloud layer), and the rate of velocity dispersion increase slows. While not universally observed among all studies to date (perhaps relating to problems of separating thin disk and IPII/thick disk stars from one another), some studies (as shown above) show a saturation effect at ages well within the domain of thin disk stars alone. In either case, given that the present observed density of giant molecular clouds is only sufficient to increase the vertical velocity dispersion σ_W to at most $15 - 25$ km s^{-1} (Villumsen 1985, Binney & Lacey 1988), it is difficult to understand how the Spitzer-Schwarzschild process can account for the velocities of the oldest thin disk stars, let alone the *IPII* stars, which appear to have $\sigma_W \sim 45$ km s^{-1}. Moreover, it is hard to understand how disk globular clusters, and the high Z-distance open clusters, could have been pumped up to such large distances above the plane with a secular heating process.

Of other proposed secular heating mechanisms, (1) the passage of spiral arm waves (Barnabis & Woltjer 1967, Carlberg & Sellwood 1985) appears to be coupled mainly to increasing velocity dispersions *within* the plane, and unable to produce significant dispersion perpendicular to it (Carlberg 1987), and (2) a vast sea of fast moving, massive ($10^6 M_\odot$) objects (supermassive black holes?) penetrating the disk and creating short-lived perturbations, while able to heat some stars to large enough velocity dispersions, is apparently unable to form a 10% mass (see Figure 12) IPII component (Ipser & Semenzato 1985, Lacey & Ostriker 1985). If they are the major constituent of the dark matter, some $10^5 - 10^6$ of these $10^6 M_\odot$ objects would be needed, and one might expect that they would have been seen in the various microlensing experiments. Indeed, the latter experiments suggest that the main contributor to the microlensing phenomenon are only of mass 0.5 $M_\odot$.

Secular heating almost certainly acts on disk stars, but is insufficient to accommodate the large velocity dispersions seen in the older disk and IPII stars. Alternatively, the disk may have experienced one or more episodes of *violent heating*, as expected during *minor mergers* of small satellite galaxies. Due to the processes of dynamical friction, the orbits of satellite galaxies in the presence of dark matter halos should decay and lead to accretion by parent galaxies. The Milky Way is in the process of just such an accretion event, as the recently discovered Sagittarius galaxy clearly is in the midst of tidal disruption (see below). Dynamical simulations show that such merger events can

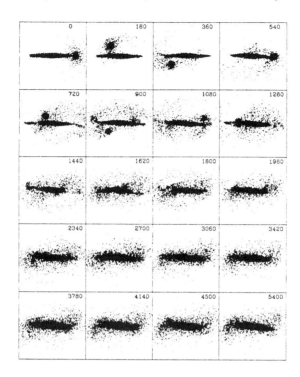

FIGURE 27. Edge-on view of a merger between a disk galaxy and a satellite galaxy with 10% the mass of the parent galaxy. The panels are each 30 kpc in size and the time steps indicated are in Myrs. Initially the satellite's orbit is inclined by 30° to the disk plane, but this orbital plane becomes aligned with the disk plane before the satellite's orbit shrinks radially. From Hernquist & Quinn (1993).

lead to the formation of thickened, heated disks that resemble the IPII (Figure 27). Because thin disks are "destroyed" in the merging process, most of the thin disk we see today must have formed subsequently to any merger event. Therefore, the thickened stars must be older than the oldest stars in the thin disk, and this may well be the case in our Milky Way. If this scenario is the explanation for the origin of the IPII, then, by the apparent age of that component, we have a relatively tight constraint that the merger event must have occurred early on in the life of the Milky Way, and, moreover, that no event of that magnitude has occurred since. It is interesting to note that since thick disks do not appear to exist in all disk galaxies (van der Kruit & Searle 1981, Morrison et al. 1994), the formation of a thick disk may not be a requisite phase of disk galaxy evolution. Thus, a stochastic process for the formation of the thick disk, like minor merger events, seems consistent with observations of external galaxies.

"Top-down" Formation

The idea that the disk kinematics we observe now may have been formed *in situ* has roots in the model of ELS. A key component of their model is the idea of "spin-up" – that as a proto-galactic cloud collapses it must preserve angular momentum and therefore it increases its rotational velocity as its radius shrinks. Another key component of the ELS model is that the proto-disk gas came from the leftover gas from the formation of the halo. While the timing of the formation of the IPII, which apparently began forming while the halo itself was still forming, may now preclude the halo-then-disk phasing of ELS†, the basic concept of spin-up, as it is a direct result of the conservation of angular momentum, must occur at some point in the formation of the Galaxy. Sandage (1990) proposes that the ELS spin-up, rather than applying to the halo and disk, may apply solely to the disk components, with the IPII representing the point at which the collapsing proto-disk gas first encountered partial pressure support (a slowing period in the collapse). Thus, the IPII might represent a transitional phase in the formation of the disk, perhaps the point at which the gas becomes predominantly supported against collapse by rotational support (any previously formed halo would be supported by the kinetic pressure provided by large anisotropic velocity dispersions of stars in predominantly radial orbits).

Given the apparent narrow age range of the IPII (as given, for example, by the disk globular clusters), its formation must have been rather rapid and early on in the Galaxy's life. Of course, this question is a function of how distinct the IPII is from the thin disk. If they are smoothly joined, then the requirement would seem to be only that the first phase of disk formation be dramatic, to produce quickly an extended population of stars *and* clusters, but not necessarily a population discrete in age. Both possibilities – a smooth, gradual formation of the entire disk (Larson 1976), or a disk formed with discrete components formed in separate bursts of star formation (Marsakov et al. 1990, Burkert et al. 1992, Katz 1992) – have been produced in chemodynamical models of disk formation without satellite merging. Pauses in the formation of stars during the collapse of the gas can come from one of several suppression mechanisms as a result of a previous star formation burst:

• *Disruptive mechanisms* in the form of tidal shocks, supernovae, and destructive cloud collisions could make it difficult for substantial star forming clouds to collapse (Gilmore 1984).

• *Temporary depletion of gas* of sufficient density to form stars might occur in a particularly vigorous starburst, and require the accumulation of more gas from larger radii to reinvigorate star formation. Initiation of star formation may require surpassing a threshold in mass density, and this threshold might be reached through replenishment from infall, possibly extragalactic or from a dying first generation of stars (Larson 1974, Gratton et al. 1996, Chiappini et al. 1997).

• *Intense heating* of the gas not used up in the previous starburst could come from the injection of energy, momentum, and metal-enriched material in the form of strong stellar winds (Marsakov & Suchkov 1977).

Thus, because of the possibility of suppression mechanisms in a dissipational collapse, *discrete* thick disk and thin disk components would seem to be an insufficient condition to discriminate between collapse and merger models for IPII formation. Note, even without a gap in *star formation activity*, it is possible to produce a rather chemodynamically distinct IPII and thin disk if the end of the IPII phase is coincident with a rapid increase in the rate of dissipation and collapse of the gas. One proposed possibility is that when the gas self-enriches to a certain metallicity ([Fe/H]~ -1), the dominant dissipation mechanism of the gas switches from the less efficient process of free-free transitions of electrons in encounters with positive ions (bremsstrahlung) to the much more efficient cooling by line radiation (Wyse & Gilmore 1988, Burkert et al. 1992). The increased

---

† Moreover, from data not presented here, it is fairly clear that the disk and halo are rather distinct from one another in a chemodynamical sense. That is, the halo and IPII are not smoothly joined, but rather, they are disjoint in their chemodynamical distributions. Combined with the near simultaneity of their age, it is difficult to see how their formations could have been part of a smooth, contiguous process. It is more likely that the leftover halo gas proceeded to make up the bulge of the Galaxy (Wyse & Gilmore 1993).

cooling rate causes rapid dissipation of energy and the cloud is able to collapse more quickly. The quick transition to higher densities probably induces an increased rate of star formation.

In summary, two formation scenarios for the formation of the old, thin disk and/or IPII seem to be viable at the present time: (1) some kind of merger event at early times, a process that we would expect to create (from the "bottom-up") a distinct, likely disjoint IPII component, with likely little kinematical gradient perpendicular to the disk, or (2) some form of "top-down" formation with ELS spin-up during a dissipational collapse, which may or may not form an IPII that is discrete and disjoint from the thin disk, but which might show a global kinematic gradient.

In my opinion, despite, and *because of*, several recent claims each way, the jury is still out on the discreteness of the thin and thick disk populations. Without a clear idea of whether and how the disk should be divided into thin/thick disk populations, it is too early to discriminate between models, and it is certainly premature to attempt to construct a true *birth* dynamical population box for the disk populations. The best we can do at present is formulate what the *observed* distribution of thin and IPII thick disk dynamics are like (Figure 26).

3.4. *The Halo*

3.4.1. *Age Characteristics of the Halo*

A great deal of what we know about the age of the halo comes from the halo globular cluster system. Indeed, the halo globulars provide critical constraints on the age of the universe, and, thereby, the size of the Hubble Constant. A few constraints on the age of the halo come from halo stars, but generally this work is less well developed compared to the long history associated with age-dating globular clusters:

- The *white dwarf luminosity function* for halo stars obviously would provide a very useful constraint on age, or at least a convenient relative chronometer to the disk. However, as shown in Figure 17 above, data on field halo white dwarfs is extremely scant, so the techniques described for age-dating the disk with this technique cannot yet be brought to bear on the halo field. But, because halo white dwarfs are considered the "least unlikely" explanation for the MACHO microlensing events (Méra et al. 1998), a number of new searches for halo white dwarfs are now underway. In the meantime, deep HST imaging on halo globular clusters now makes possible the production of CMDs reaching faint enough to see the cluster white dwarf sequences (Richer et al. 1997; see King and Castellani contributions to this volume). Thus, we are on the verge of bringing a new technique to verify the cluster age scale as determined from isochrone fitting, or, inversely, we can use the latter as a check on white dwarf cooling theory.

- The *main sequence turn-off color* for the most metal-poor subdwarfs is near $B - V = 0.36$, which is similar to the main sequence turn off of the metal-poor clusters M15 and M92 (Sandage & Kowal 1986, VandenBerg 1991). This confirms that the age for the bulk of the youngest halo field subdwarfs is similar to that of the oldest globular clusters.

- *Strömgren photometry* of metal-poor, halo stars by Marquez & Schuster (1994) does show that there is an age scatter of several Gyr in the halo field. Moreover, they find that the stars with apogalactica outside of 10 kpc have a mean age some 2 Gyr *younger* and a smaller age dispersion than the stars interior to this radius. This result is in direct contradiction to expectations from the original ELS picture, which describes a more or less ordered progression of Milky Way formation from the outside in. The Marquez & Schuster results are consistent with the latest findings concerning the order of formation of the halo globular cluster system (Zinn 1993a, Sarajedini et al. 1997).

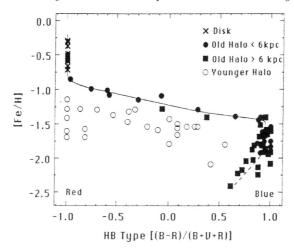

FIGURE 28. The distribution of Galactic globular clusters as a function of abundance, [Fe/H], and horizontal branch type. From Zinn (1993a).

• The *halo globular clusters* have provided us with some of the most important breakthroughs in understanding the formation of the halo in the past several decades. In principle the mean age of the halo globular clusters sets the time of halo formation, while the existence of any cluster age *range* defines the duration of the formation epoch for the halo. The former is more difficult to determine since absolute ages depend critically on accurate determinations of cluster abundances (not only [Fe/H], but [O/H], etc.), as well as distance moduli, which must be known to 0.10 magnitude accuracy to achieve a 10% precision in the age. There has been up to a factor of two range in the age scale among various groups in recent years, although the most recent results based on HIPPARCOS data (see review by Reid 1999) tend toward the lower age range (i.e., clusters more distant, on average, than previously thought). The details of absolute age determinations are beyond the scope of the present discussion, but see discussion by Castellani (also, e.g., Bergbusch & VandenBerg 1997, Gratton et al. 1997, Chaboyer et al. 1992, 1998). Here we focus on *relative* cluster ages, where some of the most fruitful results are found from the point of view of Galactic structure and formation studies.

The availability of high precision color-magnitude diagrams and accurately determined, spectroscopic metallicities and radial velocities has made a substantial impact in the study of differential ages in globular clusters, to where relative ages can in most cases be determined to 1 Gyr or better (modulo, of course, the age scale based on the adopted distances and isochrones) between appropriately chosen pairs of clusters. Most revealing are studies of "second parameter pairs" – pairs of clusters with the same abundance but very different looking horizontal branches: one cluster with a predominantly red HB and the other one with a blue HB. The HB morphology can be quantified as $(B - R)/(B + V + R)$ – where B is the number of horizontal branch stars blueward of the RR Lyrae gap, R is the number of HB stars redward of the gap, and V is the number of RR Lyrae stars. The "primary parameter" driving the nature of the horizontal branch is mean abundance, [Fe/H]. However, the distribution of cluster HB morphologies against [Fe/H] reveals a spread of HB morphologies for the abundance range $-2 < [Fe/H] < -1$ (Figure 28). Some examples of different HB morphologies in clusters are shown in King's contribution to this volume. The main dependence of HB Type on [Fe/H] is obvious in Figure 28. The *second parameter* that causes the spread in HB Type at any given

[Fe/H], and which may actually be a combination of effects (i.e., "second parameter" = "second parameter" + "third parameter" + "fourth parameter" + ...), has been variously ascribed to, or partly to, a number of causes, including (see Fusi Pecci et al. 1996 for a summary of how some of these effects operate on the position of the HB):

- age
- cluster density
- rotation
- α-capture element abundances
- oxygen abundances affecting mass loss rates
- helium abundances and "deep mixing" of helium in highly convective stars
- helium core mass at helium flash
- differences in the number of planets swallowed up in the red giant phase.

While many details need to be worked out, and there is certainly no consensus on this, evidence seems to be pointing to age as at least one (though possibly not the sole) second parameter dictating HB morphology, after [Fe/H] (Sarajedini et al. 1997). An age range in the halo globulars, suspected since the seminal study by Searle & Zinn (1978, "SZ" hereafter) discussed the strong second parameter effect in the outer halo, is now generally accepted: Even groups that strongly disagree on the source of the second parameter and the amount of age spread in the halo globular clusters admit to finding at least *some* globulars that are clearly younger (by 1 or more Gyr) than the bulk of the halo system, and the usual culprits (the likes of Arp 2, IC 4449, Rup 106) have unusually red HBs for their metallicities (though some others – Pal 12 and Ter 7 – are so metal rich that their red HBs are expected by the primary parameter). The list of identified "younger" halo globular clusters encompass almost the full range of halo cluster metallicities. Surveying these results, an age spread of 1 Gyr seems certain, and this is generally found by groups in the compressed cluster age scale camp (e.g., Reid 1998). At the other extreme, and generally claimed by groups in the large cluster age scale camp, a halo cluster age spread has been given as large as 5 Gyr (Chaboyer et al. 1992, Sarajedini et al. 1997).

Most intriguing are *relative* age rankings of objects in *different* cluster populations. The comparison of the CMDs of several halo clusters (Pal 12, Terzan 7) to that of the archetypal disk globular 47 Tuc reveals these particular halo clusters to be about 30% *younger* (Buonnano et al. 1998, Rosenberg et al. 1998); clearly, some clusters in the halo were still forming long after the disk began to form, as first suggested by SZ. According to the MAI analysis in Figure 19, there are even *open clusters* older than the halo globulars clusters Terzan 7, Arp 2, Pal 12 and IC4449.

Detailed analysis of the abundance and kinematical properties of groups of clusters selected via their distribution in the [Fe/H]-HB Type plane has led to important breakthroughs in our paradigm for the structure and origin of the globular cluster system and its subsystems since the simpler picture presented in Figure 16a. It was noted by SZ that the second parameter effect was most prominent in the *outer* halo, while at the same time there is no radial abundance gradient in the $R_{GC} > 8$ kpc globulars; on the assumption that age was the second parameter, these observations prompted SZ to conclude that the outer part of the halo formed slowly. But, on the basis of the large abundance spread *among* the clusters in the outer halo, coupled with the fact that there is little abundance spread *within* globulars, the outer halo clusters must have formed within larger structures (they used the term "fragments") that were independently evolving, self-enriching, and mixed internally. Later these transient fragments could be destroyed by supernovae or tidal effects, and the débris continued to fall into dynamical equilibrium with the Milky Way well after any main collapse of the Galaxy had begun to form the interior regions.

A slow, SZ-like halo formation (spanning $3 - 5$ Gyr according to some studies) requires repositories of gas in the halo ("fragments") and with a range in enrichment levels well after any global collapse would have collected mass to the central regions. Natural candidates for such repositories would be either (1) smaller primordial density fluctuations left behind by a main collapse (presumably at large Galactocentric radius) and on a "slow path" to initial dwarf galaxy star formation, or (2) larger, satellite galaxies capable of sustaining multiple star formation epochs and enrichment. As sites of prolonged star, and presumably cluster, formation (see the example of the Carina dwarf, in Figure 5), satellite galaxies could contribute "young halo" clusters (as well as, of course, older clusters) to the Galaxy after being tidally disrupted. As if on cue, Nature and Fortune provided us with the recently discovered satellite galaxy of the Milky Way in Sagittarius (Ibata et al. 1995), which is now observed to be in the process of tidal disruption. In the process of this tidal disruption, Sgr is contributing second parameter clusters (Arp 2, Terzan 7) to the Galaxy, as well as two other clusters (M54, Ter 8) with "normal" HB types.

Other evidence suggests an association of second parameter clusters to dwarf satellite galaxies of the Milky Way. Two second parameter globular clusters, Pal 12 and Rup 106 have been shown (Lin & Richer 1992) to have radial velocities consistent with being tidal débris from the Magellanic Clouds. More detailed statistical analysis of the locations and possible orbits of all of the second parameter clusters show them to have a high degree of association with the positions and orbits of the dwarf satellite galaxies of the Milky Way (Majewski 1994b, Palma & Majewski 1999), while the globular clusters *still attached* to dwarf satellites of the Milky Way are almost exclusively of the Zinn "young halo" type with second parameter, redder HBs (Zinn 1993b). The weight of evidence favors a picture in which the second parameter clusters are derived from the tidal stripping of dwarf satellite cluster families by the Milky Way.

How does this affect our previous "division" of the globular cluster system into "disk" and "halo" systems? Zinn (1993a) demonstrates that there are actually two halo globular cluster systems that show distinctly separable chemodynamical properties (see also van den Bergh, 1993). The basis of the separation is the second parameter effect. Clusters with the bluest HB at any given [Fe/H] (those clusters along the ridge line in Figure 28), i.e., those that show no second parameter effect, define Zinn's "BHB" sample. If age is the second parameter, these objects are presumably the *oldest* clusters at every [Fe/H], and define the "old halo" in Zinn's (1993) analysis†. All clusters falling away from this ridge line, the "RHB" sample, show a second parameter effect, and are termed the "young halo" clusters by Zinn (1993a). With this division into halo subpopulations, we find the properties shown in Table 2. As noted previously by Searle & Zinn, the outer halo has a dominant fraction of second parameter, "young clusters", so that the Galactic density distribution of RHB clusters is essentially spherical. In contrast, the BHB clusters are in a more oblate configuration, with generally smaller Z-heights. Note too that, as found by SZ in the outer halo, the RHB clusters at large R_{GC} show no abundance gradient, while the BHB clusters, which are predominantly at smaller R_{GC}, do show an abundance gradient. Finally, the kinematics of the RHB clusters are vastly more extreme than those of the BHB clusters, with "hotter" orbits (Dinescu et al. 1999) and a mean rotation that is marginally (with respect to the uncertainty) *retrograde*. If true, the mean retrograde rotation rules out the possibility that the RHB could have

---

† In a more recent discussion, Zinn (1996) further divides this "non-second parameter" sample into two groups, the "BHB" clusters with [Fe/H]≥ -1.8 and the "metal-poor" ("MP") clusters with [Fe/H]< -1.8. The "second parameter" halo clusters are called the "RHB" sample in Zinn (1996)

	BHB/"Old Halo"	RHB/"Young Halo"	All Halo
V_{rot}(km s^{-1})	70 ± 22	-64 ± 74	44 ± 25
σ_{los}(km s^{-1})	89 ± 9	149 ± 24	113 ± 12
distribution	oblate	spherical	flattens with small R
metallicity	~ 0.5 dex from $R < 6$ to 40 kpc	no gradient	no gradient $R > 8$ kpc or $|Z| > 4$ kpc

TABLE 2.

formed in a grand collapse of the Galaxy, since the direction of angular momentum of a collapsing cloud cannot change directions. Rather, the properties of the RHB cluster system, taken as a whole, support the idea that these objects were *accreted*.

In contrast, the properties of the BHB, "old halo" cluster system are consistent with a general collapse with ELS spin-up. Zinn (1993a) suggests that the BHB globulars may simply represent a more extended, metal-poor, kinematically hotter extension of the disk globular system (Figure 29). Indeed, several rather metal poor, BHB clusters (NGC 6254, NGC 6626 and NGC 6752) are found to have rather IPII-like orbits (Rees & Cudworth 1991, Dinescu et al. 1999). Taken together, the *disk* globular clusters and the *old halo, BHB* globulars seem to be consistent with formation in a single dissipational, collapsing, metal-enriching entity. Such a progression is suggested by the distribution of Z-max distances derived from orbits determined for clusters with derived proper motions (Figure 16b). A new paradigm for the division of the Galactic globular cluster system, based on Zinn's (1993a) analysis, is shown in Figure 16c.

3.4.2. *Dual Halo Scenarios*

The need for two separate, metal-poor, halo *globular cluster* populations in our description of the Galaxy is apparently mirrored in the halo *field star* population. "Dual halo" models for the distribution of metal-poor stars in the Galaxy have become increasingly popular since Hartwick's (1987) study of the spatial distribution of metal-poor RR Lyrae stars (see examples in Table 1). In these models, the space distribution of "halo" stars incorporates both a spherical component as well as a component that is significantly flattened (with exponential distributions perpendicular to the disk of approximately 2 kpc scaleheight). Such a dual component distribution has been used to describe the spatial properties of metal-poor, field RR Lyraes (Hartwick 1987), blue horizontal branch stars (Kinman et al. 1994), and ordinary main sequence stars, whether selected by low metallicity (Allen et al. 1991, Sommer-Larsen & Zhen 1990) or by large proper motion (Carney et al. 1996).

Curiously, while studies of metal-poor field populations have moved towards "dual halo" descriptions, studies of the Galactic IPII thick disk have pushed the metallicity limit of this population to as low as [Fe/H]= -1.6 and, in some cases, as low as [Fe/H]= -2.05 or less (see discussion in Section 3.3.2). If one accepts that such metal-poor IPII stars exist, then the ability to distinguish these stars from similar metallicity stars in any "flattened halo" becomes extremely difficult: The scaleheight of some studies of the IPII (Figure 12) is similar to that proposed for the "flattened halo" — but with a higher density normalization so that the IPII dominates the flattened halo. Moreover, at large distances from the Galactic plane the kinematics of the IPII become much more extreme in some studies (see Majewski 1993) and less separable from those of halo stars. These

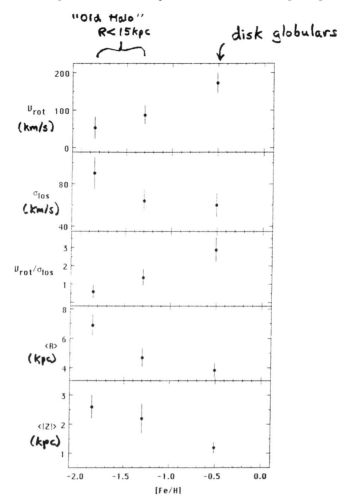

FIGURE 29. The evidence suggesting that the disk globular cluster system and the BHB, "old halo" globular clusters may represent a single continuous population formed in a dissipational, ELS-like collapse. Note the smooth progressions of spatial, chemical and kinematical properties of BHB, "old halo" clusters and disk globular clusters. From Zinn (personal communication).

sources of confusion may be part of the reason for the discrepancies in the determined relative numbers of stars in the flattened and spherical halo components: For example, the Kinman et al. (1994) HB stars yield a relative apportionment of flat to spheroidal halo components locally as 4:1, but Sommer-Larsen & Zhen (1990) quote a balance of 2:3 for their metal poor stars. Indeed, the difficulty in distinguishing between stars in the IPII thick disk and those in a "flattened halo" calls into question the *need* to distinguish between them (Majewski 1995), at least until real evidence exists that such a division is warranted. If, as has been suggested above, the combination of the BHB/"old halo" and disk globular clusters may represent one dissipational entity, might not a similar picture work satisfactorily for the field stars (particularly if the field stars are partly derived from the break-up of globular clusters)?

In such a model, the field star analogy to the "disk + old halo" cluster system might be the Population I thin disk + IPII + "flattened halo", where the latter represents the most extreme tail of a more or less continuously distributed, Norris-like (1987) "extended disk". Thus, in this view, as has been proposed for the globular clusters, we have two origin scenarios for the metal weak field star populations: one dissipational and one by accretion of satellites. Such a picture weds the Eggen, Lynden-Bell & Sandage (1962) collapse model − applied to the Population I thin disk + IPII + flattened halo − with the accretion scenario of Searle & Zinn (1978) for the more spherical halo; this marriage has been discussed variously by Sandage (1990), Majewski (1993), Norris (1994) and others. Such a unified picture, of course, must be held with caution in the face of some evidence that the IPII thick disk may have formed via violent dynamical heating. However, even in this case the ELS + SZ picture may need only to be modified: Perhaps the "flattened halo" + thin disk were formed originally as a dissipational entity, and the IPII was created soon after by the merger of an early satellite into the disk. As can be seen, studies of the thin disk, IPII and inner halo populations are ripe for more work before a comprehensive origin scenario can be proffered with more certainty.

3.4.3. *Other Evidence for Accretion in the Halo*

In contrast to the impasse just described for the more flattened, inner components of the Milky Way, there is a growing cumulation of evidence for Galactic accretion events, particularly in the outer halo. Indications of accretion of globular clusters from satellite galaxies and into the outer halo of the Milky Way have already been discussed. The same tidal stripping processes that would lead to the accretion of clusters by the Milky Way should also lead to the accretion of stars that are the tidal débris of disintegrating parent objects, presumably dwarf galaxies like Sagittarius. Theoretical modeling of satellite encounters with massive, Milky Way type galaxies (McGlynn 1990, Moore & Davis 1994, Velázquez & White 1995, Johnston et al. 1996a,b) show that stellar systems with relatively small (≤ 10 km s^{-1}) velocity dispersions (globular clusters or dwarf spheroidal satellite galaxies) may be tidally pulled into long and long-lived streams of stars strung out along the orbit of the disintegrating parent body (as modeled by Toomre & Toomre 1972). † The tidal arms of Sagittarius galaxy have now been mapped to a stretch of more than 40° of the sky (Siegel et al. 1997, Mateo et al. 1998). The destruction of Sgr may provide us with a useful paradigm for both the structure and origin of the halo, for, unless we are at some privileged epoch, events like the destruction and dispersal of Sgr must either be reasonably common or extremely long-lived.

To unwitting observers that happen to be studying a particular direction of the sky along which one of these tidal streams lies, the coherent débris trails would be observed as *moving groups* of halo stars. Indeed, it was as such a moving group that the Sagittarius galaxy first became known (Ibata et al. 1995). Other evidence for "moving groups" of halo stars has occasionally been alluded to by *in situ* surveys of the Galactic halo. Typically (and often as a "by the way" discussion) authors have pointed out possible halo moving groups manifested as small numbers of stars in a particular survey field having the same distance and radial velocity (see review by Majewski et al. 1996a). However, the possibility, postulated by Oort (1965), that tidal streamers may be a pervasive, even dominant, component of the field halo star population is suggested by the clumpy halo velocity distribution in the magnitude-limited survey of Majewski et al. (1996b). More recent searches (Harding et al. 1998, Majewski et al. 1998) specifically directed at finding this phase space substructure in the halo have shown that halo moving groups

† Because of stronger phase mixing in the less spherical potential of the inner halo, *spatial* coherence from accretion is expected to have a shorter lifetime closer to the Galactic center, although coherence in *velocity* should remain, even increase, with time (see Helmi & White 1999).

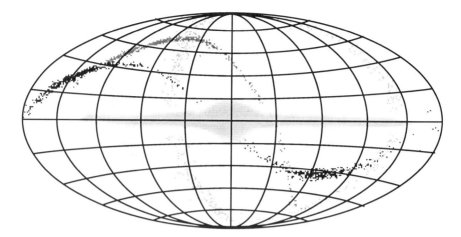

FIGURE 30. Model map of a Milky Way with a halo made from a (small) family of disrupted satellites. Modified from Johnston (1996a).

are relatively common. It may well be that the outer Milky Way contains a network of crossing tidal débris streams like that shown schematically in Figure 30.

That the halo field star population contains the remnants of disrupted satellite galaxies has long been suspected from the existence of A type stars with high velocity (Rodgers et al. 1981, Lance 1988). Main sequence stars of this spectral type must be relatively young, but their velocities and distances above the Galactic plane are difficult to reconcile with their age if the stars were formed in the Population I disk. Other examples of kinematically hot, yet apparently young stars can be seen in the left panels of Figure 25 (see also Soderblom 1990). These stars could be explained if their formation sites were within objects already having halo-like kinematics, i.e., within one or several recently disrupted satellite galaxies. Alternatively, these stars may have been created as a result of the impact of high velocity gas with the disk (Lance 1988). Contributing to the peculiar age distribution of dynamically "halo-like" stars is a population of "blue metal poor" stars that are apparently only 3 − 10 Gyr old (Preston et al. 1994); these stars have also been attributed to the break-up of a Galactic satellite within the last 10 Gyr.

The existence of "young" stars in the halo points to dwarf satellite galaxies, which are known to contain intermediate age (e.g., Carina; Figures 4 and 5) and even young populations (the Magellanic Clouds, Carina), as the likely progenitors of tidal débris. However, as we have seen (Section 1.2), globular clusters should also contribute to the field star population (Surdin 1995). Indeed, Oort (1965) suggested that the "pure races of the halo population II" (presumably the extended, spherically distributed field stars counterpart to the Zinn RHB/"young halo" clusters described above) might contain significant structure in the form of intermingling "tube-like swarms" from the breakup of some hundreds or even thousands of globular clusters.

Harris (1991) points out possible problems with this scenario: Globular cluster orbits appear to be more isotropically distributed than those of halo stars and globular clusters are more metal-poor than halo stars at the same R_{GC}. However, Lee & Goodman (1995) point out that these might not be problems when one considers that the orbits of *surviving* clusters are expected to be less eccentric than those of *destroyed* clusters (see Section 1.2). In the end it is likely that globular clusters do make at least some contribution to the halo field star population, as tidal tails have now been observed around some globular clusters (Grillmair 1998).

3.5. *The Bulge*

At a mass of $2 - 4 \times 10^{10} M_\odot$ (Blum 1995), the bulge is nearly half the mass of the disk and of order 30% the Galaxy's luminous mass. Yet the bulge population is nearly entirely confined within a radius of 1 kpc (compared to the 25 kpc radial extent of the disk, or the > 100 kpc extent of the stellar halo. Other defining characteristics of the bulge include *triaxiality* (the presence of a bar), relatively rapid rotation, and high mean metallicity. Unfortunately, the Galactic center happens to be at the point of highest density for *all* stellar populations, a more troublesome situation for separating Galactic populations than that already discussed with regard to separating the thin disk, IPII and halo near the Sun. The properties of the bulge population overlap with those of the inner disk to a degree that clear membership of individual objects cannot be assigned. I now turn to a brief description of the bulge as a stellar population.

Structure: It is thought that the shape of the bulge mimics that of the near infrared surface brightness distribution seen by the DIRBE experiment on the COBE satellite – that is, a triaxial shape consistent with a bar, and even better fit by a boxy shape (Dwek et al. 1995). Strong evidence for a bar, with some tilt to the Galactic plane, is cited by Blitz & Spergel (1991) on the basis of 2.4μ maps of the Galactic center. It is found that the distribution of Mira variables generally follows the near infrared light distribution, while the late M type stars drop off faster, with a power law of $n = -4.2$ (Whitelock 1993). On the other hand, the distribution of RR Lyrae stars in the bulge follow an $n = 3$ power law, which suggests that they may largely be representatives of the inner halo (Minniti 1996). This is supported by the finding that only in the innermost bulge do the RR Lyrae stars show any evidence for a bar (Alcock et al. 1998). In contrast, metal-rich population stars – red giant clump stars (Alcock et al. 1998), asymptotic giant branch stars (Weinberg 1992), and Miras (Whitelock 1993) – show a clear bar-like distribution.

Abundances: With a recent downward revision of the abundance scale for bulge giants (McWilliam & Rich 1994), a mean abundance of [Fe/H]=-0.25 is obtained in the inner kiloparsec of the bulge; however, stars as metal rich as [Fe/H]=$+0.5$ are still found. The abundance distribution function is broad (Rich 1988, Minniti et al. 1995, Harding 1996) and, unlike the disk, well fit by a closed box, chemical evolution model, so that there is apparently no G dwarf problem in the bulge (Rich 1990). It is significant if the mean metallicity of the bulge is less than the disk in the solar neighborhood, especially considering that the disk shows a trend of *increasing* abundance towards the center: Either the bulge formed *before* the disk or it has been replenished by infall of low metallicity gas. Leftover gas from the halo would naturally sink to the bulge, not the disk, based on the low angular momentum of the halo; thus it is possible that the halo preceded formation of the bulge (Carney et al. 1990, Wyse & Gilmore 1992), an idea consistent with many age determinations (see below).

While Tyson & Rich (1993) find no evidence for a metallicity gradient in the inner bulge, a point they use to argue for bulge formation via a dissipationless collapse, a summary of previous determinations of mean metallicities in a range of bulge fields by Minniti et al. (1995) shows a clear abundance trend in the inner 3 kpc. However, an unanswered question is whether this gradient is intrinsic to the bulge population itself, or the result of changing contributions by halo, disk and bulge stars in the inner Galaxy. A true metallicity gradient in the bulge would support its formation in a *dissipational* collapse. Minniti et al. (1995) argue that a comparison of the number of stars at the location of metal-rich giants in color magnitude diagrams from different bulge fields gives strong evidence for a metallicity gradient in the inner 1 kpc of the bulge.

It is interesting to note that the abundances of RR Lyrae in the bulge range across $-0.3 \geq$[Fe/H]≥ -1.65, with a mean around [Fe/H]$= -1$ (Walker & Terndrup 1991); this is significantly more metal poor than the mean abundance for other bulge stars, and is further evidence that most of the RR Lyrae may not be a true bulge population.

Kinematics: The dynamics of the gas in the inner bulge also suggests that it is strongly dominated by a bar (Binney et al. 1991). The kinematics of metal poor stars in the bulge generally are found to have hotter velocity dispersions than the metal rich stars (Harding 1996, Minniti 1996, Tiede & Terndrup 1997), although it is possible that this is not entirely a real population difference, but instead a result of projection effects along the line of sight (Tiede & Terndrup 1997). Representative bulge kinematics are given by the results of Harding (1996) for K giants in a field at $(l, b) = (-10°, -10°)$: For a metal rich sample ([Fe/H]> -0.9) he finds $V_{rot} = 113 \pm 11$ km s^{-1} and an azimuthal velocity dispersion of 59 ± 12 km s^{-1}, while for metal poor stars ([Fe/H]< -1.5) he finds $V_{rot} = 29 \pm 21$ km s^{-1} and dispersion 123 ± 18 km s^{-1}. This difference has been argued by Harding and others to reflect the presence of two populations. The latter stars have kinematics typical of the halo, while Harding's metal rich giants have kinematics similar to other bulge tracers, e.g., Miras, OH/IR stars, and planetary nebulae. The hot, halo-like kinematics of the metal poor K giants are also found to be characteristic of the bulge RR Lyrae sample (Gratton 1987, Tyson 1991).

Age: The age distribution of the bulge population is still not well resolved. Direct measurement of the age of the stellar population by use of the color of the main sequence turnoff is complicated due to uncertainties in the extreme reddening, the spatial distribution along the line of sight, and the large abundance spread in the bulge. One attempt (Holtzman et al. 1993) suggests that there are a significant number of intermediate age (< 10 Gyr) stars in the bulge. Houdashelt (1995) also finds a best fit isochrone to Baade's Window stars for an 8.0 Gyr population with [Fe/H]$=-0.3$. The relative numbers of red clump to red giant stars in the bulge also argues for a < 10 Gyr population (Paczyński et al. 1994). However, Ortolani et al. (1995) find that two near solar abundance globular clusters in the bulge, NGC 6528 and NGC 6553, have ages within a few Gyr of the age of halo globular clusters. Moreover, Ortolani et al. , after aligning the red clump (horizontal branch) of these globular clusters and the bulge field stars, note a striking similarity in the luminosity functions that suggests little age difference between the bulge field and the these clusters. Ortolani et al. also find little spread in age for the bulge MSTO stars. This analysis indicates an age for the bulge field stars perhaps similar to that of metal rich disk clusters like 47 Tuc, though a still younger age for the bulge with respect to disk clusters (a difference of 4 Gyr) has been suggested by analysis of Baade's Window MSTO stars by Fullton (1995).

Minniti (1995, 1996; see also Harris, in this volume) has argued that on the basis of their kinematics, abundance, and concentration – which reflect those of metal-rich bulge giant stars – the metal-rich globulars in the innermost 3 kpc of the Galactic center should rightfully be associated with the bulge, rather than a metal-rich extension of the disk, as suggested by the representations in Figures 16c and 29. However, as Zinn (1996) points out, the apparent match of systemic rotational velocity between K giants and the innermost clusters is strongly influenced by a few clusters in the sample at larger $|l|$ that may well be members of the true disk system. Removing these clusters from the sample, or restricting the sample to the most metal rich clusters, removes any signal of rotation. Unfortunately, these analyses of the inner cluster population are affected by the small number statistics of the known clusters. The issue of the proper population assignment of the metal-rich, inner globulars, if settled, would be of great value because of the use of the clusters for dating populations. Age determinations for clusters under debate include several (NGC 5927 and NGC 6553) for which significantly younger ages are found compared to other disk globulars (Demarque & Lee 1992, Fullton 1995; but see Ortolani et al. 1995 and above).

A population of stars with rather younger ages is implied by the presence of Miras in the bulge (Glass et al. 1995) with periods longer than those found in globular clusters. The large number of > 300 day period Miras suggests an age for these stars more like the young disk Miras found in the disk near the Sun.

On the other hand, there are also RR Lyrae stars in the bulge, which might suggest that at least some stars in the bulge are old. Lee (1992) looked at the change in the abundance distribution of these RR Lyrae and concluded that the changes were the result of a *decreasing* age with R_{GC} which suggests an inside-out formation, with the oldest stellar population in the bulge forming about one Gyr before the inner parts of the halo. However, if a majority of the RR Lyrae stars belong to the inner halo, rather than the bulge (as suggested by their density law, velocity dispersion, and metallicity), quite the opposite conclusion can be reached. By removing these stars from the mix, and noting strong trends of kinematics with abundance for bulge giants, Minniti (1996) argues that the bulge may have formed in an ELS-like, dissipational collapse, after formation of the halo. On the other hand, Rich (1996) warns that some of the abundance-kinematics trends might be a product of disk star contamination.

In summary, the age distribution for the bulge obviously spans a large range. Certainly intermediate age stars are present, but it is not clear whether the true bulge population has stars with an age as great as that of the halo, or whether the oldest bulge stars are several Gyr younger than the halo globulars, perhaps with an age similar to the disk clusters. Determining the age distribution of bulge stars is of course critical to fitting it into a sensible timeline with the formation of the other stellar populations in the Milky Way. The confusing situation regarding the mean age of the bulge is partly a result of the fact that it is a repository for much "waste product" (stars and gas) in the evolution of the Milky Way. For example, we have seen (Section 1.2) that clusters with small apogalactica orbits face the quickest demise from bulge shocking; therefore we might expect at least some fraction of the bulge to be cluster débris. In particular, the inner RR Lyrae population could be tracing the remains of a destroyed halo globular cluster population. Clearly it would be interesting to determine the kinematics of these stars to determine if they are on elongated orbits expected of primordial clusters most susceptible to destruction. Note that on such orbits, débris from these clusters would not feel the bar potential long enough to respond to it, and this is consistent with the observations of the RR Lyrae population (Alcock et al. 1998).

4. Putting It All Together: Chemodynamical Pictures of Milky Way Formation

In this section I attempt, via the "population boxes", to illustrate the history of stellar populations in the Galaxy as the data presented here suggest it. Several cautionary notes are worth mentioning. First, the quick review here must necessarily gloss over details, and even whole areas of Galactic studies (like the wealth of information now available from detailed chemical abundance studies) have been left out. It is hoped that other teachers at the school will fill in some of the blanks.

Second, it would be ill-advised to believe that the ideas presented here enjoy a complete consensus within the community. While I have tried to be fair in presenting what I believe to be the main themes and general results in each area, it is unavoidable that personal taste and "leanings" towards certain models creep in. This is always the case when the answers are still out of reach, and several viable models to explain the data exist. Much work remains to iron out a number of problems in the ripe, and presently very lively, research area of Galactic populations. This, then, is a call of encouragement to the next generation of Galactic astronomers who, armed with bigger telescopes, much larger surveys, and more realistic chemodynamical models, will no doubt soon (perhaps already?!) look on the present discussion as hopelessly naïve.

Nevertheless, let me summarize what I believe to be the general themes of our understanding of Galactic stellar populations, as presented here. The result is presented schematically as the composite Hodge chemical population box of Figure 31 and the composite dynamical population box in Figure 32, which are essentially Figures 22 and 26 with the addition of halo populations. Here the boxes are illustrated in projection from above in order to show details. The SFRs for the disk populations have been taken directly from Figure 18, and, although the reliability of chromospheric age dating of disk stars is not free from criticism, the data presented in Figure 18 at minimum give a flavor for what the disk SFR may have been like. The SFRs shown for the halo and IPII components are pure speculation. In both figures, the bulge has been left out due to uncertainties in the age-metallicity-kinematics relations and the definition of the true bulge population, but it is possible that the evolution of the bulge is qualitatively similar to (and would therefore overlap with) that of the early disk as shown.

While much larger samples of stars with full chemical and kinematical data are still needed, the present observational data seem to call for chemodynamical models incorporating multiple formation processes (collapse, accretion) just to explain what previously would have been considered the "classical" halo populations of the Milky Way. I illustrate the "dual halo" model of the Galaxy with clearly separated populations in the population boxes. In all populations, a tail to lower abundances, as suggested by observed metallicity distribution functions, is indicated. In these highly schematic figures I illustrate the flattened halo, the IPII and the oldest parts of the disk as partly contiguous, though I point out again that it is still not at all clear whether the IPII and flattened halo are parts of the same population, or whether they are clearly separable populations. Perhaps the old, flattened halo is the extreme extension of the disk whereas the IPII is a manifestation of an accretion event. In either case – the IPII as a transitional phase of a general, dissipational, ELS spin-up collapse, or the IPII as the result of violent heating of the thin disk – the resultant *observed* population boxes, shown here, might look the same. Indeed, this is part of the problem of discriminating between these models.

The dynamical population box shown here is based on the *observed* properties of the various stellar populations. If the IPII were formed from an accretion event, then the stars in that population would have been born with higher $\log(|V_{rot}|/\sigma)$, perhaps closer to 0.5, and then pumped up to the presently observed value. The figure also conveys the SZ picture of the halo forming in "fragments" even while the inner parts of the Galaxy were organized into a flattened, dissipational structure with ELS spin-up. Note that the dynamics shown are for the "fragment" population *as a whole*; the *internal* dynamics of the individual fragments (dwarf galaxies? globular clusters? both?) would have smaller velocity dispersions.

Finally, the age scales of Figures 30 and 31 are meant to convey something of a middle ground between the nearly factor of two range among the various quoted values for the age of Galactic clusters. If the recent trend towards young globular cluster ages is born out, then the scale of the abscissas should, to first order, be compressed.

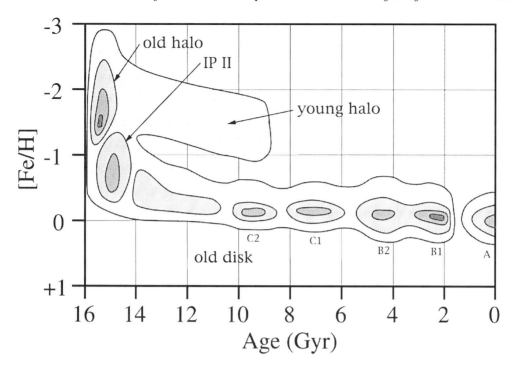

FIGURE 31. Schematic, composite Hodge population box for stellar populations in the Milky Way. The "box" is observed from above as a contour plot in order to show detail. The contour levels correspond roughly to 0.5 levels in SFR/<SFR> as shown in Figure 18 for the disk populations, with the lowest contour at 0.5. The bursts from Figure 18 are labeled.

It is a great pleasure to thank Ricky Patterson and my students, Jeff Crane, Johnny Johnson, Chris Palma, and Michael Siegel for their generous assistance with assembling this manuscript. Jeff Crane and Chris Palma went beyond the call of duty in helping to create some of the original figures. I also appreciate valuable conversations with and comments from colleagues, especially Eva Grebel, Kathryn Johnston, Bill Kunkel, Neill Reid, Bob Rood, Bill Saslaw, and Jesper Sommer-Larsen. Neill Reid and Ricky Patterson made numerous valuable suggestions for improving this manuscript.

REFERENCES

AGUILAR, L. HUT, P. & OSTRIKER, J. P. 1988 *ApJ* **335**, 720.

ALCOCK, C., ET AL. 1998 *ApJ* **492**, 190.

ALLEN, C., POVEDA, A. & SCHUSTER, W. J. 1991 *A&A* **244**, 280.

AMBARTSUMIAN, V. A. 1938 *Uch. Zap. L. G. U.* **22**, 19; English translation in 1985, *IAU Symposium 113: Dynamics of Star Clusters*, eds. J. Goodman & P. Hut, p. 521. Reidel.

ARMANDROFF. T. E. 1989 *AJ* **97**, 375.

ARMANDROFF, T. E., DAVIES, J. E. & JACOBY, G. H. 1998 *AJ* **116**, 2287.

BAADE, W. 1944 *ApJ* **100**, 137.

BAHCALL, J. N. 1986 *ARAA* **24**, 577.

BAHCALL, J. N. & SONEIRA, R. M. 1980 *ApJSuppl* **44**, 73.

BARNABIS, B. & WOLTJER, L. 1967 *ApJ* **150**, 461.

BARRY, D. C. 1988 *ApJ* **334**, 446.

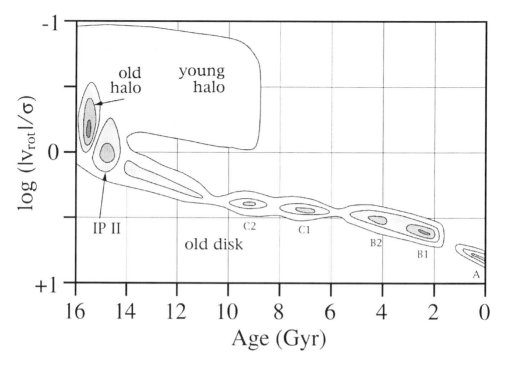

FIGURE 32. Schematic, composite dynamical population box for stellar populations in the Milky Way, *with properties as observed today.* As in the previous figure, the "box" is observed from above as a contour plot in order to show detail. The contour levels are the same as those shown in the previous figure. Note the use of $|V_{rot}|$ to accommodate the possibility of a retrograde "young halo".

BEERS, T. C. & SOMMER–LARSEN, J. 1995 *ApJSuppl* **96**, 175.

BERENDZEN, R. & HART, R. 1973 *JHA* **iv**, 73.

BERGBUSCH, P. A. & VANDENBERG, D. A. 1997 *AJ* **114**, 2604.

BINNEY, J., GERHARD, O. E., STARK, A. A., BALLY, J. & UCHIDA, K. 1991 *MNRAS* **252**, 210.

BINNEY, J. & LACEY, C. 1988 *MNRAS* **230**, 597.

BLITZ, L. & SPERGEL, D. N. 1991 *ApJ* **379**, 631.

BLUM, R.D. 1995 *ApJ* **444**, L9.

BROOKS, K. 1981 *Ph.D. Thesis.* Univ. California (Berkeley).

BUONANNO, R., CORSI, C. E., PULONE, L. FUSI PECCI, F. & BELLAZZINI, M. 1998 *A&A*, **333**, 505.

BURBIDGE, E. M., BURBIDGE, G. R., FOWLER, W. A. & HOYLE, W. A. 1952 *Rev. Mod. Phys.* **29**, 547.

BURKERT, A., TRURAN, J. W. & HENSLER, G. 1992 *ApJ* **391**, 651.

BURSTEIN, D. 1979 *ApJ* **234**. 829.

BURSTEIN, D. & HEILES, C. 1978 *ApJ* **225**, 40.

BURSTEIN, D. & HEILES, C. 1982 *AJ* **87**, 1165.

BUTCHER, H. R. 1987 *Nature* **328**, 127.

CARLBERG, R. G. 1987 *ApJ* **322**, 59.

CARLBERG, R. G. & SELLWOOD, J. A. 1985 *ApJ* **292**, 79.

CARNEY, B. W. 1993 In *Galaxy Evolution: The Milky Way Perspective*, ed. S. R. Majewski,

Astron. Soc. Pac. Conf. Ser., vol. 49, p. 83. ASP.

CARNEY, B. W., LAIRD, J. B., LATHAM, D. W. & AGUILAR, L. A. 1996 *AJ* **112**, 668.

CARNEY, B. W., LATHAM, D. W. & LAIRD, J. B. 1989 *AJ* **97**, 423.

CARNEY, B. W., LATHAM, D. W. & LAIRD, J. B. 1990 *AJ* **99**, 752.

CASERTANO, S., RATNATUNGA, K. & BAHCALL, J. N. 1990 *ApJ* **357**, 435.

CHABOYER, B., DEMARQUE, P., KERNAN, P. J. & KRAUSS, L. M. 1998 *ApJ* **494**, 96.

CHABOYER, B., DEMARQUE, P. & SARAJEDINI, A. 1996 *ApJ* **459**, 558.

CHABOYER, B., SARAJEDINI, A. & DEMARQUE, P. 1992 *ApJ* **394**, 515.

CHAMBERLAIN, J. W. & ALLER, L. H. 1951 *ApJ* **114**, 52.

CHANDRASEKHAR, S. 1943a *ApJ* **97**, 255.

CHANDRASEKHAR, S. 1943b *ApJ* **97**, 263.

CHERNOFF, D. F., KOCHANEK, C. S. & SHAPIRO, S. L. 1986 *ApJ* **309**, 183.

CHIAPPINI, C., MATTEUCCI, F. & GRATTON, R. 1997 *ApJ* **477**, 765.

CHIBA, M. & YOSHII, Y. 1998 *AJ* **115**, 168.

CLAYTON, D. 1989 In *Fourteenth Texas Symposium on Relativistic Astrophysics*, New York Acad. Sci. Ann., vol. 571, p. 79.

CROSWELL, K., LATHAM, D. W., CARNEY, B. W., SCHUSTER, W. & AGUILAR, L. 1991 *AJ* **101**, 2078.

CURTIS, H. D. 1915 *Popular Astronomy* **xxiii**, 599.

CURTIS, H. D. 1921 *Bull. Nat. Res. Council* **2**, 194.

D'ANTONA, F. & MAZZITELLI, I. 1979 *A&A* **74**, 161.

DE VAUCOULEURS, G. 1948 *Ann. d'Astrophys.* **11**, 247.

DEMARQUE, P. & LEE, Y.-W. 1992 *A&A* **265**, 40.

DINESCU, D. 1998 *Ph.D. Thesis*. Yale University.

DINESCU, D., GIRARD, T. M. & VAN ALTENA, W. F. 1999 *AJ, in press*.

DJORGOVSKI, S. 1993 In *Structure and Dynamics of Globular Clusters*, eds. S. G. Djorgovski & G. Meylan. Astron. Soc. Pac. Conf. Ser., vol. 50, p. 373. ASP.

DWEK, E., ET AL. 1995 *ApJ* **445**, 716.

EDVARDSSON, B., ANDERSEN, J., GUSTAFSSON, B., LAMBERT, D., NISSEN, P. & TOMKIN, J. 1993 *A&A* **275**, 101.

EGGEN, O. J., LYNDEN-BELL, D. & SANDAGE, A. 1962 *ApJ* **136**, 748 (ELS).

FOWLER, W. A. 1987 *QJRAS* **28**, 87.

FRIEL, E. D. 1995 *ARA&A* **33**, 381.

FULLTON, L. K. 1995 *Ph.D. Thesis*. Univ. North Carolina.

FUSI PECCI, F., BELLAZZINI, M., FERRARO, F. R., BUONANNO, R. & CORSI, C. E. 1996 In *Formation of the Halo ... Inside and Out*, eds. H. Morrison & A. Sarajedini, Astron. Soc. Pac. Conf. Ser., vol. 92, p. 221. ASP.

GEISLER, D., CLARIA, J. J. & MINNITI, D. 1992 *AJ* **104**, 1892.

GILMORE, G. F. 1984 *MNRAS* **207**, 223.

GILMORE, G. & REID, N. 1983 *MNRAS* **202**, 1025.

GILMORE, G., WYSE, R. F. G. & JONES, J. B. 1995 *AJ* **109**, 1095.

GLASS, I. S., WHITELOCK, P. A., CATCHPOLE, R. M. & FEAST, M. W. 1995 *MNRAS* **273**, 383.

GNEDIN, O. Y. & OSTRIKER, J. P. 1997 *ApJ* **474**, 223.

GOULD, A., GUHATHAKURTA, P., RICHSTONE, D. & FLYNN, C. 1992 *ApJ* **388**, 345.

GRATTON, R. G. 1987 *MNRAS* **224**, 175.

GRATTON, R. G., CARETTA, E., MATTEUCCI, F. & SNEDEN, C. 1996 In *Formation of the Halo...Inside and Out*, eds. H. Morrison & A. Sarajedini, Astron. Soc. Pac. Conf. Ser., vol.

92, p. 307. ASP.

GRATTON, R. G., FUSI PECCI, F., CARETTA, E., CLEMENTINI, G., CORSI, C. E. & LATTANZI, M. 1997 *ApJ* **491**, 749.

GREBEL, E. K. 1998 In *Dwarf Galaxies and Cosmology*, eds. T. X. Thuan et al., *in press*. Editions Frontieres.

GRILLMAIR, C. J. 1998 In *Galactic Halos*, ed. D. Zaritsky, Astron. Soc. Pac. Conf. Ser., vol. 136, p. 45. ASP.

GRUNDAHL, F. 1996 In *Formation of the Halo...Inside and Out*, eds. H. Morrison & A. Sarajedini, Astron. Soc. Pac. Conf. Ser., vol. 92, p. 273. ASP.

HANSON, R. B. 1989 *BAAS* **21**, 1107.

HARDING, P. 1996 In *Formation of the Halo...Inside and Out*, eds. H. Morrison & A. Sarajedini, Astron. Soc. Pac. Conf. Ser., vol. 92, p. 151. ASP.

HARDING, P., MATEO, M., MORRISON, H., OLSZEWSKI, E., FREEMAN, K. C. & NORRIS, J. E. 1998 In *The Galactic Halo*, Astron. Soc. Pac. Conf. Ser., *in press*.

HARRIS, W. E. 1991 *ARAA* **29**, 543.

HARRIS, W. E. 1996 *AJ* **112**, 1487.

HARTWICK, F. D. A. 1987 In *The Galaxy*, eds. G. Gilmore & B. Carswell, p. 281. Reidel.

HEATHERINGTON, N. S. 1975 *JHA* **vi**, 115.

HELMI, A. & WHITE, S. D. M. 1999 *MNRAS, in press*.

HENRY, T. J., SODERBLOM, D. R., DONAHUE, R. A. & BALIUNAS, S. L. 1996 *AJ* **111**, 439.

HERNQUIST, L. & QUINN, P. J. 1993 In *Galaxy Evolution: The Milky Way Perspective*, ed. S. R. Majewski, Astron. Soc. Pac. Conf. Ser., vol. 49, p. 187. ASP.

HODDER, P.J.C., NEMEC, J.M. RICHER, H. B., FAHLMAN, G. G. 1992 *AJ* **103**, 460.

HODGE, P. 1989 *ARAA* **27**, 139.

HOLTZMAN, J. A. ET AL. 1993 *AJ* **106**, 1826.

HOUDASHELT, M. L. 1995 *Ph.D. Thesis*. Ohio State University.

HUBBLE, E. P. 1935 *ApJ* **81**, 334.

IBATA, R. A., GILMORE, G. & IRWIN, M. J. 1995 *MNRAS* **277**, 781.

IPSER, J. R. & SEMENZATO, R. 1985 *A&A* **149**, 408.

ISERN, J., HERNANZ, M., GARCÍA-BERRO, E., MOCHKOVITCH, R. & BURKERT, A. 1997 In *The History of the Milky Way and Its Satellite System*, eds. A. Burkert, D. Hartmann & S. R. Majewski, Astron. Soc. Pac. Conf. Ser., vol. 112, p. 181. ASP.

JACOBY, G. H., ARMANDROFF, T. E. & DAVIES, J. E. 1998 *BAAS* **30**, 1258.

JANES, K. A. & PHELPS, R. L. 1994 *AJ* **108**, 1773.

JANES, K. A., TILLEY, C. & LYNGÅ, G. 1988 *AJ* **95**, 771.

JIMENEZ, R., FLYNN, C. & KOTONEVA, E. 1998 *MNRAS* **299**, 515.

JOHNSTON, K. V., HERNQUIST, L. & BOLTE, M. 1996a *AJ* **465**, 278.

JOHNSTON, K. V., SPERGEL, D. N. & HERNQUIST, L. 1996b *AJ* **451**, 598.

JOHNSTONE, D. 1993 *AJ* **105**, 155.

KAŁUŻNY, J. 1997 *AcA* **44**, 247.

KAPTEYN, J. C. 1906 *Plan of Selected Areas*, Hoitsema Brothers.

KARACHENTSEV, I. D. & KARACHENTSEVA, V. E. 1998 *A&A* **341**, 355.

KATZ, N. 1992 *ApJ* **391**, 502.

KINMAN, T. D., SUNTZEFF, N. B. & KRAFT, R. P. 1994 *AJ* **108**, 1722.

KRAFT, R. P. 1994 *PASP* **106**, 553.

KRON, R. G.. 1980 *ApJSuppl* **43**, 305.

KUIJKEN, K. & GILMORE, G. 1989 *MNRAS* **239**, 605.

KUNDIĆ, T. & OSTRIKER, J. P. 1995 *ApJ* **438**, 702.

LACEY, C. G. & OSTRIKER, J. P. 1985 *ApJ* **299**, 633.

LANCE, C. 1988 *ApJ* **334**, 927.

LANZETTA, K. M., BOWEN, D. B., TYTLER, D. & WEBB, J. K. 1995 *ApJ* **442**, 538.

LARSON, R. B. 1974 *MNRAS* **169**, 229.

LARSON, R. B. 1976 *MNRAS* **170**, 31.

LATHAM, D. W., ET AL. 1992, *AJ* **104**, 774.

LEE, H. K. & GOODMAN, J. 1995 *ApJ* **443**, 109.

LEE, M. G., FREEDMAN, W., MATEO, M., THOMPSON, I., ROTH, M. & RUIZ, M.-T. 1993 *AJ* **106**, 1420.

LEE, Y.-W. 1992 *AJ* **104**, 1780.

LEGGETT, S. K., RUIZ, M. T. & BERGERON, P. 1998 *ApJ* **497**, 294.

LIEBERT, J., DAHN, C. C., GRESHAM, M. & STRITTMATTER, P. A. 1979 *ApJ* **233**, 226.

LIN, D. N. C. & RICHER, H. B. 1992 *ApJLett* **388**, L57.

LINDBLAD, B. 1927 *MNRAS* **87**, 553.

LONG, K., OSTRIKER, J. P. & AGUILAR, L. 1992 *ApJ* **388**, 839.

LONGARETTI, P.-Y. & LAGOUTE, C. 1997 *A&A* **319**, 839.

LYNGÅ, G. 1987 *Catalogue of Open Cluster Data*, Lund Observatory.

MAJEWSKI, S. R. 1992 *ApJSuppl* **78**, 87.

MAJEWSKI, S. R. 1993 *ARAA* **31**, 575.

MAJEWSKI, S. R. 1994a In *Astronomy from Wide Field Imaging*, eds. H. T. MacGillivray et al., p. 425. Kluwer.

MAJEWSKI, S. R. 1994b *ApJLett* **431**, L17.

MAJEWSKI, S. R. 1995 In *The Formation of the Milky Way*, eds. E. J. Alfaro & A. J. Delgado, p. 199. CUP.

MAJEWSKI, S. R., HAWLEY, S. L. & MUNN, J. A. 1996a In *Formation of the Halo...Inside and Out*, eds. H. Morrison & A. Sarajedini, Astron. Soc. Pac. Conf. Ser., vol. 92, p. 119. ASP.

MAJEWSKI, S. R., MUNN, J. A. & HAWLEY, S. L. 1996b *ApJLett* **459**, L73.

MAJEWSKI, S. R., OSTHEIMER, J. C., KUNKEL, W. E., JOHNSTON, K. V., PATTERSON, R. J. & PALMA, C. 1998 In *IAU Symposium 190, New Views of the Magellanic Clouds, in press*.

MAJEWSKI, S. R., PHELPS, R. L. & RICH, R. M. 1997 In *The History of the Milky Way and Its Satellite System*, eds. A. Burkert, D. Hartmann & S. R. Majewski, Astron. Soc. Pac. Conf. Ser., vol. 112, p. 1. ASP.

MARQUEZ, A. & SCHUSTER, W. E. 1994 *A&A* **108**, 341.

MARSAKOV, V.A., SHEVELEV, Iu. G. & SUCHKOV, A. A. 1990 *Ap&SS* **172**, 51.

MARSAKOV, V.A. & SUCHKOV, A. A. 1977 *Sov. Astron. Lett.* **21(6)**, 700.

MARTIN, J. C. & MORRISON, H. L. 1998 *AJ* **116**, 1742.

MATEO, M. OLSZEWSKI, E. W. & MORRISON, H. L. 1998 *ApJLett* **508**, L55.

McGLYNN, T. 1990 *ApJ* **348**, 515.

McWILLIAM, A. & RICH, M. 1994 *ApJS* **91**, 749.

MÉNDEZ, R. A. 1995 *Ph. D. Thesis*. Yale University.

MÉNDEZ, R. H., GUERRERO, M. A., FREEMAN, K. C., ARNABOLDI, M., KUDRITZKI, R. P., HOPP, U., CAPACCIOLI, M. & FORD, H. 1997 *ApJLett* **491**, L23.

MÉNDEZ, R. H. & VAN ALTENA, W. F. 1996 *AJ* **112**, 655.

MÉRA, D., CHABRIER, G. & SCHAEFFER, R. 1998 *A&A* **330**, 937.

MEUSINGER, H. 1991 *Astrophys. Spa. Sci.* **182**, 19.

MINNITI, D. 1995 *AJ* **109**, 1663.

MINNITI, D. 1996 *ApJ* **459**, 175.

MINNITI, D., OLSZEWSKI, E. W., LIEBERT, J., WHITE, S. D. M., HILL, J. & IRWIN, M. 1995 *MNRAS* **277**, 1293.

MOORE, B. & DAVIS, M. 1994 *MNRAS* **270**, 209.

MORRISON, H. L. 1993 *AJ* **105**, 539.

MORRISON, H. L., BOROSON, T.A. & HARDING, P. 1994 *AJ* **108**, 1191.

MORRISON, H. L., FLYNN, C. & FREEMAN, K. C. 1990 *AJ* **100**, 1191.

MURRAY, C. A. 1986 *MNRAS* **223**, 649.

NEMEC, J. & NEMEC, A. F. L. 1991 *PASP* **103**, 95.

NEMEC, J. & NEMEC, A. F. L. 1993 *AJ* **105**, 1455.

NORRIS, J. 1986 *ApJSuppl* **61**, 667.

NORRIS, J. 1987 *AJ* **93**, 616.

NORRIS, J. 1994 *ApJ* **431**, 645.

NORRIS, J, E. & GREEN, E. M. 1989 *ApJ*, **337**, 272.

O' CONNELL, D. J. K. (ed.) 1958 *Stellar Populations*, Interscience.

OLLING, R. P. & MERRIFIELD, M. R. 1998 *MNRAS* **297**, 943.

OORT, J. 1965 In *Galactic Structure*, eds. A. Blauuw & M. Schmidt, p. 455. University of Chicago.

ORTOLANI, S., RENZINI, A., GILMOZZI, R., MARCONI, G., BARBUY, B., BICA, E. & RICH, R. M. 1995 *Nature* **377**, 701.

PACZYŃSKI, B., STANEK, K. Z., UDALSKI, A., SZYMANSKI, M., KAŁUŻNY, J., KUBIAK, M. & MATEO, M. 1994 *AJ* **107**, 2060.

PAGEL, B. E. J. & PATCHETT, B. E. 1975 *MNRAS* **172**, 13.

PALMA, C. & MAJEWSKI, S. R. 1999 *in preparation.*

PAUL, E. R. 1981 *JHA* **12**, 77.

PAUL, E. R. 1993 The Milky Way Galaxy and Statistical Cosmology, 1890–1924, CUP.

PENG, W. & WEISHEIT, J. C. 1992 *MNRAS* **258**, 476.

PHELPS, R. L. 1997 *ApJ* **483**, 826.

PHELPS, R. L., JANES, K. A. & MONTGOMERY, K. A. 1994 *AJ* **107**, 1079.

PRESTON, G. W., BEERS, T. C. & SCHECTMAN, S. A. 1994 *AJ* **108**, 538.

REES, R. F. & CUDWORTH, K. M. 1991 *AJ* **102**, 152.

REID, I. N. 1990 *MNRAS* **247**, 70.

REID, N. 1993 In *Galaxy Evolution: The Milky Way Perspective*, ed. S. R. Majewski, Astron. Soc. Pac. Conf. Ser., vol. 49, p. 37. ASP.

REID, I. N. 1998 *AJ* **115**, 204.

REID, I. N. 1999 *ARAA, in press.*

REID, I. N. & GIZAS, J. E. 1998 *AJ* **116**, 2929.

REID, N. & MAJEWSKI, S. R. 1993 *ApJ* **409**, 635.

RENZINI, A. & BUZZONI, A. 1986 In *Spectral Evolution of Galaxies*, eds. C. Chiosi & A. Renzini. Astrophysics and Space Science Library, vol. 122, p. 195. Reidel.

RICH, R. M. 1988 *AJ* **95**, 828.

RICH, R. M. 1990 *ApJ* **362**, 604.

RICH, R. M. 1996 In *Formation of the Halo ... Inside and Out*, eds. H. Morrison & A. Sarajedini, Astron. Soc. Pac. Conf. Ser., vol. 92, p. 24. ASP.

RICHER, H. B., ET AL. 1997 *ApJ* **484**, 741.

RICHSTONE, D., GOULD, A., GUHATHAKURTA, P. & FLYNN, C. 1992 *ApJ* **388**, 354.

ROBIN, A. & CRÉZÉ, M. 1986 *A&AS* **64**, 53.

ROCHA-PINTO, H. J. & MACIEL, W. J. 1996 *MNRAS* **279**, 447.

ROCHA-PINTO, H. J., MACIEL, W. J., SCALO, J. M. & FLYNN, C. 1998 In *Galaxy Evolution: Connecting the High Redshift Universe with the Local Fossil Record, in preparation.*

RODGERS, A. W., HARDING, P. & SADLER, E. 1981 *ApJ* **244**, 912.

RODGERS, A. W. 1991 In *Dynamics of Disc Galaxies*, ed. B. Sundelius, p. 29. Göteborg Univ.

ROMAN, N. G. 1954 *AJ* **59**, 307.

ROSE, J. A. & AGOSTINHO, R. 1991 *AJ* **101**, 950.

ROSENBERG, A., SAVIANE, I, PIOTTO, G. & HELD, E. V. 1998 *A&A* **339**, 61

RUSSELL, H. N., DUGAN, R. S. & STEWART, J. Q. 1927 *Astronomy* **2**, 910.

SALARIS, M. & WEISS, A. 1997 *A&A* **327**, 107.

SALARIS, M. & WEISS, A. 1998 *A&A* **335**, 943.

SANDAGE, A. R. 1987 *AJ* **93**, 610.

SANDAGE, A. R. 1990 *JRASC* **84**, 70.

SANDAGE, A. R. 1995 In *The Deep Universe*, eds. B. Binggelli & R. Buser, p. 1. Springer.

SANDAGE, A. & KOWAL, C. 1986 *AJ* **91**, 1140.

SANDAGE, A. R. & SCHWARZSCHILD, M 1952 *ApJ* **116**, 463.

SANDAGE, A. & WALKER, M. F. 1955 *AJ* **60**, 230.

SARAJEDINI, A., CHABOYER, B. & DEMARQUE, P. 1997 *PASP* **109**, 1321.

SASLAW, W. C. 1985 *Gravitational Physics of Stellar and Galactic Systems*, pp. 74-76. Cambridge Univ. Press.

SCHLEGEL, D. J., FINKBEINER, D. P. & DAVIS, M. 1998 *ApJ* **500**, 525.

SEARES, F. H. 1922 *PASP* **34**, 233.

SEARLE, L. 1993 In *Galaxy Evolution: The Milky Way Perspective*, ed. S. R. Majewski. Astron. Soc. Pac. Conf. Ser., vol. 49, p. 3. ASP.

SEARLE, L. & ZINN, R. 1978 *ApJ* **225**, 357 (SZ).

SHAPLEY, H. 1918 *ApJ* **48**, 154.

SHAPLEY, H. 1921 *Bull. Nat. Res. Council* **2**, 171.

SIEGEL, M. H., MAJEWSKI, S. R., REID, I. N., THOMPSON, I., LANDOLT, A. U. & KUNKEL, W. E. 1997 *BAAS* **29**, 1341.

SLIPHER, V. M. 1913 *Lowell Observatory Bull.* **ii**, 56.

SLIPHER, V. M. 1914 *Lowell Observatory Bull.* **ii**, 66.

SMECKER-HANE, T. A., STETSON, P. B., HESSER, J. E. & LEHNERT, M. D. 1994 *AJ* **108**, 507.

SNEDEN, C., KRAFT, R. P., SHETRONE, M. D., SMITH, G. H., LANGER, G. E. & PROSSER, C. F. 1997 *AJ* **114**, 1964.

SODERBLOM, D. R. 1985 *AJ* **90**, 2103.

SODERBLOM, D. R. 1990 *AJ* **100**, 204.

SOMMER-LARSEN, J. & ZHEN, C. 1990 *MNRAS* **242**, 10.

SOUBIRAN, C. 1993 *A&A* **274**, 181.

SPAENHAUER, A. 1989 *Contrib. Van Vleck Obs.* **9**, 45.

SPITZER, L. 1940 *MNRAS* **100**, 397.

SPITZER, L. 1958 *ApJ* **127**, 544.

SPITZER, L. 1969 *ApJLett* **158**, L139.

SPITZER, L. 1987 *Dynamical Evolution of Globular Clusters*, Princeton Univ. Press.

SPITZER, L. & SCHWARZSCHILD, M. 1953 *ApJ* **118**, 106.

STRUVE, F. G. W. 1847 *Etudes d'astronomie stellaire*, St. Petersburg Imperial Acad. Sci.

SUNTZEFF, N. 1993 In *The Globular Cluster-Galaxy Connection*, eds. G.H. Smith & J. P.

Brodie, Astron. Soc. Pac. Conf. Ser., vol. 48, p. 167. ASP.

SURDIN, V. G. 1995 *AstL* **21**, 508.

SURDIN, V. G. 1997 *AstL* **23**, 234.

THEUNS, T. & WARREN, S. J. 1997 *MNRAS* **284**, L11.

TIEDE, G. P. & TERNDRUP, D. M. 1997 *AJ* **113**, 321.

TOOMRE, A. & TOOMRE, J. 1972 *ApJ* **178**, 623.

TREMAINE, S. D., OSTRIKER, J. P. & SPITZER, L. 1975 *ApJ* **196**, 407.

TSIKOUDI, V. 1979 *ApJ* **234**, 842.

TWAROG, B.A. 1980 *ApJ* **242**, 242.

TYSON, N. D. 1991 *Ph.D. Thesis.* Columbia Univ.

TYSON, N. D. & RICH, R. M. 1993 In *IAU Symp. 153: Galactic Bulges*, eds. H. DeJonghe & H. Habing, p. 333. Kluwer.

VAN DE RYDT, F., DEMERS, S. & KUNKEL, W. E. 1991 *AJ* **102**, 130.

VAN DEN BERGH, S. 1993 *AJ* **105**, 971.

VAN DER KRUIT & SEARLE, L. 1981 *A&A* **95**, 105.

VAN MAANEN, A. 1916 *ApJ* **44**, 210.

VAN MAANEN, A. 1925 *ApJ* **61**, 130.

VAN MAANEN, A. 1927 *ApJ* **66**, 89.

VANDENBERG, D. A. 1985 *ApJSuppl* **58**, 711.

VANDENBERG, D. A. 1991 In *The Formation and Evolution of Star Clusters*, ed. K. A. Janes Astron. Soc. Pac. Conf. Ser., vol. 13, p. 183. ASP.

VAUGHAN, A. H. & PRESTON, G. W. 1980 *PASP* **92**, 385.

VELAZQUEZ, H. & WHITE, S. D. M. 1995 *MNRAS* **275**, L3.

VILLUMSEN, J. V. 1985 *ApJ* **290**, 75.

VOGT, H. 1926 *Astronomische Nachrichten* **226**, 301.

WALKER, A. & TERNDRUP, D. M. 1991 *ApJ* **378**, 119.

WEINBERG, M. D. 1992 *ApJ* **384**, 81.

WEINBERG, M. D. 1993 In *The Globular Cluster-Galaxy Connection*, eds. G.H. Smith & J. P. Brodie, Astron. Soc. Pac. Conf. Ser., vol. 48, p. 689. ASP.

WHITELOCK, P. A. 1993 In *IAU Symp. 153: Galactic Bulges*, eds. H. DeJonghe & H. Habing, p. 39. Kluwer.

WHITING, A. B., IRWIN, M. J. & HAU, G. K. T. 1997 *AJ* **114**, 996.

WIELEN, R. 1977 *A&A* **60**, 263.

WIELEN, R., JAHREISS, H. & KRÜGER, R. 1983 In *IAU Colloq. 76: The Nearby Stars and the Stellar Luminosity Function*, eds. A. G. Davis Philip & A. Upgren, p. 163. L. Davis Press.

WILDEY, R. L., BURBIDGE, E. M., SANDAGE, A. R. & BURBIDGE, G. R. 1962 *ApJ* **135**, 94.

WOLF, M. 1914 *Astronomische Gesellschaft* **xlix**, 162.

WOOD, M. A. & OSWALT, T. D. 1998 *ApJ* **497**, 870.

WYSE, R. F. G. & GILMORE, G. 1988 *AJ* **95**, 1404.

WYSE, R. F. G. & GILMORE, G. 1992 *AJ* **104**, 144.

WYSE, R. F. G. & GILMORE, G. 1993 In *Galaxy Evolution: The Milky Way Perspective*, ed. S. R. Majewski. Astron. Soc. Pac. Conf. Ser., vol. 49, p. 209. ASP.

WYSE, R. F. G. & GILMORE, G. 1995 *AJ* **110**, 2771.

YOSHII, Y. 1982 *PASJ* **34**, 365.

YOSS, K. M., NEESE, C. L. & HARTKOPF, W. I. 1987 *AJ* **94**, 1600.

YOUNG, P. J. 1976 *AJ* **81**, 807.

ZAHN, J.-P. 1977 *A&A* **57**, 383.

ZINN, R. 1985 *ApJ* **293**, 424.

ZINN, R. 1993a In *The Globular Cluster-Galaxy Connection*, eds. G.H. Smith & J. P. Brodie, Astron. Soc. Pac. Conf. Ser., vol. 48, p. 38. ASP.

ZINN, R. 1993b In *The Globular Cluster-Galaxy Connection*, eds. G.H. Smith & J. P. Brodie, Astron. Soc. Pac. Conf. Ser., vol. 48, p. 302. ASP.

ZINN, R. 1996 In *Formation of the Halo ... Inside and Out*, eds. H. Morrison & A. Sarajedini, Astron. Soc. Pac. Conf. Ser., vol. 92, p. 211. ASP.

VITTORIO CASTELLANI was born on 13 March 1937. He graduated in Physics at the University "La Sapienza" in Rome, with a thesis on "annichilation e+e- in the Frascati storage ring AdA". Shortly after the thesis he joined in Frascati the astrophysics group of proof. Livio Gratton, where four young people (he, V. Caloi, C. Firmani and Z. Renzini) decided to start from the ground an activity in theoretical stellar models. Just at that time he fell in love with globular clusters, looking at the impressive structure of these wonderful stellar systems.

All along the years, the investigation of Pop. II (and Pop. III) stars has remained the main target of the investigations he has developed with a growing group of students and coworkers. Today one finds that these investigations have covered all the evolutionary phases expected in globular clusters, from the very low mass MS stars to the final stage of cooling White Dwarfs, with several relevant outcomes.

More recently, these investigations have been also extended to metal rich and supermetal rich stars, disclosing some unexpected and relevant evolutionary features.

One has finally to quote the several papers devoted to the problem of solar (and stellar) neutrinos and Solar Standard Models.

In 1980 he was appointed to the Chair of Astrophysics at the University "La Sapienza", passing afterwards to the Chairs of Theoretical Astrophysics and, recently, of Stellar Physics at the University of Pisa.

He has written two books (in italian) "An introduction to nuclear astrophysics" and "Stellar Astrophysics". Beside this activity, he has been engaged in the investigation of natural and artificial undergrounds, participating in several missions (Moroccan Sahara, Greece, Central Anatolia, West China... and Antarctica) mainly for archaeological pourposes. On this subject he is author of several papers and of a book (The civilization of water) now appearing in Italy and which is being translated into English.

He has been President of the Italian Astronomical Society and of the Italian Spaelological Society, Director of the Laboratory for Space Astrophysics (Frascati), of the Teramo Astronomical Observatory and of the Institute of Astronomy of Pisa University.

Like Galileo, he is a member of the Accademia Nazionale dei Lincei.

Globular Clusters as a test for Stellar Evolution

By VITTORIO CASTELLANI

Department of Physics, University of Pisa, 56100 Pisa, Italy

Current theoretical predictions concerning the evolution of old, metal-poor stars in galactic globulars are revisited in the light of recent improvements in the input physics. After a short introduction, the role of fundamental physics in constraining stellar structure all along the various evolutionary phases is shortly recalled, together with the additional role played by the evaluation of some macroscopic mechanisms, like convection and diffusion (Sect. 2). Theoretical predictions concerning the CM-diagram location of the best-known evolutionary phases are discussed in some detail, with particular regard to the existing uncertainties in modeling stellar structure as well as in handling observational data (Sect. 3). This discussion is thus extended to the faint stars recently revealed by HST, either at the faint end of the MS or along the WD cooling sequence (Sect. 4). Additional theoretical constraints given by the pulsational properties of RR Lyrae pulsators are recalled (Sect. 5) and the case of extragalactic globulars in the Local Group are briefly discussed (Sect. 6). Some general and methodological considerations close the paper.

1. The CM diagram: an introduction

The birth of modern physics dates back to the time when Galileo Galilei stated that any attempt to understand the world around us must—first of all—*save the phenomena* ("salvare i fenomeni"). In modern words, we say that physics is studying relations between *observable quantities*, so that the identification of suitable "observables" is a first-priority step in any physical investigation. Bearing in mind such a priority, when dealing with the evidence for stars, one has first to identify the relevant observables to be used as the "starting" information on the matter.

In this respect, one has no choices: observing stars is just collecting the photons emitted by these luminous sources. Therefore the number of photons and their distribution in energy are the only natural observables one is dealing with, since these two quantities give an exhaustive description of the phenomena collected by our telescopes. This alone shows that the well-known Hertzsprung–Russell (HR) diagram is nothing peculiar, but the natural way to organize information collected from stellar objects.

The CM diagrams of Galactic stellar clusters have been early interpreted as evidence for stellar evolution. As well known, globular clusters populate the Galactic halo where recent star formation is inhibited by the lack of diffuse interstellar matter: therefore Galactic globulars must have been be born in the past. This is not the case for open clusters, which populate the Galactic disc and for which one has additional dynamical evidence for a relatively recent formation. In this context, the striking difference between the CM diagrams of open and globular clusters has been early interpreted as due to the older ages of the latter.

These simple observations drive us into the scenario into which stellar evolutionary theory is moving. The history of this theory, as it developed mainly in the second half of this century, is the history of our progressive understanding of the observational features of cluster CM diagrams in terms of the nuclear evolution of stellar structures and —thus—in terms of the cluster age and chemical composition. The chief motivation for such an investigation is—at least in my mind—the Galilean need for "saving the phenomena", i.e., for understanding observation in terms of the physical laws governing the behavior of matter in this Universe. As we will see, this means testing many physical laws well beyond their experimental limits in terrestrial laboratories.

However, one also has to bear in mind that a precise knowledge of the evolution of stars provides us with an invaluable tool for investigating the evolutionary history of the Universe as a whole, far beyond the investigation of the evolutionary status of Galactic clusters. To give some examples, one may recall here that theoretical predictions about stellar luminosity can provide useful "standard candles" to constrain the distance of far-away stellar systems and, thus, to achieve information about the actual value of the Hubble constant H_0. Or, more generallly, that theoretical predictions about the evolution of stellar populations represent the key ingredient for understanding the light from even remote galaxies in terms of the evolutionary history of those stellar aggregates.

In the following I will try to give a picture of the present situation, presenting a rather large number of theoretical predictions that allow an observational test, but also discussing the existing uncertainties in the theoretical scenario. Attention will be primarily focused on the investigation of the old globulars populating our Galactic halo. On this ground, we will also discuss the main theoretical problems concerning globulars in external galaxies. An overview of the connections between evolutionary and pulsational theories will close the paper. All through the paper, the main ingredients playing a role in the evolutionary scenario will be presented and discussed. However, this cannot be done in excessive detail here. The interested reader can find more detailed discussions of several topics in the many books that have already appeared in thid field (as, e.g., the classical book by Cox & Giuli 1968 or the more recent book by Kippenhahn & Weigert 1990), as well as in more recent review papers such as the excellent ones recently published by vandenBerg, Stetson & Bolte (1996) or Chiosi (1997). References given in this paper will allow a deeper look at the more recent problems.

2. Stellar evolution investigations: the ingredients

To approach a meaningful discussion of current evolutionary results, this section will be devoted to a preliminary review of the various theoretical ingredients entering into the manufacture of evolutionary predictions. On this basis we will try to shed light on the development as well as on the uncertainties of evolutionary theories, advancing from now on a rather firm warning against the uncritical use of evolutionary results, which sometimes causes substantial suspicions about the reliability of the theory as a whole, uselessly increasing the level of noise in the literature.

2.1. *Cluster isochrones.*

On very general grounds, one expects that the evolutionary status of the structure of a star depends, at least, on the star's age, mass and chemical composition. Figure 1 shows the typical CM diagram for a well observed Galactic globular cluster. The evidence for stars arranging themselves along a well defined single-parameter sequence was understod long ago understood as evidence that the clusters are populated by a "simple stellar population", i.e., by stars having with good approximation a common age and original chemical composition, while only one parameter, "the original mass", governs the location of the individuals along the observed sequence. Therefore a stellar cluster appears as a natural celestial laboratory where stellar evolution shows itself in its simplest form. Note that the evidence for a one-parameter sequence also excludes the possibility that further structural parameters, such as, e.g., stellar rotation, can have a macroscopic influence on the location in the CM diagram.

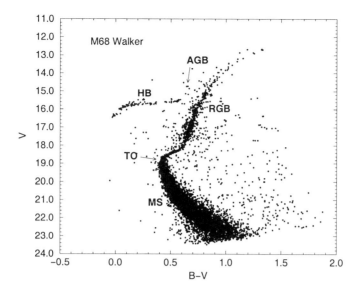

FIGURE 1. The CM diagram of the Galactic globular cluster M68 presented by A. Walker (1994) on the basis of CCD observations at the 4-m telescope of Cerro Tololo (Chile) with the various evolutionary phases labeled.

In this context, Galactic globulars play a role of particular relevance, since they give particular and statistically significant evidence for stars in advanced evolutionary phases. In a Galactic globular one finds indeed hundreds of stars which have already evolved off their Main Sequence (MS), compared with the few (if any) found in young open clusters. The reason for such an occurrence is twofold: the intrinsic richness of stars forming these clusters and the old cluster ages, which push off the MS low-mass stars which in any stellar population represent the most abundant component.

In Figure 1 one recognizes the traditional evolutionary phases characterizing a globular cluster, as labeled in the same Figure. These experimental data (i.e., the cluster color–magnitude (CM) diagram) can be quantitatively predicted by theory only if the theory is able to predict for each given stellar mass, age and chemical composition the CM location of a stellar object. Accordingly, the first goal for stellar evolutionary theory is to produce "evolutionary tracks", i.e., to predict the CM location as a function of time for stars with a given mass and chemical composition. When a suitable set of evolutionary tracks is available, one can easily derive the predicted location of cluster stars, as given by the so called "cluster isochrone".

However, to allow a comparison with observational data one needs a further step, as given by suitable model atmospheres transforming the theoretical parameters luminosity (L) and effective temperature (T_e) in the observational magnitudes and colors, as in the widely adopted $V, B - V$ or $V, V - I$ CM diagrams. This step joins together the main ingredients that compose the scenario for stellar-evolution investigations, namely, the CM diagram on the observational side and stellar models plus model atmospheres on the theoretical side. The goal of stellar evolution is the match between observation and theory, primarily to test the reliability of the adopted theoretical scenario. On this basis, one can finally attain relevant information about the cluster and, in particular, about the cluster age.

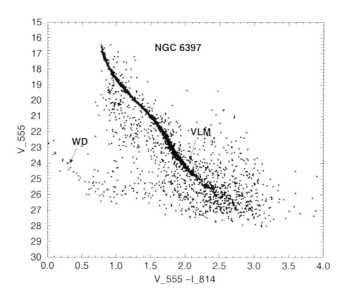

FIGURE 2. HST V, $V - I$ diagram of faint stars in the Galactic globular NGC 6397 (Cool et al. 1996)

Before leaving this topic, one has finally to notice that till now we referred to the "classical" evolutionary phases observed in Galactic globulars, as given by the more luminous members of the cluster. However, the Hubble Space Telescope (HST) is now allowing us to complete this sample, opening for the first time to observation new and relevant evolutionary phases. This is shown in Figure 2, where one can recognize the well developed sequence of Very Low Mass (VLM) Main Sequence stars, together with a portion of the expected White Dwarf (WD) cooling sequence.

2.2. *Theoretical models: the input physics.*

The physical relations constraining a stellar structure are known since a long time. Let me briefly recall these relations, as they can be found in all even elementary books on the matter:

$$\frac{dP}{dr} = -\frac{GM_r\rho}{r^2} \qquad \text{(hydrostatic equilibrium)} \tag{2.1}$$

$$\frac{dM_r}{dr} = 4\pi\rho r^2 \qquad \text{(mass conservation)} \tag{2.2}$$

$$P = [P(\rho, T)] \qquad \text{(equation of state)} \tag{2.3}$$

$$\frac{dT}{dr} = -\frac{3}{4ac}\frac{[\kappa]\rho}{T^3}\frac{L_r}{4\pi r^2} \qquad \text{(no convection)} \quad \text{or} \tag{2.4}$$

$$= \frac{dT}{dP}_{\text{eff}}\frac{dP}{dr} \qquad \text{(convective layers)}$$

$$\frac{dL_r}{dr} = 4\pi\rho r^2 [\varepsilon] \qquad \text{(conservation of energy)} \tag{2.5}$$

with the usual meaning of the symbols. These relations, which hold for the stellar interiors, need to be completed by boundary conditions at the stellar atmosphere, as normally given by suitable assumptions about the dependence of the temperature on the optical depth, derived under the so-called Eddington approximation (but see the case of VLM stars further on in this paper).

The occurrence of convection, and the choice of the local gradient of temperature, are governed by the well-known Schwarzschild criterion, for which a stellar layer is unstable against convection any time that the radiative gradient becomes larger than the adiabatic one: $(dT/dP)_{rad} > (dT/dP)_{ad}$.

One is facing a set of mathematical relations which, however, require the knowledge of some physical ingredients, as highlighted within square brackets in the previous relations. It may be convenient to have a better look at such an "input physics", just to give some rough indication of the complex physics one is dealing with:

$P(\rho, T)$ = *Equation of State* or EOS, which governs the behavior of the mixture of ions, electrons and photons in a stellar interior. Even in the simplest case of a perfect gas this requires a quantitative evaluation of the degree of ionization of all the various kinds of atoms composing the matter. Moreover, in cool dense structures one has also to account for the relevance of Coulomb corrections and/or of quantum contributions to pressure by fermions (electron degeneracy).

$\kappa(\rho, T)$ = *radiative opacity*, as related to the mean free path of photons, governing the heat transfer. Note that just to start the evaluation of such a parameter one needs to evaluate not only the degree of ionization but also the population of all the excited levels in the various atoms which interact with the radiation. At higher densities one has also to evaluate the contribution to the heat transfer by quantum-degenerate electrons.

$\varepsilon(\rho, T)$ = *energy generation per gram of matter*, which governs the balance of energy. This comes from three different contributions. The first one is the thermodynamic balance as given by the first law of thermodynamics, which now requires the evaluation of the internal energy of each ionic species. The second is the contribution by nuclear reactions, with the evaluation of the various cross-sections for the strong reactions that occur. At high densities and temperatures one has also to evaluate the additional (negative) contribution given by the production of neutrinos by weak interactions.

From such a quick look, one easily understands why the production of quantitative stellar predictions is a rather recent job. As a matter of the fact, even the simplest stellar structure, such as that of our Sun, requires the knowledge of the most advanced physics. As an example, we cannot have a reliable physical picture of the Sun without a detailed knowledge of quantum mechanics together with strong and weak interactions, which all govern the behavior of matter and/or the nuclear energy production in the solar interior. Therefore the theory of stellar structure has of necessity followed the progress of physics, waiting to apply to the celestial laboratories the knowledge born in the terrestrial laboratories. Without going deeper into this matter, let us note here that according to such a scenario one understands that stars can also contribute to our physical knowledge, as we will shortly recall further along in this paper.

2.3. *Theoretical models: macroscopic mechanisms*

Before producing an evolutionary sequence, one has however to make some additional choice about the efficiency of several macroscopic mechanisms that can affect the structure of stars. Let me here give a short list with only a few general comments:

External convection: Cool stars have convective envelopes, as a consequence of the large opacity and, thus, the large radiative gradient of partially ionized hydrogen. In stellar interiors we know that the convective gradient $(dT/dP)_{\text{eff}}$ is to all practical purposes given by the adiabatic one. This is not the case in the less dense stellar envelopes, where we have only the firm theoretical constraint $(dT/dP)_{\text{rad}} > (dT/dP)_{\text{eff}} > (dT/dP)_{\text{ad}}$. In these stellar envelopes $(dT/dP)_{\text{eff}}$ is largely evaluated through a theoretical approach which contains the free parameter "mixing length". As a consequence, the effective temperature of stars with cool envelopes (roughly below $\log T_e = 4.0$) has no firm theoretical constraints. More recent approaches to the evaluation of this "superadiabatic" temperature gradient will be presented in the final section.

Diffusion: The diffusion of elements throughout the structure of stars has often been neglected in stellar-evolution computations. However, Solar Standard Models can be put in agreement with helioseismological constraints about the solar interior only when accounting for the efficiency of such a mechanism (Bahcall & Pinsonneault 1995), which should also be at work in solar-like old globular-cluster stars, with non-negligible observational consequences (see, e.g., Castellani et al. 1997).

Core Overshooting: Canonical models assume that convective stellar cores are bounded right at the point where the radiative gradient reach the adiabatic one. However, strictly speaking this condition marks the point where convective elements stop being accelerated, and the elements must obviously go a bit further in the "radiative" region before coming to rest, mixing such an "overshooting" region. Canonical models thus assume that such a region has a negligible thickness. However the suggestion has been advanced for greater efficiency of such a mechanism, thus mixing stellar interior sensibly beyond the point stated by the neutrality criterion (see, e.g., Bertelli et al. 1986).

Breathing pulses: Stars burning He at the center, in the phase just before the exhaustion of He in the central convective core, may suffer a pulsating instability of the convection, which should drive fresh He into the core, thus increasing the He-burning lifetime. Caputo et al(1989) discussed this point, showing that—at that time—the comparison between observation and evolutionary predictions were suggesting (but only suggesting!) a negligible efficiency of such a mechanism. It is not clear if such a suggestion will survive the recent improvement in the evolutionary computations.

2.4. *The theoretical scenario*

Once a suitable formulation of the input physics has been adopted and once the preferred assumptions on the efficiency of the macroscopic mechanisms have been taken, one can write down an evolutionary code solving the five equations of stellar structure to produce an evolutionary sequence for each given assumption about the stellar mass and original chemical composition. The first computed model is taken during the early phase of gravitational contraction, before the onset of nuclear reactions, where one can assume that matter has a homogeneous composition throughout all the structure of the star. The solution of the equations gives the physical quantities P, ρ, T, M_r, L_R in each point of the star and, on this basis, a prediction of the distribution of chemical composition after a time step Δt. A new model can be thus computed, and the iteration of such a procedure produces at the end the required evolutionary sequence.

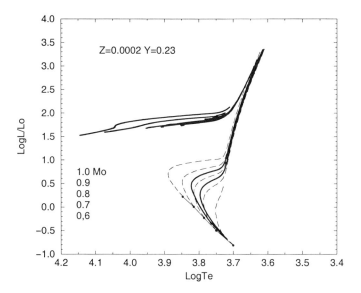

FIGURE 3. Evolutionary tracks (dashed lines) for the labeled values of the stellar mass and for the given chemical composition (from Cassisi et al. 1998a) with the predicted isochrones superposed for the ages 10 and 15 Gyr.

To do that, all the evolutionary codes adopt the algorithm devised by Henyey et al. (1965), as based on a relaxation technique over a set of difference equations. According to such a procedure the solution, when attained, is independent of the procedure adopted to reach the final convergence. This forces us to the relevant conclusion that different codes reach the same solution when the same input physics is adopted. In other words, the reliability of stellar models depends only on the reliability of the adopted input physics, provided that no trivial errors have been made in writing down the equilibrium equations. Therefore the recurrent question of which is "the better" result appears largely to be a meaningless question if it refers to the performance of the code. Moreover, one has just to look at the input physics to have all the information on the consistency of the theoretical scenario.

Given the input physics one can derive evolutionary tracks and from evolutionary tracks one obtains predictions of cluster isochrones. This is shown in Figure 3, which presents selected evolutionary tracks for low-mass metal-poor stars as recently presented by Cassisi et al. (1998a), with superimposed cluster isochrones, as derived from these tracks for the stated assumptions about the cluster ages.

As expected, and as we will discuss in detail later on, now we understand the distribution of stars along an isochrone in terms of a distribution of stellar masses. Low-mass stars are still in their Main Sequence (MS) phase, burning H at the center of their structure. Increasing the mass, one finds Turn Off (TO) stars in the phase of central H exhaustion, RGB stars burning H in a shell surrounding a He core, Horizontal Branch (HB) stars burning He at the center and H in a shell and, finally, Asymptotic Giant Branch (AGB) stars with a double (He and H) shell burning.

Inspection of Figure 3 shows some additional relevant features to be borne in mind all through this paper. One finds that a time variation moves the location of the isochrone essentially in the region between the MS and the base of the RGB. Stars in that region

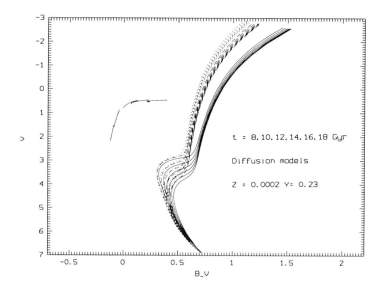

FIGURE 4. Theoretical cluster isochrones for the labeled assumptions about the cluster age and for three different choices for the mixing length parameter (left to right) $l = 2.5$, 2.2 and 1.6 H_p (from Brocato et al. 1997).

represent indeed the theoretical clock to be used to investigate cluster ages. At the lower luminosities, low-mass stars are not appreciably evolved from their MS location, and the MS will be therefore populated according to the so called Initial Mass Function (IMF), i.e., according to the original distribution of masses. On the contrary, at the larger luminosities one finds relatively rapid evolutionary phases; as a consequence one expects that the RGB (and the HB) are populated by stars with quite a restricted range of original masses. In such a case we will find that the distribution of stars along these evolved evolutionary phases is independent of the IMF but proportional to the time spent in each given portion of these phases.

The run of the isochrones in the HR diagram, as given in Figure 3, obviously depends on the adopted theoretical scenario, which in that case includes all the updating to the input physics that has been presented in the literature. As for the macroscopic mechanisms, element diffusion has been taken into account with the efficiency calibrated on the Sun. Neither core overshooting nor breathing pulses have been taken into account: however, as we will discuss later on, in low-mass stars these mechanisms could affect only the lifetimes of He burning stars but not their HR location. Finally, the efficiency of convective transport in the external layers of the structure has been fixed by the assumption of a mixing length $l = 1.6 \, H_p$, where H_p represents the local pressure scale height.

Bearing in mind that the value of the mixing length is a free theoretical parameter, Figure 4 illustrates the effect of varying the assumptions regarding the value of l, after transformation of the isochrones in the V, $B - V$ diagram according to the model atmospheres by Kurucz (1993a, b). As already noted, one finds that cool stars are all affected by this assumption: if we increase the mixing length the convective transport increases, the local gradient decreases and the effective temperature of the star increases.

Figure 4 shows that this effect becomes larger as the star's luminosity (and thus the

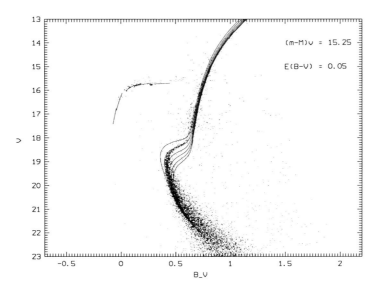

FIGURE 5. Best fitting of theoretical isochrones as in Figure 4 to the CM diagram of Figure 1 (Brocato et al. 1997), for the assumed ages (top to bottom) 8, 10, 12, 14, 16 and 18 Gyr.

stellar radius) increases. The simple reason is that as the radius increases, the density of the stellar envelope decreases and convection becomes less and less efficient, i.e., more and more superadiabatic, increasing the sensitivity to the assumptions about l. In this context, note that low luminosity MS are less and less affected by l, since in the dense external layers the convection is becoming more and more adiabatic. We will come back to this point later on. Here we conclude that the theoretical fitting of the RGB is largely a matter of a cosmetic intervention into the theory, just tuning l to match the observed temperatures.

The relevance of such a theoretical scenario is shown in Figure 5, where these theoretical predictions are used to best fit the CM diagram already given in Figure 1, adopting for the mixing length $l = 2.2\ H_p$.

One finds that theoretical predictions can nicely fit the observational data all along the various evolutionary phases, indicating a cluster age of the order of 10 Gyr. How far can we be confident of such a result? We have found the evidence for a fitting between theory and observation, but the uniqueness of such a result is not a trivial question. The problem is thus: how good is the physics we are using in modeling stars? This is really not a trivial question. One has to notice that a decade ago the adopted input physics was already producing models allowing a reasonable fitting of globular-cluster evolutionary features. According to such evidence, people dealing with evolutionary theory felt confident that the adopted physics was "good enough" to believe in the quantitative results, even if well aware of the difficult physics one is dealing with.

Such a belief has passed as a dream. It happened that the results of helioseismology came to show that the solar-interior models produced at that time were challenged by such experimental evidence. This has stimulated a large effort in improving our knowledge about the physics of stellar matter, so that in these last ten years all the various branches of such knowledge have been revisited and, hopefully, improved. So we are now dealing with new EOS, new opacities, new neutrino productions ...and so on.

This has produced a continuous improvement not only of solar models but also of the whole evolutionary scenario, with non-negligible variations in the quantitative predictions concerning, in particular, the age of Galactic globulars. A discussion of the relevance of such improvements in the input physics can be found in Cassisi et al. (1998a). We have already shown that the most updated physics gives for a metal-poor Galactic globular such as M68 an age of the order of 10 Gyr, sizably decreasing previous estimates. However the age of Galactic globulars is still open to debate (see, e.g., Pont et al. 1998). In order to have a meaningful approach to such a debate, let me here recall once again that stellar models in the literature are only the result of the adopted physics and, thus, that users should investigate this point rather than the name or the nationality of the authors.

3. What we know and what we do not know

Bearing in mind the scenario presented in the previous section, in this section we will go a bit deeper in discussing the predictions of theory. As a first point, one has to recall that in recent years there have been several improvements in the input physics and, in turn, in theoretical evolutionary models. A rather detailed investigation of this issue can be found in Cassisi et al. (1998a,b). In the following let me refer to the most updated evolutionary scenario as presented in that paper to discuss the various evolutionary phases, with particular regard to the predictions of observables quantities as well as to the related theoretical uncertainties.

3.1. *H-burning structures*

In this section we will discuss theoretical predictions concerning the portion of the CM diagram in Figure 1, running from the MS through the subgiant branch till the tip of the RGB. We will discuss in particular the use of MS stars as standard candles, the cluster TO as a clock marking the cluster ages and the several observational tests concerning RGB evolution.

3.1.1. *Main Sequence*

Long ago, the evidence that high velocity, metal-poor stars in the solar neighborhood are less luminous than solar-like stars was one of the first pieces of evidence for Population II stars in our Galaxy. Theories of MS stars confirmed such an occurrence quite early. Figure 6 gives the results of a recent investigation of the matter (Castellani, Degl'Innocenti & Marconi 1998), showing theoretical expectations for a large range of metallicities, both in the theoretical ($M_{bol}, \log T_e$) and in the observational $V, B - V$ HR diagram.

Contrary to what has been often assumed, one finds that the magnitude of the MS at a given $B - V$ color is far from being a linear function of the metal content Z. Assuming a He content $Y = 0.23$ as the common suitable assumption for Galactic globulars one finds (Cassisi et al. 1998b) that the magnitude of the MS at $B - V = 0.6$ goes with metallicity according to:

$$Mv_{(B-V=0.6)} = 3.889 - 2.160[\text{Fe/H}] - 0.509[\text{Fe/H}]^2 \qquad (3.6)$$

Similar predictions can be tested on the sample of field subdwarfs with trigonometrical parallaxes from Hipparcos. This is shown in Figure 7, where the absolute magnitudes of dwarfs in selected classes of metallicity are compared with theoretical predictions. Contrary to previous claims that have appeared in the literature, we found no evidence against the theoretical predictions quoted above.

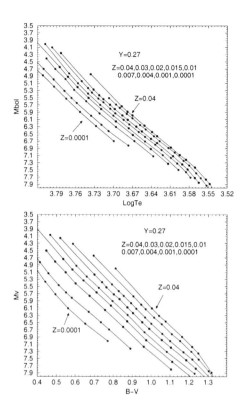

FIGURE 6. The predicted location of MS structures for the labeled assumptions about the star metallicity and original He content in the theoretical (upper panel) or in the observational (lower panel) V, $B - V$ HR diagram (from Castellani, Degl'Innocenti & Marconi 1998, but see also Pagel & Portinari 1998).

By relying on theoretical prediction, one could use MS stars as standard candles, since theory predicts a luminosity and, thus, an absolute magnitude for each given stellar temperature and/or for each given color. However, on observational ground one has to notice that it could be quite a risky procedure, being seriously affected by errors in the estimated cluster reddening. As a matter of fact, from the predicted MS slope one easily finds that an uncertainty of $\Delta(B - V) = \pm0.02$ gives a variation in the predicted theoretical magnitudes by $\Delta M_V \simeq \mp0.1$ m.

On the theoretical side the question arises of how firm these predictions are. As for the input physics, one finds that the estimated uncertainties in the p-p reaction rate play a rather negligible role (Brocato et al. 1998). Moreover, numerical experiments show that a variation of the opacity by $\pm5\%$ moves the models essentially along the MS, with the new models departing from the previous MS locus by no more than few hundreds of magnitude. As a tentative estimate of uncertainties connected with EOS, one can finally take the predicted difference in magnitude (at a given $B - V$) between MS models computed according to advanced EOS evaluations, as given by Straniero (1988) or Roger

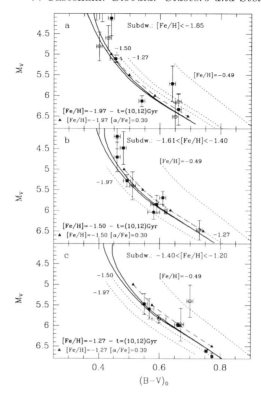

FIGURE 7. Comparison of theoretical isochrones with Hipparcos subdwarf absolute magnitudes from Gratton et al. (1997). Each panel shows the run of the 10 and 12 Gyr isochrones (full lines) for the labeled choice of [Fe/H]. Dotted lines show the run of the 10 Gyr isochrones with different [Fe/H], as labeled. The dashed line with triangles shows the shift of the 10 Gyr isochrone caused by an enhancement [α/Fe] = 0.3. Filled squares and open squares indicate single stars or detected or suspected binaries, respectively (from Cassisi et al. 1998b).

et al. (1996). This would give an uncertainty of the order of ±0.1 m (Brocato, Cassisi & Castellani 1996).

As for the role of macroscopic mechanisms, neither element diffusion nor, obviously, core overshooting or breathing pulses can affect MS structures. However, in low-mass MS stars one finds a convective envelope and, thus, uncertainties about superadiabatic convection. An estimate of such an uncertainty is given in Figure 8, which shows selected cluster isochrones as computed under various (reasonable) assumptions about the value of the mixing-length parameter and for selected cluster ages in the (reasonable) interval 9 to 15 Gyr. One finds (Cassisi et al. 1998b) that varying the mixing length between 1.3 and 2.0 H_p at $B - V = 0.55$ (i.e., around theoretical predictions for $M_V = 6.0$) moves theoretical predictions by ±0.09 m. Moreover, one finds that the MS magnitude at the given color is still affected by an age effect, which can be estimated of the order of ±0.05 m. As a whole, one should conclude that at the present state of the art, theory can hardly give very stringent constraints on the magnitude of MS stars, whose uncertainty is certainly no lower than ± 0.1 m.

3.1.2. *TO stars*

The magnitude of the TO, defined as the bluest point of the H-burning isochrone, is the well-known clock marking the cluster age. However, this appears to be a difficult

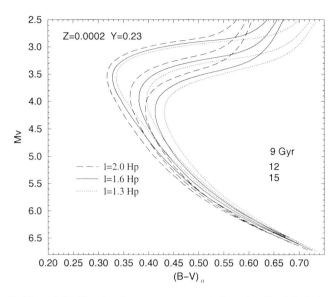

FIGURE 8. The 9, 12 and 15 Gyr isochrones as computed under the labeled assumptions about the mixing-length parameter and translated into the CM diagram according to model atmospheres by Castelli et al. (1997a,b) (from Cassisi et al. 1998b).

observational parameter, in particular in the most metal-poor globulars, where theory predicts quite a small variation in the stellar effective temperature over a rather large range of luminosity around the nominal TO. Therefore the most reliable way to "read" a cluster age is given by isochrone fitting in the whole TO region rather than relying on a difficult and often subjective determination of the observational TO magnitude. Alternative approaches to the use of TO luminosities have been suggested by Chaboyer et al. (1996) and Buonanno et al. (1998). However, here we will discuss this parameter as a useful indicator of the behavior of the isochrone with time.

As a warning, and to avoid possible misunderstanding, let me first recall that the TO of the evolutionary track *is not* the TO of the isochrone, as well as that the stars with the higher effective temperature (i.e., the TO of the theoretical $\log L$, $\log T_e$ isochrone) *are not* the stars with the smallest predicted $(B-V)$. Here we will discuss theoretical predictions concerning the magnitude of the isochrone TO, bearing in mind that a variation of 1 Gyr in the (globular) cluster age is expected to produce a decrease by about 0.1 m in such a TO luminosity.

An exemplary analysis of the dependence of TO magnitudes on the assumptions about input physics has been presented by Chaboyer (1995, but see also Brocato et al. 1998). As a result, one can estimate that theory predicts the magnitude of the TO (the cluster age) with an uncertainty of some tenths of magnitude (of some Gyr). The color of the TO is sensitive to the assumptions on the mixing length parameter, which—however—only marginally affects the TO luminosity in metal-poor clusters.

However, if the Sun is affected by element sedimentation, this should also be the case for globular-cluster stars, which have similar masses and much greater ages. The effect of such a mechanism on the isochrone TO is shown in Figure 9, where we present a cluster isochrone of 11 Gyr as computed with or without allowing for the efficiency of element sedimentation. For each given TO magnitude, such a macroscopic mechanism tends to reduce the predicted cluster ages by about 1 Gyr. However, diffusion is a rather complex

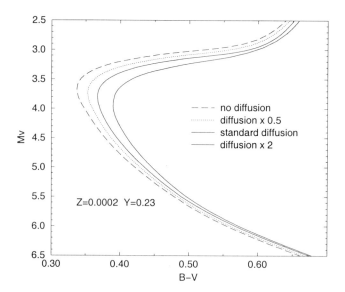

FIGURE 9. Cluster isochrones for a cluster age $t = 11$ Gyr for the various labeled assumptions about the efficiency of element sedimentation (from Castellani & Degl'Innocenti 1999).

mechanism, difficult to evaluate firmly. Pont et al. (1998) argue that in globular cluster stars somethingh could act to prevent the gravitational settling whereas Fiorentini et al. (1998) estimated that helioseismology constrains the efficiency of such a mechanism in the Sun no better than within ±30%. However, if diffusion is at the work, this uncertainty increases greatly when passing from the Sun to stars with different masses and different original chemical composition. Therefore one can safely estimate that the efficiency of diffusion is probably known to no better than a factor of two.

The same Figure 9 shows the effect of varying the efficiency of sedimentation between the two quoted limits. One finds that for ages of the order of 10 Gyr the uncertainty on such a mechanism moves—for each given cluster age—both the TO and the MS magnitudes but in opposite directions. Over the whole range of assumptions, i.e., passing from no sedimentation to twice the solar efficiency, one finds that the MS magnitude (at $Mv \sim 6.0$) *decreases* by about 0.12 m, whereas the magnitude of TO *increases* by ~ 0.34 m (!). By adding the further uncertainties on the MS luminosity discussed in the previous section, one reaches the depressing conclusion that, even assuming that theory is based on a perfect physics and that model atmospheres allowing the evaluation of observational quantities are assumed also to be perfect, *if the distance modulus of a cluster is derived by theoretical predictions about the MS luminosities* the TO magnitude cannot be predicted better than ±0.37 m, which gives an uncertainty on the absolute cluster age of about ±4 Gyr. This uncertainty is reduced to less than a half if the cluster distance can be evaluated in some alternative way.

3.1.3. *Subgiant and Red Giant Branch*

When going beyond the TO one is leaving the "best founded" stellar models, as given by computations using the "minimum" and simplest input physics that one deals with in stellar structure. As a matter of fact, after the exhaustion of central H (which occurs around the track TO) a central He core develops, where the pressure by quantum-degenerate electrons slows down the contraction, delaying the ignition of central He burning. Such a delay is further increased by the cooling deriving from the efficiency of both electron conduction in degenerate matter and of neutrino production through weak interactions. As a result, low-mass stars in Galactic globular clusters cannot rapidly ignite He after the exhaustion of central H, developing the RGB which characterize these clusters. The interested reader can find detailed discussions about such a *Red Giant Transition* in Sweigart, Greggio &Renzini (1989, 1990).

We have already found that the CM location of cool giant stars depends increasingly on the assumptions about the efficiency of superadiabatic convection. Therefore, at the present state of the art, one can only empirically tune the mixing length to overlap the observed colors of these stars (see, e.g., Salaris & Cassisi 1996, Brocato et al. 1997). However, the mixing length only affects the radius (and, thus, the effective temperature) of these stars, and the behavior of luminosity with time remains a well-founded theoretical prediction. This allow a series of quite relevant checks of evolutionary theories. To enter into the subject, let us show in Figure 10 the theoretical predictions concerning the distribution of star luminosity all along the post-MS evolution, as presented for the labeled chemical composition and for different assumptions about the cluster age (lower panel) or the exponent of the Initial Mass Function (IMF) (upper panel), assumed to be a power law. In the figure n represents the expected number of stars in a given interval of luminosity ΔL for each given luminosity L and for a given number (N_{tot}) of stars along the RGB.

A quick look at the figure shows a feature that is already well known but is relevant: the luminosity function of stars moving from the MS toward the RG branch depends both on the IMF and on the assumed cluster age. This is not the case for stars along the RGB, whose distribution appears independent of both these parameters. Note that these distributions do depend on the assumptions about N_{tot}. However, thanks to the chosen representation $\log N$, $\log L$ (Castellani 1976), one immediately understands that a variation of N_{tot} only shifts the curve in Figure 10 along the y-axis, so that shape of these curves has to be regarded as a firm and unequivocal prediction of evolutionary computations, for each given assumption about the cluster age and original IMF.

One has in particular to notice that the LF appears independent of the IMF before the TO luminosity. In this case, theory tells us that the expected number of giants in a given interval of luminosity is proportional to the time spent by these stars in the interval. This is a particular case of a general rule by which all along the fast evolutionary phases following the central H burning the number of stars expected in a given selected phase is in all cases proportional to the time spent by stars in that phase. Therefore star counts in selected portions of the CM diagram give additional constraints on the evolution of stars, adding a "third dimension" to the constraints given by the run of the isochrones in that diagram.

In the case of the RGB, theory predicts that $\log N$ should depend linearly on $\log L$, with only a bump occurring along this distribution. The first occurrence (the linearity) is a straightforward consequence of the well-known existence of a relation between the RG luminosity and the mass of the electro-degenerate He core (Castellani, Chieffi & Norci 1989). The second one, the bump, is the predicted consequence of the H burning shell reaching the chemical discontinuity at the bottom of the maximum extent of the convective envelope (first *dredge up*).

As for the general linear distribution of stars along the RGB, we can thus use star counts to test the velocity of evolution of the structure along this sequence, testing in this way our prediction about the efficiency of electron conduction and neutrino production. As shown by Figure 11, comparison of observational data with theoretical predictions, as given by several authors (see, e.g., Paczynski 1984, Ratcliff 1987, Bolte

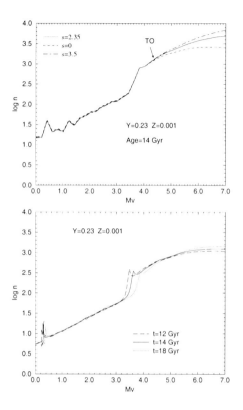

FIGURE 10. The predicted distribution of cluster stars along the MS, the subgiant and the RG branch for the labeled assumptions about the exponent of the IMF (top panel) or for the assumed cluster age (bottom panel). The arrows mark the luminosity of the cluster TO (from Degl'Innocenti, Weiss & Leone 1997).

1994), reveals in general a beautiful agreement, therefore supporting our adopted input physics (but see VandenBerg et al. 1998). However, to shed light on the relevance of such a theoretical constraint, one has to recall that Faulkner & Swenson (1988, 1993) have suggested the occurrence of a mismatch between observed and predicted LF, to be connected with an additional cooling of the stellar interior by cosmic Weak Interacting Massive Particles (WIMPS). Even though the reality of this mismatch has been challenged by Degl'Innocenti, Weiss & Leone (1997), the above suggestion clearly shows the relevance of such an observational constraint.

As for the bump, after the pioneering papers by Thomas (1967) and Iben (1968), this subject was first investigated by Sweigart & Gross (1978). More recently both the predicted luminosity and the population of the RG bump have been further investigated by Castellani, Chieffi & Norci (1989) (see also Bono & Castellani 1992). Summarizing the results, one finds that the luminosity of the bump decreases when the cluster age and/or the metallicity increase, whereas the signal-to-noise ratio (where the "signal" is the bump and the "noise" is given by the statistical fluctuation of RG counts) has a more complicated behavior. The level of agreement between observation and theoretical predictions has been widely debated in the past (see, e.g., Fusi Pecci et al. 1990). The matter has

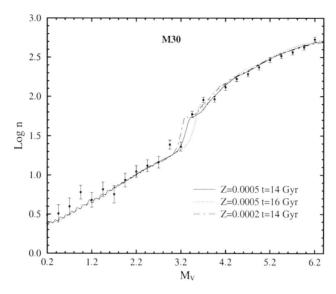

FIGURE 11. The observed luminosity function for H-burning stars in the globular cluster M30 as compared with theoretical predictions for the labeled assumptions about the cluster age. Error bars give the 1 σ statistical error of the stellar counts.

been recently reviewed by Cassisi, Degl'Innocenti & Salaris (1997), who conclude that there is a reasonable agreement between observation and current theoretical scenarios, as shown in Figure 12. If this is the case, this appears to be a relevant result, since the predicted luminosity of the bump marks the evolutionary phases when the H-burning shell (i.e., the boundary of the central He core) crosses the chemical discontinuity at the maximum depth previously reached by surface convection. The agreement quoted above would thus imply that the adopted computations give a realistic evaluation of the extension of the convective envelope, leaving little room for additional mechanisms like the already quoted overshooting (in this case: undershooting!) extending the mixing beyond the boundaries defined by Schwarzschild's criterion.

A further prediction of stellar models allowing a comparison with observational data is given by the maximum luminosity reached by a RGB star before igniting the He flash to start the following HB evolutionary phases. However, the detailed theoretical predictions about such a luminosity face the observational difficulty of a precise determination of such a parameter. As a matter of fact, we have already found that the population of the RGB is strongly decreased when the luminosity increases. As a result, in the final portion of the branch one finds in general only few stars randomly scattered below the maximum theoretical luminosity. This point has been early discussed in a pioneering paper by Frogel, Cohen & Persson (1983) who estimated that the most luminous giant observed in a cluster should approach the luminosity of the RG tip better than 0.1 mag. The connection between observational and theoretical tip luminosities has been further discussed in Castellani, Degl'Innocenti & Luridiana (1993).

A similar approach has already prompted several investigations, aiming either to test the predictions of stellar evolution or to use such very luminous stars as standard candles to constrain the distance of extragalactic stellar systems (see Da Costa & Armandrof 1990, Lee, Freedman & Madore 1993). As for the test of evolutionary predictions, Figure 13 shows the most recent results on the matter, as given by Salaris & Cassisi (1998).

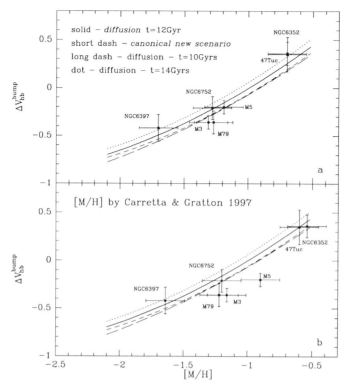

FIGURE 12. Comparison between predicted and observed luminosities of the RG bump as a function of the cluster metallicities for the labeled assumptions about the cluster age and adopting the cluster metallicities from the compilation by Salaris & Cassisi 1996 (upper panel) or from Carretta & Gratton 1997 (lower panel). The luminosities predicted by no-diffusion models are also reported for comparison.

The reasonable agreement between observation and theory at least gives support to the adopted physics governing the size of the He core in a RG stars and, in particular, the adopted treatment of neutrino cooling and electron conduction, which both contribute to the cooling of the core, delaying the onset of the He flash.

Once again, this offers us the opportunity of emphasizing the strict connection between stellar models and fundamental physics. As a matter of fact, one finds that the luminosity of the RGB was demonstrating the efficiency of neutral currents in weak interactions well before ground-based physics was able to prove such an efficiency. This comes about because in the absence of neutral currents one should have less efficient neutrino production and, in turn, lower size of the He core at the flash and sensitively lower luminosity of the RG tip. Much more recently, the evolution of cluster RG has been used to put severe observational (experimental!) upper limits to the neutrino magnetic moment (Raffelt & Weiss 1992; Castellani & Degl' Innocenti 1993).

3.2. *HB and beyond*

In this section we will discuss theoretical predictions concerning stars all through their He burning phases, from the initial Zero Age HB models through the phases of central and shell He burning till the exhaustion of nuclear fuels and the entry into the final cooling phase as a White Dwarf. We will find several tests for theoretical predictions

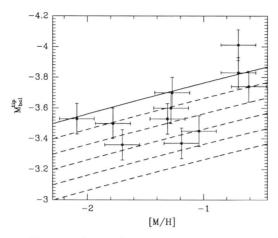

FIGURE 13. Observed (crosses) and predicted luminosities (full line) of the RG tip as a function of the cluster metallicity.

and, in particular, two relevant predictions concerning the luminosity of both the HB and the AGB.

3.2.1. *The Zero Age Horizontal Branch*

After the He flash a stars readjusts its structure quickly (i.e., on a gravitational time scale), moving to its Zero Age Horizontal Branch location, where the structure is supported by both central He and H shell burnings. It has been suggested that the first HB model powered by nuclear energy generation only, should experience a rapid excursion toward larger temperatures (a "blue nose"), following the readjustment of CNO elements in the H-burning shell (Caputo, Castellani & Tornambè 1978). Such an occurrence would give only marginal evidence in the CM diagram, but could affect the pulsational behavior of the RR Lyrae pulsators that we will discuss later on. Detailed computations, as recently given by A. Chieffi (private communication) seem to support such an occurrence.

As early recognized, the CM location of a low-mass, He-burning model depends on the mass of the central He core as well as on the mass of the H-rich envelope surrounding this core. Roughly speaking, one can say that the first one governs the luminosity of the structure whereas the second one governs its effective temperature. In order to handle the issue concerning the HB one has to bear in mind some general results, namely:

i) The range of temperatures covered by observed HB can be understood only by assuming a differential amount of mass loss during the previous RGB evolution.

ii) In old globulars the mass of the He core is predicted to be rather independent of either the original stellar mass (and, thus, of the cluster age: see the discussion in Cassisi et al. 1998a) or the amount of mass loss (Castellani & Castellani 1993). It decreases when the envelope helium content (Y_{env}) increases and, to a lesser extent, when the metallicity increases.

iii) Not-too-hot new born HB stars from progenitors with a larger amount of original He have smaller He cores BUT larger luminosity, since the increased efficiency of the H shell burning overcomes the decreased efficiency of He burning in the core. This is not the case for hot HB stars where the thin H-rich envelope do not support the efficiency of

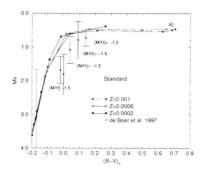

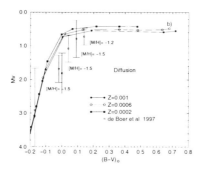

FIGURE 14. The run in the CM diagram of the predicted ZAHB for the given values of the metallicity and for alternative assumptions about the occurrence of element diffusion.

the H shell, and the 3α luminosity is governed by the size of the He core (see, e.g., Caloi, Castellani & Tornambè 1978)

As for macroscopic mechanisms, the radius of cool HB stars is obviously affected by the treatment of superadiabatic convection, whereas overshooting from the central convective cores would affect only models evolving off their ZAHB phase. Due to the (relatively) short HB evolutionary times, the influence of diffusion on the structure of He-burning stars is the result of the influence on the H-burning progenitors. As already predicted by Proffitt and Vandenberg (1991) and confirmed by evolutionary calculations (Caloi et al. 1997, Castellani et al. 1997) when diffusion is at work helium is ignited within a slightly larger He core (M_c^{He}) and with a lower He abundance in the envelope (Y_{env}).

Figure 14 shows the predicted run in the CM diagram of selected ZAHBs for the labeled values of chemical composition from Cassisi et al. (1998a). In the same paper one can find a discussion of the comparison with previous results that have already appeared in the literature. As repeatedly stated in this paper, one has to read such differences in terms of the adopted input physics. As an example, the evidence that our values for the mass of He cores at the He flash are larger by about 0.004 $M_\odot$ with respect to the evaluation by Straniero, Chieffi & Limongi (1997) is the simple consequence of our adoption of neutrino losses by Haft et al. (1994).

Bearing in mind that the off-ZAHB evolution in all cases runs toward larger luminosities, one finds that the ZAHB defines the lower envelope of the predicted distribution of HB luminosities. In Figure 5 we have already shown that theoretical predictions, when transferred into the observational $B - V$, V plane, appear in excellent agreement with

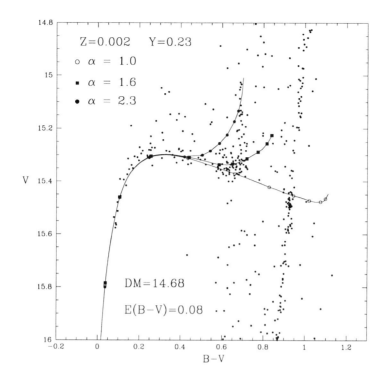

FIGURE 15. The CM diagram of HB stars in the moderately metal-rich globular NGC 6362 as compared with the predicted distribution of ZAHB structures for the labeled choices of the mixing length parameter.

observation. However, it is worth noticing that the beautiful turn-down of the HB luminosity at the blue side of the branch is essentially due to the growing of the bolometric correction rather than to the intrinsic behavior of the true luminosity of the stellar models. On these grounds, such an observational feature has been already used as an useful tool to derive the reddening of globulars with well populated blue portion of the HB (Brocato et al. 1998b).

Again concerning the HB morphology, one has to notice that recent observations have revealed in some moderately metal-rich globulars the occurrence of "tilted" HB, i.e., of an HB whose luminosity is not smoothly decreasing from the red to the blue side of the branch, showing on the contrary a maximum just before the turn-down we were speaking of above (Piotto et al. 1997, Rich et al. 1997) . Such evidence has suggested the occurrence of non-canonical mechanisms affecting HB stars, such as a differential enrichment in surface He or an increase of the mass of the He core in a flashing Red Giant due to stellar rotation (Sweigart 1997, 1998, Sweigart & Catelan 1998) . However, here we notice that tilted HBs are also the natural prediction of canonical theories when the cluster metallicity increases. This can be found by looking at the data for $Z = 0.001$ presented eight years ago in the paper by Castellani, Chieffi and Pulone (1991). However, the evidence for tilted canonical HBs can be much better realized looking at Figure 15, where we show a recent CM diagram of stars in the globular NGC 6362 with superimposed theoretical ZAHBs for selected assumptions about the mixing length parameter (from Brocato et al. 1999).

Coming back to the data in Figure 14, over the range of explored metallicities one would derive a quadratic relation between the magnitude of the ZAHB at $\log T_e$=3.85 (i.e., at the RR Lyrae gap) and [Fe/H], as given by:

$$M_V = 0.993 + 0.461[Fe/H] + 0.097[Fe/H]^2,$$

which, at the lower metallicities, can be approximated by the linear relation:

$$Mv = 0.18[Fe/H] + 0.77$$

On the observational side, this dependence of HB luminosities on metallicity has already deserved quite a large number of investigations, since a firm knowledge of such a dependence would allow the use of HB stars as standard candle to derive the cluster distance moduli and, in turn, to constrain the TO luminosity and—thus—the cluster age. By looking in the current literature one finds several results, as based on various different approaches to the problem. Here let me quote, in chronological order:

$$Mv = 0.15[Fe/H] + 0.73 \qquad \text{(Walker 1992)}$$
$$Mv = 0.30[Fe/H] + 0.94 \qquad \text{(Sandage 1993)}$$
$$Mv = 0.19[Fe/H] + 0.97 \qquad \text{(Clementini et al. 1995)}$$
$$Mv = 0.23[Fe/H] + 0.83 \qquad \text{(Chaboyer et al. 1998)}$$
$$Mv = 0.18[Fe/H] + 0.74 \qquad \text{(Gratton et al. 1998)}$$

One has however to notice that such relations refer to the *mean magnitude* of HB stars rather than to the ZAHB luminosity, which represents the lower envelope of the observed luminosities. Therefore the zero point of the previous relation should be increased by some hundredths of a magnitude, as discussed by Caputo et al. (1993). Also taking into account such a difference one can only say that theoretical predictions appear in rather good agreement with *some* observational results, though the question is far from being settled.

Our problem is now: how far can we be confident in the theoretical predictions? In this respect one has indeed to bear in mind that theoretical predictions about He-burning stars are more uncertain than predictions on central H-burning structures. In fact theoretical evaluations concerning He-burning structures require the intervention of additional physical mechanisms which do not affect MS and TO stars, such as neutrino production by weak interactions and electron conduction in electron-degenerate matter. A further source of uncertainty on the HB structures comes from the cross section of the 3α reaction, which governs the onset of the He flash and, thus, the He core mass of the ZAHB structure.

To look into such problems we have performed some numerical experiments to investigate the sensitivity of HB luminosities to the above quoted input physics (Castellani & Degl'Innocenti 1999). As in the previous section, the efficiency of sedimentation has been varied by a factor of two. Moreover, we took from Haft et al. (1994) the estimated uncertainty of plasma neutrino energy losses, tentatively adopting a similar uncertainty in the radiative opacity. The uncertainty in the 3α reaction rate is $\pm15\%$ (see, e.g., Rolfs & Rodney 1988). Finally, electron conduction deserves a bit more discussion. According to Catelan, de Freitas Pacheco & Horvath (1996) a completely satisfactory estimate of conductive opacities for RGB stars is still not available; in particular Catelan et al. (1996) pointed out that in RG evolutionary computation the widely adopted Itoh et al. (1983) formulation is used beyond its range of validity, suggesting the alternative adoption of the Hubbard & Lampe (1969) evaluations.

TABLE 1. The luminosity and the visual magnitude of ZAHB stars in the RR Lyrae region (at $\log T_e = 3.85$) and the luminosty at the RG tip (last column) under the labeled assumptions about the physical inputs (see text). The third column gives the difference in the HB visual magnitude with respect to the model with standard diffusion. Chemical composition as labeled. H&L (69) indicates the adoption of Hubbard & Lampe (1969) coefficients for the conductive opacity.

	$\log L$(HB)	M_V(HB)	ΔM_V(HB)	$\log L$(tip)
Z=0.0002 Y=0.23				
no diffusion	1.748	0.386	$-$ 0.033	3.349
std. diffusion	1.735	0.419	—	3.352
diffusion x 0.5	1.743	0.398	$-$ 0.021	3.350
diffusion x 2	1.719	0.460	$+$ 0.041	3.356
plasma ν +5%	1.739	0.408	$-$ 0.011	3.358
plasma ν $-$5%	1.732	0.429	$+$ 0.010	3.346
3α +15%	1.732	0.426	$+$ 0.008	3.347
3α $-$15%	1.740	0.408	$-$ 0.011	3.357
opacity +5%	1.733	0.415	$-$ 0.004	3.337
opacity $-$5%	1.737	0.424	$+$ 0.005	3.368
H&L (69) conduction	1.718	0.461	$+$ 0.042	3.330

Table 1 summarizes the variations of the ZAHB luminosity (at the RR Lyrae gap) due to the uncertainties of the physical parameters considered; one clearly sees that the uncertainties in the efficiency of element diffusion and the conductive opacity are the main source of variation. If one decides to be very conservative, one can simply add the errors given in the table to derive that theory gives the ZAHB luminosity $Mv_{-0.059}^{+0.106}$ mag. As a result, one finds that in order to constrain the cluster distance, theoretical ZAHB work better than theoretical MS, whose uncertainty has been already estimated ± 0.20 mag as given by the uncertainty on macroscopic mechanisms alone.

The last column in the same Table 1 gives the predicted luminosity of a RG star at the He flash, showing that such a parameter can be affected by rather similar uncertainties, and allowing an evaluation of uncertainties in the often discussed difference in luminosity between the RG tip and the HB.

3.2.2. *Horizontal branches*

During the main phase of central He burning, stars are burning He into C (and O) at the center of a convective core. In the canonical scenario (i.e., no overshooting) that we are moving in, the increased opacity of the C-enriched core causes a growth of the central convective instability and eventually the onset of a semiconvective region surrounding the convective core (Castellani, Giannone & Renzini 1971). The relevant consequence of such an occurrence is that more fuel is driven into the burning region and the HB lifetime is appreciably increased, thus increasing theoretical expectations about the number of stars to be found along the HB.

During this phase, a low-mass star moves slowly from its ZAHB location, with a path which depends on the assumed chemical composition, as early shown in the seminal paper by Sweigart & Gross (1976). As a result, one expects that HB stars cover a rather restricted range of luminosities above their ZAHB. Sandage (1990) first indicated that such a luminosity dispersion appears to increase when the cluster metallicity increases. Such an occurrence is beautifully predicted by theory, as discussed by Bono et al. (1995). As already mentioned, such a dispersion does play a relevant role in the evolutionary scenario, since many observational results refer to the *mean luminosity* of HB stars, which has to be connected with theoretical predictions about the ZAHB luminosity through the above quoted evaluation of the luminosity dispersion (see, on this matter, Caputo et al. 1993).

A problem arises in the last phases of this central He-burning phase, when stars are burning the small residual amount of He into the C-enriched convective core. Even a small amount of He brought into the core by semiconvection can cause a sudden increase of the luminosity generated in the core, which in turn drives a growth of the radiative gradient (see the role of L_r in the equation for the gradient!) and a series of sudden increases of the convective core known as "breathing pulses" (Sweigart & Demarque 1973). The effect of this occurrence is easy to understand: more fuel is driven into the burning region, and the lifetimes of these last phases of HB evolution increase.

However, owing to the complexity of the phenomenon, the reality of these breathing pulses has been largely debated. Caputo et al. (1989) on the basis of star counts on the HB and the RGB suggested (but it was only a suggestion) that breathing pulses should be neglected in modeling HB evolution, and this is the recipe adopted in all more recent evolutionary computations. However, it is not clear if such a suggestion will survive the several variations in the evolutionary scenario that have occurred since that time.

Now we are in the position, at least in principle, of using the theory to derive relevant information not directly accessible to observation, namely the amount of original He in globular-cluster stars. This is a quite relevant parameter, since it puts an obvious upper limit to the amount of cosmological He produced during the big bang. However, He lines can be observed only in hot HB stars, which have been found to be deeply depleted in He. As early suggested by Greenstein, Truran & Cameron (1967), this is due to diffusion, which does not affect the structure as a whole, but which succeeds in depleting He from the very thin atmospheres (in radiative equilibrium) of these hot stars. The best evidence for such an efficiency is probably the investigation by Heber et al. (1986), who mapped the surface He along the hot tail of the globular NGC 6752. As shown in Figure 16, one finds that He anticorrelates with the temperature and, thus, with the gravity of the stars. Therefore surface He in these stars cannot be used to investigate the original amount of He.

However, the evolutionary scenario is providing us with a quite simple way to investigate such a parameter. We have already found that the HB luminosity is expected to increase when the original He increases. On this basis Iben (1968) early suggested that the number ratio (R) of HB to RGB stars brighter than the HB can be used as an indicator of the original He, for the very simple reason that increasing Y decreases the portion of RGB contributing to the counts. The method has been carefully calibrated by Buzzoni et al. (1986) and, more recently, by Bono et al. (1995), who discussed the dependence of this parameter on the cluster metallicity. On these bases, one finds in the literature several estimates of Y, normally ranging around $Y \simeq 0.23$, thus in agreement with evaluation of primeval He in extragalactic HII regions. However, Cassisi et al. (1998a) have recently shown that HB lifetimes and, thus, theoretical calibrations of R are strongly affected by uncertainties on the $^{12}C + \alpha$ reactions which govern the burning near the exhaustion of central He, concluding that at present time the uncertainty in the calibration of R does not really constrain the value of Y with a reasonable accuracy.

Before closing this section one has to notice that the above uncertainty in HB lifetimes

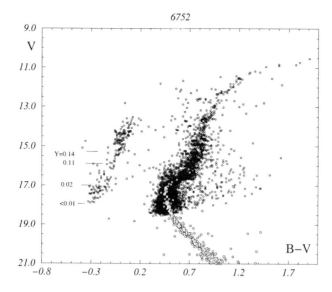

FIGURE 16. Photographic photometry of luminous stars in NGC 6752, supplemented at the fainter magnitudes by CCD data, from Buonanno et al. (1986), with the evaluation of the atmospheric abundance of He as given by Heber et al. 1986 for stars along the hot tail of the HB.

may mask the increase of this lifetime that is due to the breathing pulses described above. Moreover, if one prefers an overshooting scenario, one is avoiding the occurrence of breathing pulses, but HB lifetimes keep being uncertain since the amount of overshooting appears as a free parameter to be adjusted to fit these lifetimes.

3.2.3. *He shell-burning phases*

As already noted, observational evidence tells us that the distribution of stars along the ZAHB is mainly governed by the amount of mass lost by their Red Giant progenitors. Since the effective temperature of ZAHB stars is very sensitive to the amount of envelope surrounding the H-exhausted core, a dispersion of only few hundredths of a solar mass can account for the dispersion of temperatures observed in a cluster like M3 or M5. However, in some other clusters, like, e.g., NGC 6752 in Figure 16, the long blue tail of the HB has to be taken as evidence that several stars have lost practically all the H-rich envelope. The origin of such behavior is beyond the scope of these lectures. Here let me say only that, at least in my opinion, we are probably looking at the evidence of atmospheric stripping in dense (post-core-collapse) clusters. Some general constraints on the hot end of the branch are discussed in Castellani, Degl'Innocenti & Pulone (1995), D'Cruz et al. (1996).

Here we notice that in clusters with very hot HB tails it is also possible that some giants fail to reach the He flash, leaving the RG branch directly for the WD cooling sequence. Evolutionary constraints on similar structures are discussed in Castellani, Luridiana & Romaniello (1994). Interesting enough, some RG—failing to ignite He at the tip of the RGB—can undergo the flash when cooling down along the WD sequence, reviving as extremely hot HB stars (Castellani & Castellani 1993).

Without entering into further details, now we will face the evidence that the post-HB evolution depends basically on the mass of the HB structure. After the exhaustion of central He, not-too-hot HB stars run toward their Hayashi track to ignite shell He

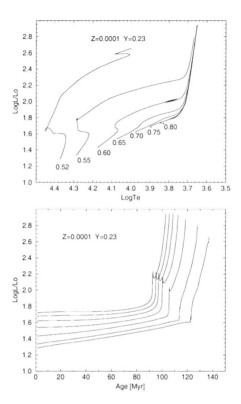

FIGURE 17. Upper panel: The run in the CM diagram of He-burning stars all along the HB and AGB phases. Lower panel: the evolution with time of the luminosity of stars in the upper panel (from Bono et al. 1995).

burning at the base of their Asymptotic Giant Branch, whereas hot HB structures miss such an approach. A detailed discussion of the approach to the AGB can be found, e.g., in Castellani, Chieffi & Pulone (1990), whereas the structural parameters governing the post HB evolution are discussed in Castellani et al. (1994). Figure 17 shows a detailed picture of the evolutionary behavior of stars, starting from points on the ZAHB, for the given values of chemical composition and for the labelled values of stellar masses.

One expects that the more massive stars reach their AGB phase, climbing along the branch to eventually experience the onset of thermal pulses before exhausting nuclear fuel to cool down as White Dwarfs. Less massive stars reach the branch, but leave it before the onset of thermal pulses; and, finally, stars of even lower mass do not reach the branch at all, crossing the HR diagram to reach the final WD cooling sequence. According to the classification given by Greggio & Renzini (1990) the stars approaching their final cooling can be thus divided in the three corresponding classes of i) post-AGB, ii) post-Early AGB and, iii) AGB-manqué stars. This behavior has been largely discussed in connection with the problem of integrated UV fluxes from external galaxies (see, e.g., Dorman, O'Connell & Rood 1995, Yi, Demarque & Oemler 1998); this matter is however beyond the limits of the present lectures.

Here we will focus our attention on stars which reach the AGB. The lower panel in Figure 17 shows the time evolution of a selected sample of HB stars throughout the central and shell He burning. One see that the slow increase of luminosity during the HB (central He burning) phases ends with an abrupt increase of the luminosity at the exhaustion of central He (the star is crossing the HR diagram to reach the AGB) which is followed by a small plateau after which the luminosity starts again to rise rapidly. In terms of the HR diagram this implies that we expect a small clump of stars well above the HB magnitudes and some few stars scattered along the branch at the higher luminosities. And, at least qualitatively, this is just what one see in the clusters.

Quantitatively one has two theoretical predictions which can be compared with observations, as given by the luminosity of the AGB clump and the relative abundance of stars in the AGB phases. Castellani, Chieffi & Pulone (1991) have shown that their predictions about the luminosity of the AGB clump appear in good agreement with observations. As for the abundance, they predict (in HBs that are not too blue!) a number ratio of AGB to HB stars as given by $R2=N(\text{AGB})/N(\text{HB})\simeq0.13$ in reasonable agreement with the value 0.15 ± 0.01 estimated by observations (Buzzoni et al. 1983, Buonanno et al. 1985). However, more recent models by Cassisi et al. (1998a) with improved input physics give now $R2=N(\text{AGB})/N(\text{HB})=0.15$! This agreement appears too perfect, taking into account the various sources of errors already discussed in this paper. One has finally to notice that the same value can be obtained also by tuning the efficiency of overshooting within reasonable values (Bressan et al. 1986, Chiosi 1986)

4. HST: enlarging the evolutionary scenario

In the previous sections we have discussed theoretical constraints concerning the evolutionary scenario covering the evolutionary phases which have for a long time characterized the CM diagram of Galactic globulars. However, at the beginning of these lectures we have already focused attention on the new evidence brought to the light by HST. In this section we will approach this problem, discussing current theoretical constraints on that matter.

4.1. *Very Low Mass MS stars*

In Figure 2 we have already shown the beautiful sequence of VLM disclosed in the cluster NGC 6397 by HST. It appears that such photometry reveals MS stars down to about the limit for the ignition of H, i.e., down to about 0.1 $M_\odot$. Lower masses will become degenerate during the contraction, failing the ignition of H and continuing to cool down as Brown Dwarfs. Similar results have recently given new impetus to the investigation of Very Low Mass (VLM) MS structures, whose evidence was for a long time confined to the stars in the solar neighborhood with known trigonometric parallaxes.

From the theoretical point of view, VLM structures have several advantages but, at the same time, one is also facing new and difficult problems. As a general overview, let me recall that we expect that the globular cluster TO are populated by stars with masses around 0.8 $M_\odot$, whose envelopes are affected by convection. Going down along the MS, the stars become cooler and cooler and, thus, external convection sinks down until, around 0.3 $M_\odot$, stars become fully convective. However, the envelopes are becoming increasingly dense, and the convection more and more adiabatic. As a result, below about 0.6 $M_\odot$ the models are no longer affected by assumptions about the mixing length. This is of course a great advantage.

However, as discussed by Burrows et al. (1993), stellar atmospheres are largely affected by convection and the Eddington approximation is no longer valid; thus the boundary conditions have to be derived from a proper evaluation of stellar atmospheric models. This is not an easy task, since the EOS has to take into account the coexistence of electron

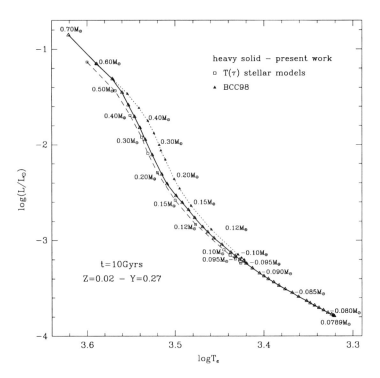

FIGURE 18. The HR diagram location of VLM stars as computed adopting either Brett's (Brocato et al. 1998: BCC98) or Allard & Hauschildt's (1998) ("present work") model atmospheres.

degeneracy and strong coulomb interactions, whereas both the EOS and the opacity of such cool matter have to take into account grains and molecules. As a consequence, a theory for VLM structures has been a long job, which started with the pioneering papers by Limber (1958), followed by several research papers scattered in time, as recently reviewed, e.g., in the introduction to the paper by Brocato, Cassisi & Castellani (1998). To be brief, for a long time one was facing the tantalizing evidence that theoretical models were too blue to reproduce the observed sequence of metal-rich VLM stars in the solar neighborhood.

VLM models with suitable assumptions about the external boundary conditions have been presented only by a few groups. One has to quote here the pioneering series of paper given by Baraffe, Chabrier and coworkers (see, among the most recent ones, Baraffe & Chabrier 1996, Baraffe et al. 1997, and references therein). Figure 18 gives theoretical expectations in the $\log L$, $\log T_e$ HR diagram according to quite recent computations by Cassisi (private communication). One can take the difference between models computed alternatively adopting Brett (1995a,b) or Allard & Hauschildt (1998) model atmospheres as an indication of the residual uncertainties in these models. Moreover, one can appreciate the curious occurrence that in adopting Allard & Hauschildt atmospheres one is coming back very near the prediction of canonical atmospheres as computed under the Eddington approximation.

The left panel in Figure 19 shows that HST observations of VLM stars in the metal-poor globular NGC 6397 (see Fig. 2) are indeed rather well reproduced by these models. Moreover, the right panel in the same Figure shows that the discrepancy for solar-metallicity VLM stars is, at least, noticeably reduced. There is still some work to do on

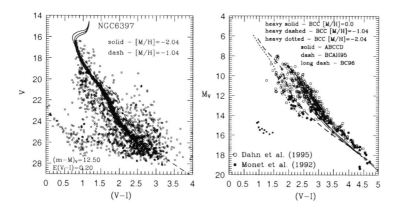

FIGURE 19. Left panel: the best fitting of theoretical isochrones with a suitable value of metallicity ([Fe/H] = −2.04) to the NGC 6397 MS observed by HST. For the sake of comparison, the dashed line shows the predicted cluster MS when the metallicity is increased to [Fe/H] = −1.04. Right panel: The comparison of theoretical predictions for selected assumptions of stellar metallicity (from Brocato, Cassisi & Castellani 1998: BCC) with the observed distribution of dwarfs with known parallaxes. Solar-metallicity models by Baraffe et al. (1995: BCAH95), Baraffe & Chabrier (1996: BC96) and Alexander et al. (1997: ABCCD) are also shown for comparison.

the matter, even if it is not clear, at least to me, whether the remaining discrepancy originates from the inadequacy of models or, on the contrary, of the adopted color-temperature relations.

4.2. *Cluster White Dwarfs.*

To complete the discussion of the evolutionary phases populating Galactic globulars we finally have to discuss theoretical predictions concerning the final cooling of cluster WD. To introduce the subject, one has to recall that any star that develops a degenerate C-O core and in which mass loss succeeds in reducing the mass below the Chandrasekhar limit must end its life as a C-O WD. The upper limit for masses of WD progenitors is a widely debated topic in the literature, together with the relation between the mass of the progenitor and the mass of the final cooling structure. Weidemann & Koester (1983, 1984) estimated an upper mass limit of WD progenitors of $\simeq 8\ M_\odot$, whereas an estimate of the relation between original and final mass can be found, e.g., in Wood (1992).

By relying on the above quoted estimates one finds that a cluster takes only 30 Myr before starting to produce WD. Therefore one is facing the exciting fact that a cluster WD sequence contains the "frozen" evidence about practically the whole cluster history. As for theoretical predictions, one has to notice that cooling times are so long that they can exceed the current estimates of the Hubble time. Therefore in these last phases one must again produce "isochrones" on the basis of suitable evaluations of the evolutionary tracks (as given by the cooling laws along the constant-radius line) for the various masses, whereas predictions about the luminosity distribution of WD can be obtained by *synthetic*

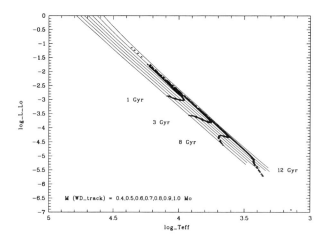

FIGURE 20. WD cooling isochrones for the various labeled assumptions about the cluster age. Cooling sequences for the various labeled WD masses are also shown (Brocato, Castellani & Romaniello 1999).

cooling sequences, i.e., populating the isochrones according to suitable assumptions about the IMF.

However, as discussed by Wood (1992), cooling times depend on the composition of the C-O core, which in turn depends on the uncertain cross-section for ^{12}C + α reactions. Moreover, Castellani, Degl'Innocenti & Romaniello (1994) have shown that nuclear sources can appreciably slow down the cooling in the first phases ($\log L \geq -4$) of cooling (see also Iben & Tutukov 1984; Koester & Schoenberner 1986). Bearing in mind these uncertainties, Figure 20 presents selected WD isochrones as computed assuming pure carbon models and He envelopes as thin as 10^{-4} $M_{\odot}$. One finds that in all cases a large portion of the sequence follows with good approximation the cooling line of low-mass (0.5–0.6 $M_{\odot}$) WD. At the bottom of the sequence, the most massive WD, which originated earlier from the more massive progenitors, leave the sequence to move toward higher effective temperature.

As a relevant point, the luminosity of WD clumping in the "hook" at the base of the sequence can be calibrated in terms of the cluster age. One finds :

$$\log t \; (\mathrm{Gyr}) = -1.41 - 0.51 \log L,$$

which, for each given $\log L$, gives a safe lower limit for the age of a cluster older than, about, 3Gyr, with an uncertainty of about 20%. For shorter ages, the contribution by nuclear sources could however decrease the above predicted age up to, about, 1 Gyr. As a relevant point, one finds that the population of the WD sequence can give information about the cluster IMF. This can be easily understood from the data in Table 2, where we give the expected number of WD per each surviving HB star for selected assumptions about the cluster age and the exponent of the IMF.

As a final point, Figure 21 closes the general problem of reproducing the CM diagram, showing the satisfactory agreement between predictions and observation. Note that, on

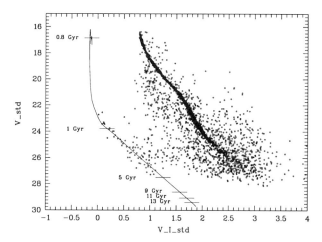

FIGURE 21. HST observation of faint stars in NGC 6397 (Cool, Piotto & King 1997) compared with the cooling sequence of a 0.5 $M_\odot$ WD. The predicted luminosity of the faint end of the sequence is indicated for the labeled assumptions on the cluster age.

TABLE 2. Expected number ratios $N_{\rm WD}/N_{\rm HB}$ for different assumptions for the cluster age and the exponent of the IMF.

t (Gyr)	$\alpha = 0.0$	$\alpha = 2.35$	$\alpha = 3.35$
1	82	22	11
3	490	65	42
8	2080	187	121
12	4166	295	180

this basis, the cooling sequence can be used as a standard candle constraining the cluster distance modulus, a procedure already applied by Renzini et al. (1996) to constrain the distance modulus of the globular NGC 6752.

5. The pulsational connection

Throughout the previous sections we have seen that the CM diagram provides several tests for theoretical predictions as given either by the CM location of the stellar models or by the relative abundance of these models along selected evolutionary phases. In this respect, we have already spoken of a "third dimension" of the HR diagram. Now our trip into the world of stellar evolution will enter into a "fourth dimension", opening a new and additional source of experimental data concerning the structure and the evolution of cluster stars, as given by the well known occurrence of RR Lyrae variables along the HB of several Galactic globulars.

The evidence for a periodic variability of some HB stars, with period typically shorter than 1 day, was among the first observational phenomena which drew attention to Galactic globulars. The investigation of these variables started as early as 1885 (!) when Bailey discovered a group of these variables in the Galactic globular ω Cen. On observational grounds, one has to notice that the phenomenon of variability has its equivalent of the CM diagram in a diagram where one can arrange the two direct observables characterizing the pulsation, as given by the period and by the amplitude of the light curve. This is the well known Bailey diagram. In contrast with the CM diagrams, and this is quite a relevant point, note that the Bailey diagram appears largely unaffected by the many uncertainties in the link between theoretical predictions and the observed star magnitudes and colors, since periods and amplitudes are affected neither by cluster reddening nor by the distance modulus. Therefore we are happily dealing with firm observational constraints, easy to link with the predictions of theory.

According to the distribution of data in this diagram, Bailey (1902) found that within RR Lyrae pulsators one finds the occurrence of two distinct classes of variability: RRab, with asymmetric lightcurves and large amplitudes (of the order of 1 mag in B) decreasing for longer periods, and RRc with short periods and symmetric, smaller-amplitude lightcurves. Now we know that we are in the presence of an instability for radial pulsations, where the hotter RRc structures are pulsating in the first overtone (Schwarzschild 1941), whereas the cooler RRab structures experience the instability of the fundamental mode.

The theoretical approach to this variability has slowly developed throughout this century. To look at the basic nature of the problem, one has to consider that the stability of a stellar structure can be tested with a linear approach, whereas the growing of the pulsation can be followed only with the adoption of non-linear algorithms. Moreover, the problem is further complicated by the predicted occurrence in the stellar envelopes of more or less extended layers of convective instability. Not surprisingly, the theory has thus developed starting from linear adiabatic models (Eddington 1918), to linear non-adiabatic models (van Albada & Baker 1971), to non-linear radiative models (Christy 1966), and, eventually, to non-linear convective models, as given by Deupree (1977) and Stellingwerf (1975, 1984). To fix our ideas, let me recall that linear adiabatic models can give a reasonable approximation only of the periods, but not of the modal stability; linear non-adiabatic models can predict the blue edge of the instability strip but not the red edge, which is the result of the non-linear coupling between pulsation and convection; non-linear radiative models are able to provide theoretical light curves but again cannot predict the red edge; finally non-linear convective models are able to reproduce both the detailed topology of the instability strip (blue and red boundaries for each studied mode) and the pulsation amplitudes, as well as the morphological details of light curves resulting from the interplay of dynamical and convective motions.

Each step of this investigation has produced relevant constraints concerning the evolutionary status of HB stars, as we will separately discuss here in the following.

5.1. *Periods*

As early as 1972 vanAlbada & Baker predicted that the period of a RR Lyrae fundamental pulsator is connected to the structural parameter of the stars according to the relation:

$$log P_{\mathrm{F}} = 11.497 + 0.84 \log L - 0.68 \log M - 3.48 \log T_{\mathrm{e}}.$$

which can be written also in the form:

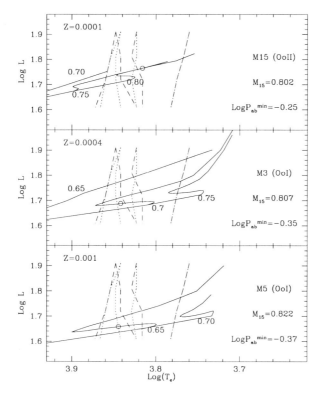

FIGURE 22. The HR-diagram location of the ZAHB for the three labeled assumptions for the cluster metallicity, together with the evolutionary path of selected He-burning models for the labeled values of the masses. Dashed and dotted lines give the boundaries of instability for the masses 0.75 and 0.65 $M_\odot$, respectively. For each given metallicity the name of the typical cluster, the observed minimum fundamental period and the maximum mass allowed in the strip are also given.

$$log P_{\rm F} = 11.497 + 0.84A - 3.48 \log T_{\rm e}.$$

where $A = \log L - 0.81 \log M$ depends on the mass-to-luminosity ratio of the structure. For the first-overtone pulsators they found:

$$log(P_{\rm H}/P_{\rm F}) = 0.438 + 0.14 \log L - 0.032 \log M - 0.09 \log T_{\rm e}.$$

Recent non-linear computations (Bono et al. 1997) have substantially confirmed this scenario, which will be used here to discuss the link between the pulsation period and stellar evolution. This link appears clearly in Figure 22, which shows the predicted evolutionary tracks of stars spending a substantial amount of their HB life in the instability region. One finds that theory gives rather severe constraints on the mass and the luminosity of these stars. Thus for each observed RR Lyrae, if the effective temperature is derived from the observed color, one finds a period to be compared with the actual one, testing the consistency of the evolutionary scenario. Of course, there are problems, as given by the relation between colors and temperatures as well as by possible uncertainties in the cluster reddening. However, this is at least in principle a very powerful method constraining the HB luminosity (and, thus, the amount of original helium!).

As a matter of fact, one finds that, for each given temperature, the range of allowed masses has little influence on periods. This is also the case if we allow for an enhancement of α-elements in the Pop. II mixture (see, e.g., Gratton 1998 and references therein), an occurrence which would moderately decrease the predicted masses at any given effective temperature. Therefore periods are mainly constraining the HB luminosity, without the uncertainties connected with the $^{12}C + \alpha$ cross section which affects the alternative R-method we described earlier. More directly, it has been shown that the mass–luminosity parameter A is a sensitive function of the original Y only, to be calibrated through stellar models in terms of this quantity (see, e.g., Caputo 1985), providing a relevant constraint connecting fundamental periods to that evolutionary parameter of cosmological relevance.

As for first-overtone pulsators, from the previous relations one finds with reasonable accuracy:

$$log P_{\mathrm{H}} \simeq log P_{\mathrm{F}} - 0.12.$$

This relation has been early used by Stobie (1971) to get rid of the complication of the two mode periods, dealing in the analysis with either fundamental or fundamentalized periods, the last one obtained implementing the experimental overtone period according to the previous relation. However, when dealing with some peculiar variables where both modes are unstable (the so called *double mode pulsators*) a precise observational determination of the ratio $P_{\mathrm{H}}/P_{\mathrm{F}}$ can be used to derive direct information about the mass of these pulsators (Jorgensen & Petersen 1967). This point will be further discussed in the light of the result of the non-linear theory, which adds more strength to this method.

5.2. *The instability strip.*

The use of RR Lyraes as a test of the evolutionary scenario can be greatly improved when information about the edges of the instability strip is added. Theoretical results are shown in the same Figure 22 for two different assumptions about the masses of the pulsators. Left to right one finds: 1) the blue boundary for instability in the first overtone, ii) the same but for the fundamental mode, iii) the red boundary for the first overtone and, finally, iv) the red boundary for the fundamental pulsators. Therefore, again left to right, one finds first a region (i.e., a range of temperatures) where only the first overtone can occur, then the both-mode region, where a star can pulsate either in the first overtone or in the fundamental mode (if not in both!), and finally the region of fundamental pulsators. Therefore theory appears in beautiful (though qualitative) agreement with observation, which places the c-type pulsators at the blue side of the instability strip.

By relying on these theoretical edges, one can now improve the test of HB luminosity, dispensing with the RR Lyrae colors and using only the distribution of periods. In short, for each assumed HB luminosity level one predicts a distribution of periods as given by the distribution of temperature across the strip. However, if one increases the adopted luminosity this distribution moves toward longer periods as a consequence of the effects on periods of both the increased luminosity and the shift of the instability strip toward smaller effective temperatures. The agreement between the predicted and the observed distribution gives the required HB luminosity level. Such a procedure, already devised by Caputo, Castellani & Quarta (1984), has been further implemented by Caputo (1997) on the basis of recent results of the pulsational theory and by discussing observational results in the plane $\log P$, M_V, as shown in Figure 23 for the case of the globular cluster M5 (Caputo et al. 1999).

As a particular case, the predicted edges of the instability strip have been used to shed light on a question which has been chewed over for a large portion of this century, namely,

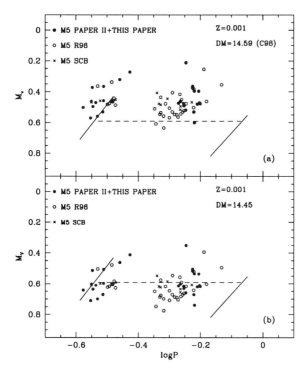

FIGURE 23. The comparison between pulsational predictions and the distribution of RR Lyrae magnitudes in the $\log P$, M_V plane under the labeled assumptions about the distance modulus of M5 suggests that the DM should be not larger than 14.45 mag.

the question of the origin of the so called Oosterhoff dichotomy in the properties of RRab stars in Galactic globulars. The problem was raised as early as 1939 by Oosterhoff, who found that the mean period of RRab populating a single cluster are clumped around two different values, above or below 0.6 day. Typically, the more metal-poor clusters (such as M15, M68 ...) have the longer mean periods (Oo.II type), whereas moderately metal-rich clusters, like M3 or M5, are Oo.I type clusters with shorter mean periods.

Sandage (1981, 1982) has suggested that such a dichotomy can be understood as an evidence for a dichotomy (not predicted by evolutionary theories) in the luminosity of the RRs, Oo.II variables being more luminous than the RRs in Oo.I clusters, according to the relation between period and luminosity. As an alternative explanation, van Albada & Baker (1973) suggested that the dichotomy originated from the fact that the "both mode" zone is populated by RRc in Oo.II clusters but by RRab (with higher temperature and thus shorter periods) in Oo.I clusters. One can test these two alternative explanations observing that RRs in both Oo. types have a rather similar distribution of the fundamentalized periods. Taking into account the behavior of the instability edges with luminosity and its dependence on stellar mass, it follows that Sandage's suggestion runs against not only evolutionary but also pulsational theoretical constraints which in that case would predict a sensitive shift of the distribution of fundamentalized periods (see Bono et al. 1995). In the next section we will find additional evidence showing, at least in my feeling, that Sandage's suggestion cannot be further maintained.

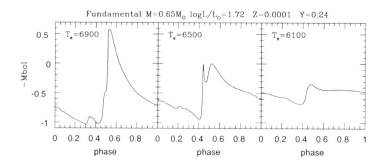

FIGURE 24. Theoretical light curves for fundamental pulsators and for the labeled values of stellar effective temperatures.

5.3. *Bailey's diagram*

As already recalled, convective, non-linear pulsational models allow a full coverage of the observational properties of radial pulsators. As an example, Figure 24 shows the predicted light curves for fundamental pulsators as obtained from recent non-linear, non-local and time-dependent convective envelope models (Bono et al. 1997). Similar computations support the already recalled behavior of these pulsators, with RRab having asymmetric curves and RRc with small amplitude and much more symmetric light curves. Interestingly enough, one finds that such an agreement is predicted only by assuming for the RRs a luminosity around the value predicted by evolutionary models; if we move the luminosity away from these predictions, this (qualitative) agreement would vanish.

Among the various uses one can made of such detailed theoretical predictions, one has to notice that non-linear results give theoretical predictions about the amplitude of the pulsators, allowing for the first time a theoretical interpretation of the behavior of RRs in that Bailey diagram which has been present throughout the history of RR Lyrae investigations. Without entering into too detailed a discussion (but see Bono et al. 1997) one finds that the Bailey diagram does contain information about both the masses and the luminosities of the variables. As an example, Figure 25 shows the Bailey diagram for RRs in the cluster M15 as compared with theoretical predictions on that matter, disclosing that the combined predictions of evolutionary and pulsational theories appear in reasonable agreement with observation. From the same Figure one can realize that Oo.II type clusters, as M15 in the Figure, lack large amplitude fundamental pulsators whereas the "descending branch" of c-type pulsators appears well populated. The contrary holds for Oo.I clusters, as M3, an occurrence which has to be interpreted as an evidence that the both-mode region is populated by RRab in Oo.I clusters only.

As an additional gift, detailed non-linear computations remove the degeneration of the Petersen diagram with luminosity, showing that double-mode pulsators can give information on both the mass and the luminosity of these stars, as shown in Figure 26. Comparison with observational data, as given in the same figure, indicates, e.g., for Oo.I pulsators $M \simeq 0.65 M_\odot$, $\log L \simeq 1.66$, in rather good agreement with evolutionary predictions. However, not to be over-confident, one has to note here that pulsational theories also suffer uncertainties, both in the input physics and in macroscopic mechanisms. As a

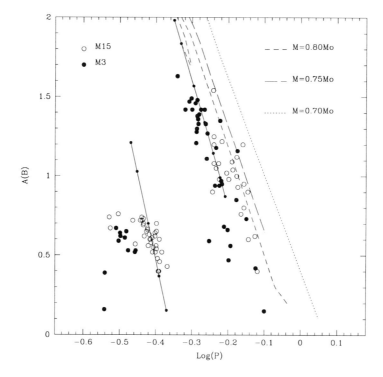

FIGURE 25. The distribution of M15 (open circles) and M3 (filled circles) RR pulsators in the Bailey diagram as compared with theoretical predictions for metal-poor stars ($Z = 0.0001$) on their ZAHB (full line) and along the HB evolutionary lines for the labeled masses.

major point, one finds that evaluations of both the instability boundaries are affected by the assumption about the mixing length. Without entering into many details, one may conclude that, e.g., the uncertainties in the "pulsational" luminosities we were dealing with is of the same order of magnitude of the uncertainty in the evolutionary luminosity, namely $\simeq 0.1$ mag (Caputo et al. 1999).

6. Leaving our Galaxy ......

Till now we have discussed evolutionary and pulsational theories concerning globular clusters in our own Galaxy. However, one knows that globulars accompany galaxies everywhere in the Universe, and it appears worth while to give at least some hints on the extension of theories to these extraGalactic objects. The main point is that in our Galaxy we are always dealing with old and more or less metal-poor systems. This is not the case for other globulars beyond our Galaxy, since one finds both young, metal-poor clusters (as, e.g., in the Magellanic Clouds) or old metal-rich systems.

Keeping the name of Pop. II for metal-poor stars born from matter with a rather short genetic distance from the big-bang nucleosynthesis, one can say that extragalactic globulars give evidences for both young Pop. II stellar systems as well as for old Pop. I stars. Of course, we cannot here give a detailed account of such a huge enlargement of the assumed evolutionary parameters. However, let me at least mention the major points one is dealing with.

As for old clusters, the extension to larger metallicity of the evolutionary and pulsa-

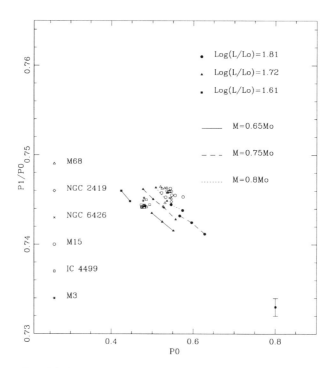

FIGURE 26. The predicted behavior in the Petersen diagram of the ratio between first overtone and fundamental periods (P1/P0) as a function of P0, for selected assumptions about star masses and metallicities.

tional scenario has been already done by several authors (see, e.g., Fagotto et al. 1994, Bono et al. 1997b and references therein). Therefore one already finds in the literature the extension to larger metallicities of the theoretical predictions we have discussed for Galactic globulars. One may here notice that the behavior of the HB luminosity level depends on the assumed ratio between the Z and Y enrichment. By assuming $\Delta Y/\Delta Z \simeq 2$ (as suggested by a linear interpolation between Pop. II stars and the Sun) one finds that the HB luminosity reaches a minimum near the solar metallicity, to increase again for larger Z values. As a relevant point, it has been found that old, metal-rich HB stars can be affected by what has been named a *gravonuclear instability*, which could have sizable effects on both the amount of UV radiation from these systems as well as on the occurrence and abundance of some classes of variable stars (Bono et al. 1997c)

When moving toward younger clusters, one has first to bear in mind that the RG transition depends on the star's chemical composition. As a consequence, one has to be cautious with observations, since the luminosity of the RG tip can even depend strongly on the cluster age and, in the case of mixed populations, the RG branch itself could be populated by selected metallicities only (Cassisi, Castellani & Castellani 1997). The rather constant luminosity of the He-burning structures disappears, and if one decreases the cluster age, one expects first a decrease and thus a further increase of this luminosity, following the balancing contribution of the He- and H-burning energy sources and the removal of electron degeneracy in the core and, thus, the growing of a convective core in MS stars. In this case, theory predicts a clump of cool central He-burning stars (as observed), and theoretical predictions become largely affected by assumptions about the mixing length, whereas sedimentation becomes less and less efficient (no time to act!).

The case of old but not-too-old clusters (i.e., of clusters still having RG branches) has been discussed in several papers (see Caputo, Castellani & Degl'Innocenti 1996 and references therein). It has in particular been found that the theoretical predictions on the clump of red He burning giants can be used to give relevant constraints in decoding the observational features. Stars in even younger clusters can be affected not only by the quoted assumptions about the mixing length but also by core overshooting, by problems about the amount of mass loss and, in the more massive stars, about the efficiency of convection developing at the exhaustion of central H (see, e.g., the discussion in Brocato & Castellani 1993). However, we are now moving well beyond the scope of these lectures, and I must stop here, referring the reader interested in similar evolutionary features to the beautiful lectures given by Chiosi (1998) last year in this school. The few points in this paper which will be found not in agreement with the above quoted lectures can be taken as a final useful indication of the debate that is still alive among theoreticians of evolution.

7. Final remarks

Even if we tried to do so, it appears difficult to give a precise estimate of the residual uncertainties in the most up-to-date input physics. Curiously enough, and to tell you the truth, we were much happier with the lower HB luminosities predicted by Castellani, Chieffi & Pulone (1991) with the old physics, since these luminosities were in much easier agreement with predicted pulsational properties of RR Lyraes. Even if agreement is still possible (Caputo et al. 1999) I suspect that theory is presently overestimating these luminosities. Discussing the uncertainties in this parameter, we have indeed found that possible errors go in that sense.

When closing this paper, such a suspicion has probably been substantiated by observational data provided by the Hipparcos satellite. According to these observations, it appears that the clump of red He-burning stars in the solar neighborhood is characterized by a mean absolute I magnitude given by $< M_I > = 0.23 \pm 0.03$ (see, e.g., Cole 1998). These stars represent the nearby counterpart of HB stars in Galactic globulars, from progenitors which were a bit more massive. However, they still undergo electron degeneracy at the end of their central H-burning phase. The evidence that "updated" evolutionary scenarios overestimate the predicted luminosity of these stars can be immediately translated into a quite probable corresponding overestimate of HB luminosities. Note that this offers for the first time the opportunity of a reliable calibration of the luminosity of He-burning stellar models to observational data, possibly reducing considerably the uncertainties we are dealing with.

As a parallel problem, one is facing the lack of firm theoretical constraints on the temperature of cool stars. It turns out that every effort at producing improved treatments of superadiabatic convection , such as the one, e.g., given by Canuto & Mazzitelli (1991), is obviously of great interest, going—at least—in the right direction (see also Canuto, Minotti & Schilling 1994, Grossman 1996 and references therein). The final observational test for similar theories will be the capability of firmly predicting the radius and, thus, the colors of the structures more affected by such a mechanism, as RGB and AGB stars are. In this context, one has to notice that throughout this paper we have largely neglected the additional uncertainties affecting the link between theory and observations, i.e., the theoretical evaluations for bolometric corrections and colors. Unfortunately, this is a point which adds further uncertainties in the evolutionary results and which should be always taken into account when comparing theories with observation. Here let me only quote that the progress of stellar interferometry, providing reliable stellar radii, can in the near future give a much more solid basis to this relevant ingredient (see, e.g., Di Benedetto 1999).

In conclusion, let me advance some recommendations, both to theoreticians (and thus also to myself) as well as to observational people. Theoreticians should always test their models with available observational constraints, possibly discussing not only what appears in agreement with their theory but also what appears in disagreement. Bear in mind also the existence of the rather severe constraints given by helioseismology. Moreover, it is probably time that theoretical results be presented together with an evaluation of errors, as common physics does require. To the observational people here I can only repeat not to take theoretical results simply at their face values, bearing in mind that theory does not give the "true", but only the predicted consequences of the adopted input physics.

As a final point, I must apologize for referring so often to papers by myself or by those around me. Of course this has been suggested by the obvious contiguity with these works, as well as by the opportunity of developing these lectures in a way that makes reference to a homogeneous theoretical scenario. I hope that this would not be taken as an understatement concerning the quite valuable work accomplished all around the world by several evolutionary groups, whose contribution I have tried to bring to light through the rather extended references accompanying the various sections of this paper.

8. Acknowledgments

To finish, I should acknowledge the many persons who all along the road shared with me the pleasure of investigating such a fascinating subject as globular clusters. However, let me here recall only the ones I am still enjoying close cooperation with, namely Giuseppe Bono, Enzo Brocato and Filippina Caputo, together with the last generation of pupils represented by Santino Cassisi, Scilla Degl'Innocenti, Marcella Marconi and Vincenzo Ripepi. All have been of invaluable help in producing this paper; and their enthusiasm for stellar investigation is daily teaching me something, stimulating me to keep on the job. Last, but not least, I gratefully acknowledge Ivan King for a critical reading of the manuscript and for the precious assistance to put this paper in its final form.

REFERENCES

Alexander D.R., Brocato E., Cassisi S., Castellani V., Degl'Innocenti S., Ciacio F. 1997, A&A 317, 90

Bahcall J.N., Pinsonneault M.H. 1995, Rev. Mod. Phys. 67, 781.

Baraffe I, Chabrier G. 1996, ApJ 461, L51.

Baraffe I., Chabrier G., Allard F., Hauschildt P.H. 1995, ApJ 446, L35.

Baraffe I., Chabrier G., Allard F., Hauschildt P.H. 1997, A&A 327, 1054.

Bertelli S., Bressan A., Chiosi C., Angerer K. 1986 A&AS 66, 191.

Bolte M. 1994, ApJ 431, 223.

Bono G., Castellani V. 1992, A&A 258, 385.

Bono G., Caputo F., Castellani V., Marconi M. 1995, ApJL 448, 115.

Bono G., Caputo F., Castellani V., Marconi M. 1997, A&A 121, 327.

Bono G., Castellani V., Degl'Innocenti, Pulone L. 1995, A&A 297, 115.

Bono G., Caputo F., Cassisi S., Castellani V., Marconi M. 1997b, ApJ 479, 279.

Bono G., Caputo F., Cassisi S., Castellani V., Marconi M. 1997c, ApJ 489, 822.

Bressan A., Bertelli G., Chiosi C. 1986, Mem. Soc. Astron,It 57, 411.

Brett J.M. 1995a, A&A 295, 736.

Brett J.M. 1995b, A&AS 109, 263.

Brocato E., Castellani V. 1993, ApJ 410, 99.

Brocato E., Cassisi S., Castellani V. 1996, in "Advances in stellar evolution", R.T. Rood, A.Renzini eds.

Brocato E., Cassisi S., Castellani V. 1998, MNRAS 295, 711.

Brocato E., Castellani V., Piersimomi A. 1997, ApJ 491, 789

Brocato E., Castellani V., Romaniello M. 1999, (in preparation)

Brocato E., Castellani V., Villante F. 1998, MNRAS 298, 557

Brocato E., Castellani V., Raimondo G., Walker A. 1999 *in preparation*

Brocato E., Castellani V., Scotti G.A., Saviane I., Piotto G., Ferraro F.R. 1998b, A&A 335, 929.

Buonanno R., Corsi C.E., Fusi Pecci F. 1985, A&A 145, 97.

Buonanno R., Corsi C.E., Fusi Pecci F., Bellazzini M., Ferraro F.R. 1998, A&A 333, 505.

Buonanno R., Caloi V., Castellani V., Corsi C.E., Fusi Pecci F., Gratton R. 1986A&AS 66, 79.

Burrows A., Hubbard W.B.,Saumon D., Lunine J.I. 1993, ApJ 406, 158.

Buzzoni A., Fusi Pecci F., Buonanno R., Corsi C.E. 1983, A&A 128, 94.

Caloi V., Castellani V., Tornambè A. 1978, A&AS 33, 169.

Caloi V., D'Antona F., Mazzitelli I., 1997, A&A 320, 823

Canuto V., Mazzitelli I 1991, ApJ 370, 295.

Canuto V., Minotti F.O., Schilling O. 1994, ApJ 425,303.

Caputo F. 1985, Rep. Prog. Phys. 48, 1235.

Caputo F. 1997, MNRAS 284, 994.

Caputo F., Castellani V., Degl'Innocenti S. 1996, A&A 309, 413.

Caputo F., Castellani V., Quarta M.L. 1984, A&A 138, 457.

Caputo F., Castellani V., Tornambè A. 1978, A&A 67, 107.

Caputo F., Castellani V., Chieffi A., Pulone L., Tornambè A. 1989, ApJ 340, 241

Caputo F., Castellani V., Marconi M., Ripepi V. 1999, MNRAS (submitted)

Caputo F., De Rinaldis A., Manteiga A., Pulone L., Quarta M.L. 1993, A&A 276, 41.

Carretta E., Gratton R.G. 1997, A&AS 121, 95.

Cassisi S., Castellani M., Castellani V. 1997, A&A 317, 108.

Cassisi S., Degl'Innocenti S., Salaris M. 1997, MNRAS 290, 515.

Cassisi S., Castellani V., Degl'Innocenti S., Weiss A. 1998a, A&AS 129, 267.

Cassisi S., Castellani V., Degl'Innocenti S., Salaris M., Weiss A. 1998b, A&AS, in press.

Castellani M., Castellani V. 1993, ApJ 407,649

Castellani V. 1976, A&A 48, 461.

Castellani V., Degl'Innocenti S. 1993, ApJ 402, 574.

Castellani V., Degl'Innocenti S. 1999, A&A (accepted)

Castellani V., Chieffi A., Norci L. 1989, A&A 216,62.

Castellani V., Chieffi A., Pulone L. 1991, ApJS 76, 911

Castellani V., Degl'Innocenti S., Luridiana V. 1993, A&A 272,442.

Castellani V., Degl'Innocenti S. & Marconi M. 1998. A&A (submitted)

Castellani V., Degl'Innocenti S., Pulone L. 1995, ApJ 446, 228.

Castellani V., Degl'Innocenti S., Romaniello M. 1994, ApJ 423, 266.

Castellani V., Luridiana V., Romaniello M. 1994, ApJ 428, 633.

Castellani V., Giannone P., Renzini A. 1971, Ap. Sp.Sci 10, 355.

Castellani V., Ciacio F., Degl'Innocenti S., Fiorentini G. 1997, A&A 322, 801.

Castelli F., Gratton R.G., Kurucz R.L., 1997a, A&A 318, 841

Castelli F., Gratton R.G., Kurucz R.L., 1997b, A&A 324, 432

Catelan M., de Freitas Pacheco J.A., Horvath J.E., 1996, ApJ 461, 231

Chaboyer B. 1995, ApJ 444, L9.

Chaboyer B., Demarque P., Kernan P.J., Krauss L.M. 1998, ApJ 494, 96.

Chaboyer B., Demarque P., Kernan P.J.. Krauss L.M., Sarajedini A. 1996, MNRAS 283, 683.

Chiosi C. 1986, in *Spectral Evolution of Galaxies* eds. C. Chiosi & A. Renzini, (Dordrecht:Reidel), p. 237

Chiosi C. 1998, in *Stellar Astrophysics for The Local group*, VIII Canary Islands Winter School, eds. A. Aparicio, A. Herrero, F. Sanchez, p1/95, Cambridge University press

Christy R.F. 1966, ApJ, 144, 108.

Clementini G., Carretta E., Gratton R.G. et al. 1995, AJ 110, 2319.

Cole A.A. 1998, ApJ 500, L137.

Cool A.M., Piotto G., King I.R. 1996, ApJ 468, 655

Cox J.P., Giuli R.T. 1968, *Principles of Stellar Structure*, Gordon & Breech, New York

Da Costa G.S., Armandrof T.E. 1990, AJ 100. 162.

D'Cruz N.L., Dorman B., Rood R.T., O'Connell R.W. 1996,ApJ 466, 359.

Degl'Innocenti S., Weiss A., Leone L., 1997, A&A 319, 487

Deupree R.G. 1977 ApJ 214, 502.

DiBenedetto P. 1999, A&A, in press.

Dorman B., O'Connell R.W., Rood R.T. 1995, ApJ 442, 105.

Eddington A.S. 1918, MNRAS, 79, 2.

Fagotto F., Bressan A., Bertelli G., Chiosi C. 1994, A&A S 105, 39.

Faulkner J., Swenson F.J., 1988, ApJ 329, L47

Faulkner J., Swenson F.J., 1993, ApJ 411, 200

Fiorentini G., Lissia M., Ricci B., 1998, A&A, in press

Frogel J.A., Cohen J.G. Persson S.E. 1983, ApJ 275,773

Fusi Pecci F., Ferraro F.R., Crocker D.A., Rood R.T., Buonanno R. 1990, A&A 238, 95.

Gratton R.G. 1998, MNRAS 296, 739.

Gratton, R.G. et al. 1997, ApJ, 491, 749

Gratton R.G., Clementini G., Fusi Pecci F., Carretta E. 1998, in "Views on distance indicators", Caputo F. (ed.)

Greenstein G.S., Truran J.W., Cameron A.G.W. 1967, Nature 213, 871.

Greggio L., Renzini A. 1990, ApJ 364,35.

Grossman S.A. 1996, MNRAS 279, 305.

Haft M., Raffelt G., Weiss A., 1994, ApJ 425, 222

Heber U., Kudritzki R.P., Caloi V., Castellani V., Danziger J., Gilmozzi R. 1986, A&A 162, 171.

Henyey L.G., Vardya M.S., Bodenheimer P.L. 1965, ApJ 142, 841.

Hubbard W.B. & Lampe M., 1969, ApJS 163, 297

Iben I. Jr. 1968, Nature 220, 143.

Iben I. Jr., Tutukov A.V. 1984, ApJ, 282, 615.

Itoh N., Mitake S., Iyetomi H., Ichimaru S., 1983, ApJ 273, 774

Jorgensen H.E., Petersen J.O. 1967, Zeit. Astrophys. 67, 377.

Kippenhan R., Weigert A. 1990, Stellar Structure and Evolution, Springer-Verlag, Berlin

Koester D., Schoenberner D. 1986, A&A, 154, 125

Kurucz R.L. 1993a, CD-ROM 13, ATLAS9 Stellar Atmospheres Programs and 2 km/s Grid (Cambridge: Smithsonian Astrophysica Obs,)

Kurucz R.L. 1993b, CD-ROM 18, SYNTHE Spectrum Synthesis Programs and Line Data (Cambridge: Smithsonian Astrophysical Obs.)

Lee M.G., Freedman W., Madore R.F. 1993, ApJ 417, 553.

Limber D.N. 1958, ApJ 127, 387.

Oosterhoff P.Th. 1939, Observatory 62, 104.

Paczynski B. 1984, ApJ 284, 670.

Pagel B.E.J., Portinari L. 1998, MNRAS 298, 747.

Piotto et al. 1997, in *Advances in Stellar Evolution*, ed. R.T.Rood & A. Renzini, Cambridge University Press, p.84

Pont F., Mayor M., Turan C., VandenBerg D.A. 1998, A&A 329, 87.

Proffitt C.R., Vandenberg D.A., 1991, ApJS 77, 473

Ratcliff S.J. 1987, ApJ 318, 196.

Raffelt G., Weiss A. 1992, A&A, 264, 536.

Rich R.M. et al. 1997, ApJ 484, L25

Rogers F.J., Swenson F.J., Iglesias C.A., 1996, ApJ 456, 902

Rolfs C.E. & Rodney W.S., 1988, "Cauldrons in the Cosmos" the University of Chicago press, Chicago and London

Salaris M., Cassisi S. 1996, A&A 305, 858.

Salaris M., Cassisi S. 1998, MNRAS 298, 166.

Sandage A. 1981, ApJ 248, 161.

Sandage A. 1982, ApJ 252, 553.

Sandage A. 1990, ApJ 350, 603.

Sandage A. 1993, ApJS 106, 703.

Schwarzschild M. 1941, ApJ 94, 245.

Stellingwerf R.F. 1975, ApJ 195, 441.

Stellingwerf R.F. 1984, ApJ 277, 322.

Stobie R.S. 1971, ApJ 168, 381.

Straniero O., 1988, A&AS 76, 157

Straniero O., Chieffi A., Limongi M. 1997, ApJ 490, 425.

Sweigart A.V. 1997, ApJ 474, L23.

Sweigart A.W. 1998, in "3rd Conference on faint blue stars" A.G.D. David Philip, J.Liebert, R.A.Saffer eds., Cambridge University Press.

Sweigart A.V., Catelan M. 1998, ApJ 501, L63

Sweigart A.V., Demarque P. 1973, in *Variables stars in globular clusters* J.D. Fernie (ed.), Reidel, Dordrecht, p.221.

Sweigart A.V., Gross P.G. 1976, ApJS 32, 367.

Sweigart A.V., Gross P.G. 1978, ApJS 36, 405.

Sweigart A.V., Greggio L., Renzini A. 1989, ApJS 69, 911.

Sweigart A.V., Greggio L., Renzini A. 1990, ApJ364, 527.

Thomas H.-C. 1967, Zeit. Astrophys. 67, 420.

vanAlbada T.S., Baker N. 1971, ApJ 169, 311.

vanAlbada T.S., Baker N. 1973, ApJ 185, 477.

vandenBerg D.A., Larson A.M., Di Propris R. 1998, PASP 110, 98.

vandenBerg D.A., Stetson P.B., Bolte M. 1996, ARA&A 34,461.

Walker A.R. 1992, ApJ 300, L81.

Walker A.R. 1994, AJ 108, 555.

Weidemann V., Koester D. 1983, A&A 121, 77.

Weidemann V., Koester D. 1984, A&A 132, 195.

Wood M.A. 1992, ApJ 386, 539.

Yi, S., Demarque P., Oemler A.Jr. 1998, ApJ 492, 480.

RAFFAELE GRATTON was born in La Plata (Argentina), on November 21, 1956.

Italian, he studied at University "La Sapienza" in Rome where he got his Degree in 1979. He worked at Cerro Tololo Inter-American Observatory, La Serena, Chile, 1980-1981, Asiago Astrophysical Observatory, Italy, 1981-1984 Astronomer; Rome Astronomical Observatory, Italy, 1984-1989, and he is presently Associate Astronomer at Padua Astronomical Observatory, Italy. Gratton is an observational astronomer, but he is now working also on technological aspects (spectrograph design and construction). His speciality is abundance analysis, and his main area of interest are globular clusters and chemical evolution of the Galaxy. He also works on surface abundances as signatures of stellar evolution. Gratton is PI of the High Resolution Spectrograph for the Galileo National Telescope, under construction at the Roque de Los Muchachos Observatory (La Palma).

Early nucleosynthesis and chemical abundances of stars in globular clusters

By R. GRATTON

Osservatorio Astronomico di Padova, Vicolo dell'Osservatorio 5, 35122, Padova, Italia

This cycle of lectures presents a self consistent sketch of current understanding about chemical composition of globular clusters and its aftermaths. The first two lectures give basic about nucleosynthesis, chemical models, and abundance determinations. Main results for globular clusters are presented in the next two lectures. In the final lecture I review various indices used to derive abundances from photometry and low dispersion spectroscopy.

1. Early Nucleosynthesis and models of galactic chemical evolution

In this first lecture I will briefly present the fundamentals of nucleosynthesis and chemical evolution. Owing to lack of time, only few sketches can be given.

The basic observation that we live in an environment rather rich in heavy elements (hereinafter metals) that could not be produced by Big Bang leads us to try to describe the mechanisms of formation of these elements. There is a close interaction between chemical and dynamical evolution of stellar systems; chemical abundances provide then a basic diagnostic for models of galactic evolution.

Figure 1 sketches the most important features to be introduced in this picture. Stars form from condensation of the most dense clouds within the interstellar medium (ISM). Metals are produced by nucleosynthesis processes within the stellar interiors. Stars lose part of their metal-enriched material either through more or less quiescent stellar winds, or through explosive events (SNe) at the end of their lives: the amount of each element produced within stars and returned to the ISM depends on the stellar masses and in some case on the presence of close companions. The material ejected from stars pollutes the ISM, so that following generations of stars form with a larger metal content. Stellar radiation, winds and SNe also deposit large amounts of energy into the ISM, affecting its physical status, and then the ability to further form stars. The effects of interaction between stars and the ISM depend on the stellar masses, which also determine time-scales of the related phenomena. Properties of the ISM are deeply influenced not only by interaction with star radiation and winds, but also by general gravitational field.

The basic ingredients to be introduced in a model for the chemical evolution then refer to:

- the physics of star formation from the ISM, which is usually described by three functions: the star formation rate (SFR), the distribution of stellar masses (initial mass function, IMF), and the binary fraction

- the physics of stellar evolution, through the timescale of stellar evolution and the efficiency with which newly produced (or destroyed) elements are given back to the ISM (usually called yields), which is determined by the nucleosynthesis processes occurring within stars, by the efficiency of mixing episodes, and by the mechanisms through which matter is returned to the ISM (quiescent winds, SN explosions, etc.); and finally

- the physics of the ISM, represented at small and intermediate scale by its thermokinetic behavior (and for some element, by nuclear reactions stimulated by high energy particles present in the ISM), and at large scale by the dynamics of gas and dust within the general gravitational field

155

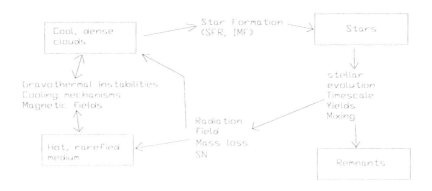

FIGURE 1. Schematic representation of a model for the galactic chemical evolution

Clearly, a detailed description of all the aspects of this complex scenario cannot be presented in a single lecture; furthermore, several important featrures are not at all well understood at present; finally, an adequate model for the galactic chemical evolution represents a formidable computational task even for the most powerful modern computers. Here, I will simply presents some terminology, and a few basic concepts. The reader interested in a more detailed description of these problems, may found extensive reviews in Wallerstein et al. (1997) and McWilliam (1997).

1.1. *Star formation*

In spite of notable progresses, the physics of star formation is still far from being well understood (for reviews see Shu et al. 1986, and Larson 1991). In general, we are presently unable to predict the rate of star formation and the IMF from first principles. Observations indicate that different phases coexist in the ISM (McCray & Snow 1979): a hot, very rarified medium (density 10^{-2} cm^{-3}, temperature 10^8 K); a warm diffuse medium (density 0.1 cm^{-3}, temperature 10^4 K); and cool dense clouds (density $\geq 10^2$ cm^{-3}, temperature < 100 K), often organized in very large systems (molecular clouds: total mass $10^8 \div 10^9$ M$_\odot$: Scoville & Hersh 1979). Simple theoretical arguments as well as observations indicate that star formation occurs in the highest density, coolest interstellar clouds. The main uncertainties relate to important details of the cooling mechanisms (Raymond et al. 1976), in particular the role of dust (and hence of metal abundance); to the role of interstellar magnetic fields (Draine 1980; Draine et al. 1983); and to the impact of previous star formation in the same region (Hartquist & Dyson 1997); or even to the presence of external ionizing UV sources (like e.g. active galactic nuclei). Mechanisms leading to formation of binary and multiple stars are also not well known.

The paradigm of star formation is that lower mass begin forming first in star forming regions, and continued forming until the high mass stars form; fast winds and SN explosions from these last shut down further star formation in the molecular cloud (see Elmegren 1983, 1985). However, the general impact of this observation on SFR and IMF is not clear: in fact, most well studied star forming regions (all those in our own Galaxy)

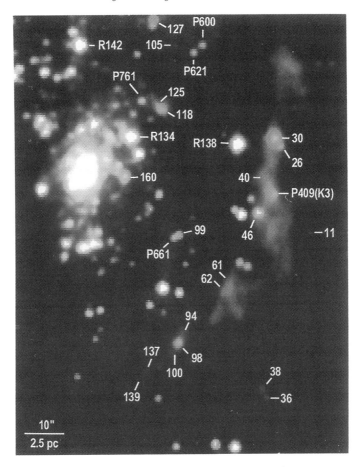

FIGURE 2. The large star forming region S Doradus in the LMC (composite of J, H, and K images from Rubio et al. 1998)

are rather small; none is known even barely as massive as a globular cluster: inference of star formation as occurred in the halo or even in the bulge from these observations appears quite risky. More data can be drawn from studies of large HII regions in the Magellanic Clouds, and in particular from the largest one (S Doradus in the LMC: see Figure 2). In this case, a very compact ($r_{0.5} \sim 1.7$ pc) bright cluster including about a hundred O stars is observed at the very center (R136, with an age ≤ 3 Myr: Hunter et al. 1995, 1996; Massey & Hunter 1998); fast winds from these very massive stars sweep the ISM generating a nearly empty bubble. Very active star formation occurs at the walls of the bubble where the shock fronts interact with the preexisting ISM (Rubio et al. 1998 and references therein); however, it is doubtful that stars forming at these large distances (~ 20 pc) from the core will remain tied to the system after the end of the large mass loss episode related to the evolution of the most massive stars: most likely they will form a loose association that will disperse in a few 10^7 yr. Low mass stars are certainly present in large number (Brandl et al. 1996; Massey & Hunter 1998), although their observation at the distance of the LMC is not easy. Furthermore, they might be spread over a much larger volume. However, observations are not clearcut: the slope of the mass function determined for R136 is not much different from the Salpeter (1955); the mass cut (if any) is however below threshold (at $M > 2.8\ M_\odot$). It should also be noticed that evolution of small mass stars is slower than that of massive stars, so that most of them are still well in the accretion phase while more massive stars are already somewhat evolved. R136 is likely the closest analogue we have of a proto-globular cluster: total mass (2×10^4 $M_{\odot}$ in stars with masses $M > 2.8$ $M_{\odot}$ within a radius of 4.7 pc: Hunter et al. 1995; the mass is about 3 times larger if the IMF continues with the same slope down to $M > 0.1$ $M_{\odot}$) and concentration place it at the low-mass, low luminosity end of the spectrum of globular clusters in the Milky Way (typical masses 10^5 $M_{\odot}$, with a range of a factor of 4, from mean luminosity of $M_V = -7.3 \pm 1.4$, Harris 1991, and mass to luminosity ratio of $M/L_V \sim 1.6$: Illingworth 1976; typical $r_{0.5} \sim 1.7$ pc with values ranging from 1 to 8 pc: van den Bergh et al. 1991). When comparing data for R136 with those of typical galactic globular clusters, it should also be reminded that the metal abundance of the LMC ([Fe/H]~ -0.3: see Wheeler et al. 1989†) is currently much higher than it was in the halo of our own Galaxy. However, there is a general agreement between this scenario and those usually considered for the evolution of proto-globular clusters, see e.g. Burkert et al. (1993) and references therein. In general such models predict formation of a first generation cluster in the core by gravitational collapse of structures above the Jeans mass; the formation of a supershell driven by the SN explosions in the center of this collapsing protocloud (for a detailed discussion see Brown et al. 1992); and the formation of a second generation of stars in the supershells on timescales of 2×10^7 yr (McCray & Kafatos 1987). Stars of this second generation are lost unless the efficiency of star formation is very high: this might however be the case of globular clusters (Brown et al. 1995). Stars of this second generation will show signatures of enrichment by type II SNe alone. Turbulent mixing will keep the supershell chemically homogeneous (Brown et al. 1991).

Given the large theoretical uncertainties and the limited usefulness of currently available observational data, estimates of SFR and IMF in environments different from those typical of the thin disk are very uncertain: the usual practice is to adopt a constant (for an observational support, see Richer & Fahlman 1996), simple IMF (either a power law similar to the Salpeter 1955 one; or that proposed by Miller & Scalo 1979), and a SFR which is a power law related either to the volume or to the surface gas density (Schmidt law: Schmidt 1959, 1963), where the constant factor and the exponent are given by (rather uncertain) fits to determinations of the current SFR the solar neighborhood and extragalactic star forming regions (see e.g. Timmes et al. 1995). Furthermore, a constant (mass independent) binary fraction is usually assumed. Note that under these assumptions star formation is assumed to be a universal phenomenon, independent of chemical composition: while this assumption might be easily criticized if the physics involved is considered, it is considered a safe conservative approach given our poor understanding of the phenomenon.

1.2. *Nucleosynthesis and Stellar Evolution Processes*

Since the discovery that nuclear burning is the energy source of the Sun and other stars by Bethe, great progresses have been made in our understanding of the nucleosynthesis processes occurring in stars. A great contribute is due to the classical work of Burbidge et al. (1957), who clarified the relative importance of the various mechanisms for the different nuclei. Presently, most processes are quite clear, although important details (in particular accurate cross sections at the relevant energies) are still missing in some cases (for review, see Wheeler et al. 1989; and Arnett 1995).

---

† Through these lectures we adopt the usual spectroscopic notations, i.e. $[X] = \log(X)_{\text{star}} - \log(X)_{\odot}$ for any abundance quantity X, and $\log n(X) = \log(N_X/N_H)$ $+ 12.0$ for absolute number density abundances

TABLE 1. Fe yields from some type II SNe

SN	Progenitor Mass $M_\odot$	^{56}Ni Mass $M_\odot$	Source
SN 1987a	20	0.075	Arnett 1989
SN 1997d	$25 \div 40$	0.002	Turatto et al. 1998
SN 1998bw	40	$0.6 \div 0.8$	Iwamoto et al. 1998

On the other side, rather large uncertainties still exist on relevant aspects concerning the astrophysical sites where the basic nucleosynthesis processes occur. I will briefly review a few cases.

The production of the lightest elements after He (Li, Be, B) is still not clear. These elements are easily destroyed in the stellar interiors; Big Bang is only able to produce significant amount of ^{7}Li: most investigators identify the so-called Spite plateau (Spite & Spite 1982) with this original Li, although a lively debate exists about the possibility that some depletion has occurred even in these stars (see for instance King et al. 1996). It is not clear where the remaining Li was produced: in order to produce significant amounts of Li, the so-called ^{7}Be-transport mechanism must be active, requiring first production of the very unstable ^{7}Be through alpha-capture on ^{3}He, and then a fast decrease of temperature (Cameron & Fowler 1971). This mechanism might occur in convective regions with H-burning: as a matter of fact, strong Li lines have been observed in the spectra of bright AGB stars (Wallerstein & Conti 1969; Smith & Lambert, 1990). Even more mysterious is the production of the remaining light elements: traditionally, these elements were explained as due to spallation events by energetic cosmic rays on C and O nuclei trapped in interstellar grains (Meneguzzi et al. 1971); however the recent discovery that the abundances of Be and B scale as the abundance of O (Boesgaard & King 1993; Duncan et al. 1997) have cast doubts over this interpretation. To explain such a behavior, alternative mechanisms have been proposed, including the symmetrical reaction (energetic C and O nuclei with H and He nuclei: Vangioni-Flam et al. 1998; Lemoine et al. 1998), as well as reactions induced by the strong neutrino fluxes in SNe environments (the ν-process: Woosley et al. 1990). Predictions from such mechanisms are currently under tests.

Of even larger impact on stellar and galactic astrophysics are uncertainties related to the production of Fe-group elements. There are two major concerns here:

(*a*) While we are pretty sure that massive stars finally explode as (type II) SNe, current (unidimensional) models are still unable to naturally reproduce these explosions (Woosley & Weaver 1995; Thielemann et al. 1996). The usual practice is to arbitrarily introduce a "piston" which should simulate the expansion shock front due to the explosion. The initial location of this piston determines the mass that is finally tied into the compact remnant, and the mass which is ejected into the outer space. This mass cut is located within the region where Fe-group elements are produced (in conditions of nearly nuclear statistical equilibrium). Since in models the location of the mass cut is arbitrary, the mass of Fe-group elements present in the ejecta is unpredictable; in practice, it is calibrated using the mass of ^{56}Ni required to power the decay of the SN light curve. However, this empirical calibration is fostered by the lack of reliable progenitor mass estimates. Pre-explosion observations are available for only a single case (SN 1987a: 20 $M_\odot$, Arnett 1989). In other cases, progenitor masses may be determined only from fits of spectra and of the early light curve; reliable data exist for only a few SNe. Recently, Nomoto (1998) presented two additional cases (see Table 1), including the anomalous (very bright, very massive progenitor) SN 1998bw, likely connected to GRB980425 (Iwamoto et al. 1998); this data seems to suggest that the Fe/O ratio for type II SNe is highly variable from object-to-object, quite indipendent of progenitor mass; this is clearly different from model predictions, which gives a strong dependence of this ratio on progenitor mass (see e.g. Woosley & Weaver 1995). More data is urgently required

(*b*) The light curves of type Ia SNe (which is powered by the decay of ^{56}Ni) indicate that a fairly large amount of Fe-group elements is produced by these objects (Nomoto et al. 1984). While there is a widespread consensus about the explosion model (C-deflagration in degenerate conditions for white dwarfs at the Chandrasekhar mass limit: Nomoto et al. 1984), there is a lively debate about the progenitors (although it is quite clear that they should be in most cases the result of the evolution of intermediate mass stars in binary systems: Whelan & Iben 1973; see Branch et al. 1995 and Nomoto et al. 1997 for recent reviews). The most favourite models (double degenerate, i.e. merging of double C+O white darfs with a combined mass exceeding the Chandrasekhar limit: Iben & Tutukov 1984, Webbink 1984; or a single degenerate scenario, i.e. accretion of H-rich matter via mass transfer from a binary companion: Nomoto et al. 1984) differentiate in lifetime and progenitor mass spectra. Furthermore, very recently Kobayashi et al. (1998) have proposed that the mass spectrum for the progenitors of type Ia SNe is a function of metallicity, no type Ia SNe being expected for metallicity below a tenth of solar. If confirmed, this result would have a deep impact on the interpretation of the run of the O/Fe ratio with overall metal abundance

Another important and not fully understood aspect of nucleosynthesis concerns the production of elements in small and intermediate mass stars (IMS). Small and IMS are likely amongst the major producer of C and N (as well as of the neutron capture nuclei produced through the s-cycle: see the reviews by Iben & Renzini 1983; Lattanzio & Boothroyd 1997). Great observational and theoretical progresses in the eighties have shown that C is mainly produced by the less massive amongst the IMS (see e.g Frogel et al. 1990). Less clear is the case of N: N production in massive stars (through CNO cycle) is secondary (Arnett 1973; Truran 1973): however, observation requires a primary mechanism to be active, at least for metallicities characteristic of the disk (see the review by Wheeler et al. 1989; note that data are highly uncertain at low metallicities). However, following an original idea of Truran & Cameron (1971), recent studies have shown that hot bottom burning efficiently produces large amounts of N amongst the most massive and brightest IMS (Boothroyd & Sackmann 1988, Wagenhuber & Groenewegen 1998; Marigo 1998). The mass range over which hot bottom burning may occur, as well as the impact of these objects on galactic chemical evolution are still not well explored. Note that this process may also be relevant to explain inhomogeneities observed in several globular clusters.

1.3. *Equations governing chemical evolution models*

In this subsection I will give the basic equations governing chemical evolution models. Let us define $G(t)$, $S(t)$, and $R(t)$ the mass in gas, stars and remnants respectively at a time t. We have:

$$G + S + R = M_{tot}, \qquad (1.1)$$

where M_{tot} is the total mass of the system. In a closed box model, we have $M_{tot} = const$; else, more in general we may write:

$$\frac{dM_{tot}}{dt} = F - W, \tag{1.2}$$

where $F(t)$ and $W(t)$ represents the inflow and outflow (winds) of mass respectively. In general, it is assumed that only inflow and outflow of gas occurs.

The total mass locked in stars is the sum of the contribution $s(m,t)$ of mass locked in stars of different masses m:

$$S = \int_{m_1}^{m_2} s(m,t) \, dm, \tag{1.3}$$

where m_1 and m_2 are lower and upper limits of the mass function.

We introduce now a function $n(m,t)$, which gives the mass in stars of mass m forming at a time t; n is given by:

$$n = \omega G^\beta \psi, \tag{1.4}$$

where ωG^β represents the SFR (which is given here by a Schmidt law with exponent β), and $\psi(m)$ is the IMF; the IMF is normalized so that:

$$\int_{m_1}^{m_2} \psi(m) \, dm = 1. \tag{1.5}$$

Let us now introduce a second function $x(m,t)$, which gives the mass given back to the ISM by stars of mass m at a time t; for simplicity, we will assume that all mass is lost by stars at the end of their lifetime $\Delta(m)$ (this is certainly a not too bad approximation: note that Δ also depends on chemical composition; we neglect here this dependence). This introduces a *delay* with which stellar material is given back to the ISM. Note that for binary systems (type Ia SNe), Δ should depend also on orbital parameters, so that it is not any more a single valued function, but rather a distribution function; in the present formalism, we may assume that Δ has an appropriate average value. x is then given by:

$$x = [1 - r(m)] \, n(m, t - \Delta), \tag{1.6}$$

where $r(m)$ is the fraction of the mass of a star tied up in the remnant. Note that $x > 0$ only for stars with masses such that $\Delta < t$.

We are now able to write down the equations governing the total mass in gas, stars, and remnants at a time t:

$$\frac{dG}{dt} = -\omega G^\beta + \int_{m_1}^{m_2} x \, dm + F - W, \tag{1.7}$$

$$\frac{ds(m)}{dt} = \omega G^\beta \, \psi - n(m, t - \Delta), \tag{1.8}$$

$$\frac{dS}{dt} = \int_{m_1}^{m_2} \frac{ds(m)}{dt} \, dm, \tag{1.9}$$

$$\frac{dR}{dt} = -\frac{dG}{dt} - \frac{dS}{dt}. \tag{1.10}$$

These equations can be integrated numerically once the initial conditions are defined; they are $G(0) = M_{tot}$, $s(m,0) = 0$, and $R(0) = 0$.

The chemical evolution may be easily introduced in this system of equations; in fact, let us define $Z_i(t)$ as the total amount of a nuclear specie i in the ISM, and $\alpha_i(t)$ its

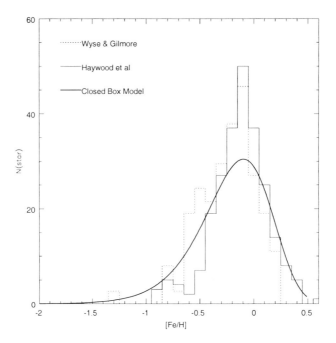

FIGURE 3. Histogram of metal abundance distribution (from Wyse & Gilmore, 1995; and Haywood et al. 1997) and comparison with predictions of simple closed-box model

fractional abundance:

$$\alpha_i = \frac{Z_i}{G}. \tag{1.11}$$

The evolution of the mass of a nuclear specie i in the ISM is then described by the equation:

$$\frac{dZ_i}{dt} = \int_{m_1}^{m_2} [\alpha_i(t - \Delta) + y_i] \, x(m, t - \Delta) \, dm - \alpha_i(\omega G^\beta + W), \tag{1.12}$$

where $y_i(m)$ are yields of newly produced nuclei i from a star of mass m, normalized to unitary mass (we have assumed here that yields are independent of chemical composition; when this assumption is relaxed, a dependence of y on $\alpha(t-\Delta)$ should also be considered).

Equation (1.12) is integrated numerically, with initial abundances α_i equal to the Big Bang abundances. Note that these equations can be easily extended to cases in which elements are destroyed rather than produced in stars (astration).

1.4. *Comparison of chemical evolution models with observations*

In order to discuss properties of chemical evolution models, it is useful to start from the closed box model ($F = W = 0$). In the following we will also assume that the IMF is constant and yields are independent of chemical composition. We will then compare the results given by this model with observations relevant to the chemical evolution of the solar neighborhood: critical approximations will be considered. We will show how they have to be released in order to reproduce observations.

It can be showed (see e.g. Pagel 1992) that for a closed box model, the distribution of

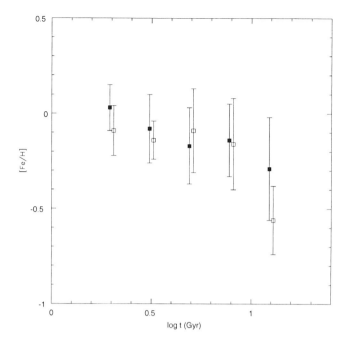

FIGURE 4. Age metallicity relation for the solar neighborhood from Edvardsson et al. (1993); filled and open squares refer to stars at galactocentric distance bins of 7-8 and 8-9 Kpc respectively. Error bars represent r.m.s. scatter of distribution (not of average value), and are much larger than expected errors in abundance determinations for individual stars

stars with metal abundance is described by the equation:

$$\frac{d\,s}{d\ln\alpha} \sim \frac{\alpha}{y} e^{-(\alpha-\alpha_0)/y} \tag{1.13}$$

where α_0 is the initial abundance. This distribution peaks at $\alpha = y$ (provided that $\alpha_0 < y$), which leads to assume that the yield is very roughly the solar metallicity, which is confirmed by calculations of nucleosynthesis yields where $y \sim 0.01$ (Matteucci 1983). The distribution of primary elements resulting from this equation predicts a too large number of metal poor stars when compared with observations for the solar neighborhood (this is called the G-dwarf problem); a modern version of this result is shown in Figure 3. There are two possible ways out of this problem:

• the IMF is not constant, and very few small mass stars were produced at low metallity; however, yields from an IMF biased toward larger masses should be larger causing a significant overproduction of metals (the observed distribution would anyway not be reproduced)

• there is a slow infall of metal-poor gas into the disk, which dilutes the ISM; also in this case it is necessary to assume slightly higher yields: however this can easily be accomodated within current uncertainties, so that these models are favoured

A second important data to be reproduced is the age-metallicity relation (see Figure 4). Models with a slow infall well succeed in reproducing the extended plateau at high metallicities (see e.g. Matteucci & François 1992).

Additional important features are the patterns observed in the element-to-element abundance ratios. The most important diagram compares the abundances of O and Fe. There are various forms of this diagram: the most useful represents the run of

TABLE 2. Mean [Fe/O] ratios in halo, thick disk, and thin disk stars

	V_{rot} (km/s)	z_{max} (kpc)	$< [Fe/O] >$	r.m.s.
Thin Disk	> 170	< 0.5	$0.33\,[O/H] + 0.03$	0.11
Thick Disk	> 100	< 1.5	-0.38	0.11
Halo	< 100	> 1.5	-0.36	0.16

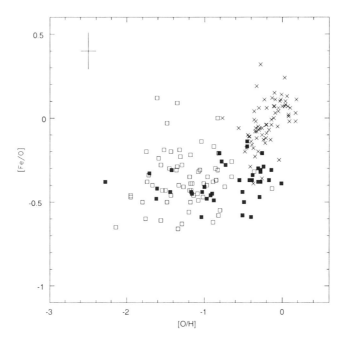

FIGURE 5. Run of the [Fe/O] ratios with metal abundance; stars belonging to the halo, thick, and thin disk (follpwing the definition of Table 2) are marked with different symbols

the [Fe/O] ratio with [O/H]. A recent version of this diagram is shown in Figure 5 (from Gratton 1998); while this diagram has been obtained from observations of stars in the solar neighorhood, it provides important constraints on three of the four major dynamical populations of our own Galaxy: the halo, the thick disk, and the thin disk (data of similar quality is still not available for stars in the bulge). These constraints are obtained by combining the chemical abundances with space velocities, which allows an appropriate definition of membership of individual stars to different stellar populations. Our definition of halo, thick and thin disk from the stellar velocities is given in Table 2; note that the present definition of the thick disk is slightly different from that adopted by other authors, and better corresponds to the dissipative component identified by Norris (1994). The main emerging features are:

• stars in the halo have a nearly constant average deficiency of Fe with respect to O. The value obtained by Gratton and coworkers is [Fe/O]~ -0.4; however, the exact value is somewhat model dependent (see e.g. Israelian et al. 1998 for a somewhat different result). An important feature is the scatter around this mean value (0.16 dex), which is

TABLE 3. Predicted scatter in [Fe/O] values in a halo formed by individually evolving fragments as a function of fragment mass

M_{star}	M_{total} ([O/H] = −1)	M_{total} ([O/H] = −2)	N(SNII)	[Fe/O] scatter	O mass
10^3	3×10^4	3×10^5	3	0.59	23
10^4	3×10^5	3×10^6	34	0.15	230
10^5	3×10^6	3×10^7	340	0.04	2300

larger than expected from observational errors alone (0.11 dex). Comparison with the scatter obtained for the other populations suggests that this is real: indeed, a few of the stars with the smallest deficiency of Fe consistently reproduce this datum in different analyses (Carney et al. 1997; King 1997; Gratton 1998; Jehin et al. 1998). It is now quite clear that an intrinsic scatter indeed exists, indicating that the ISM was not well mixed in the halo (as earlier suggested by McWilliam 1997)

• stars in the thick disk also have a constant deficiency of Fe with respect to O of about [Fe/O]~ −0.4, but the intrinsic scatter (0.11 dex) is much smaller than for the halo, and consistent with observational errors. There are thick disk stars with high (solar) O abundances. Similar results have obtained in other recent independent analyses (Nissen & Schuster 1997; Fuhrmann 1998).

• Fe deficiency is much smaller, if any, for stars in the thin disk; again, the star to star scatter is small (0.11 dex) indicating that in this case the ISM is well mixed. The transition from thick to thin disk is abrupt, with a rather large raise of the Fe abundance while O is nearly constant.

A diagram like that of Figure 5 provides a number of strong constraints on galactic evolution:

• The nearly constant average value of the [Fe/O] ratio over a couple of decades in metal abundance requires that formation of the halo and of the thick disk was fast enough so that there was no appreciable contribution from type Ia SNe. There may be a few exception (a few halo stars with small deficiencies, or even excesses of Fe): while it is possible that some of these stars were accreted from satellites (and are then substantially younger than the bulk of the halo), most of the scatter observed in the halo may be better explained by local inhomogeneities of the ISM due to random fluctuations of the mass of individual SNe, not unexpected on the basis of the low densities. Table 3 provides basic data within the hypothesis that halo individual fragments evolved individually: typical masses of fragments that could explain the observed scatter are similar to those of globular clusters. This sounds as a reasonable explanation; for the same reason, we might expect a ~ 0.15 dex cluster-to-cluster scatter in the [Fe/O] ratios.

• the transition from thick to thin disk occurs at constant O and raising Fe abundances. O is produced by massive (short living) stars, its abundance closely tracing star formation, while part of the Fe production is delayed (and is then an integral of star formation in the previous phases). This observation then indicates that there was a phase of relatively low star formation (hiatus). Gratton et al. (1996) and Chiappini et al. (1997) have presented a number of chemical models which try to reproduce this observation: a second infall phase related to the thin disk, separated from the first one which originated the halo and the thick disk, is required in order to avoid either large dilution of the metals produced in the thick disk, or a strong burst of star formation at the end of the hiatus.

From the former discussion, it is clear that a single zone model like the one insofar considered requires several ad hoc hypothesis in order to explain the rather complex observational picture. However, as noticed by Gratton et al. (1996) these hypotheses may raise more naturally in a scenario where the most important driving mechanisms of galactic evolution (dissipational collapse: Eggen et al. 1963, Larson 1974; and accretion of fragments or satellites: Toomre 1977, Searle & Zinn 1978) are considered simultaneously. An adequate model should then consider both chemical and dynamical features. Attempts to develop such models are in course (Samland et al. 1997; Raiteri et al. 1998), and in spite of computational complexities, important progresses are expected in the near future.

2. High dispersion analysis: methods

Abundances of metals in stars may be measured by exploiting the dependence of various observables on chemical composition. However, practically all indices devised insofar do not depend uniquely on chemical composition: in most cases, metallicity is a second parameter, the first one being effective temperature; often also surface gravity plays an important role. Additional factors like He content, velocity fields in stellar atmospheres, or even stellar activity may be important. Furthermore, broad band colours, or even most features observed in low dispersion spectra, measure the integrated effects of a variety of atomic species: the appropriate weights to be given to each element are not easily determined, and the real meaning of derived abundances is often not clear.

Most direct results are obtained when individual lines are observed: in this case at least atoms or molecules responsible for the absorptions/emissions can be identified. The spectrum of normal stars is dominated by absorption lines; a few emission lines form in the most external atmospheric layers. and are observable in the UV (where photospheric continuum is weak). However, physics of chromosphere and corona is poorly understood, so that its use for abundance analysis is forcely limited, and will be insofar neglected in this talk. Absorption line spectra are very rich in the visible and UV; lines are broadened by both motions of atoms in the atmosphere (both thermal and due to turbulence at various scales), that have typical velocities v of a few km/s (the sound velocity in typical atmospheric conditions), and by pressure effects. Additionally, natural broadening due to the finite lifetime of levels involved in the transitions should also be considered; line profiles are resolved at a resolution of few 10^4 (depending on atmospheric conditions). Hence, a pretty high spectral resolution is required in abundance analysis of individual lines. At lower resolution, the instrumental profile smears out the profile of individual lines, so that blends due to several lines can only be observed, confusing the physical interpretation of the data.

Analysis of high dispersion spectra provides then the most direct route to photospheric abundances, through either interpretation of equivalent widths (EWs) of individual lines or by comparison of observed profiles with synthesized spectra. However, also in these cases correct interpretation of results requires an understanding of the physical status of gas and radiation field in the stellar atmospheres. It is then clear that accurate and reliable abundance analyses require observational material of good quality, and appropriate analysis methods.

In this lecture I discuss the observational requirements (mainly resolution and S/N ratio) and critically review the most important steps of data interpretation (measure of line profiles and equivalent widths and the related issues of determination of the continuum level and blend contamination). I will then briefly examine the analysis methods employed (fine analysis, spectral synthesis), and the most important steps involved in the abundance derivation (basic atomic and molecular data, determination of the atmospheric parameters). Finally I will shortly comment on the present status of knowledge of stellar atmospheres with emphasis on solar-type stars and K-giants, considering the impact of line opacity, subatmospheric convection and departures from LTE.

Through this discussion, it should be reminded that spectroscopic abundance analyses measure present surface chemical composition (this also applies to most photometric abundance indices), while in general (but not always) we are more interested in the original stellar composition. In most cases it may be assumed that the two values coincide: however, there are at least four mechanisms that may alter the surface composition of a star:

(a) diffusion and radiation pressure may be important on a long ($\geq 10^8$ yr) timescale if the outer stellar layers are in very quiet conditions (radiative equilibrium); this is the case of warm stars (blue horizontal branch stars, upper main sequence stars, warm white dwarfs). Recently, it has been proposed that inward diffusion at the base of the outer convective envelope may significantly alter the surface composition of solar mass stars, selectively depleting this region of part of their metals (see e.g. Chaboyer et al. 1992)

(b) Inward penetration of convection (or rotation induced currents) may bring to the surface material which has been previously processed by nuclear reactions: this mechanism is important for the lighter elements (Li and Be), for nuclei participating in the CNO cycle, and for the neutron capture $s-$elements. We will discuss this point quite extensively in Section 4

(c) If a considerable fraction of the stellar mass is lost, nuclearly processed material may be exposed at the surface. This mechanism is important for very massive stars, causing the appearence of Wolf Rayet type spectra, and in very evolved low mass stars (proto-planetary nebulae and non-DA White Dwarfs). Hereinafter, we will limit ourselves to stars where mass loss does not affect the surface abundances

(d) Finally, pollution by material transferred from a companion may be important in binary systems, and perhaps in the high density cores of globular clusters. This point also will be commented in more detail in Section 4

2.1. *Analysis techniques*

In general, abundances are derived by comparing the observed spectrum with expectations from synthetic spectra, that can be obtained by solving the (monochromatic) transfer equation through a model stellar atmosphere for different points along the profile.

The simplest technique is the **fine analysis** of individual lines, which has superseded old methods based on rough representations of stellar atmospheres. Both the entire line profile and the equivalent widths alone can be used for this comparison.

Fine analysis yields accurate abundances if three basic conditions are satisfied:

(a) Reliable EWs for weak, unsaturated absorption lines are measured

(b) An accurate set of basic parameters (partition functions, oscillator strengths, broadening parameters, continuum opacities at the relevant wavelengths) is used

(c) A realistic model atmosphere exists for the star under scrutiny

Spectral synthesis is a natural extension of modern fine analysis codes, from synthesis of a single line profile, to that of an arbitrarily large number of lines. In principle, spectral synthesis may be used to interpret spectra of whatsoever resolution (once the instrumental profile is known), and it has been in fact used in variety of applications, ranging from predictions of stellar colours (see e.g. Bell & Gustafsson 1978), to the discussion of very high resolution (crowded) spectra (see e.g. Gratton & Sneden 1990).

Spectral synthesis in principle is a very powerful technique, since it fully exploits informations contained in the observed spectrum. However, it is often difficult to reproduce results obtained with spectral synthesis due to lack of details (line lists are rarely published). Furthermore, reliability of results depend on resolution and S/N, mainly for two reasons:

• the accuracy of results obtained from comparisons with synthesized spectra depend on the quality of the line list used in the computations: a poor spectral synthesis analysis may be misleading. Unfortunately, there is not a good enough line list available in the literature for any purpose: the very extensive list by Kurucz (1993, 1994, 1995) is intended to be used for statistical purposes, and its accuracy must be checked in every single application. The usual procedure is to adjust the line parameters (wavelengths, oscillator strengths, damping constants) in order to fit a template spectrum (in most cases the solar one); however this procedure is not reliable if lines vanishingly weak in the Sun are present in the spectra of the program stars (as it occurs in very cool or metal-rich stars; but some time also in warmer, metal-poor stars if the lines are weak contaminants in the wings of very strong lines). Additionally, unidentified lines are still observed in the solar spectrum

• velocity fields (in particular microturbulent velocity) cannot be determined when weak lines are unobservable.

These two sources of concern may generate important systematic errors on abundances derived from comparisons with synthetic spectra. Even if they are checked, only a few global parameters may be derived from low dispersion spectra, where most features are due to blends of lines due to different species. At resolution $\sim 1,000$ (typical of most applications), the strongest atomic lines and molecular bands may be used to deduce a mean overall metal abundance and the abundances of C and N. The obvious advantage of low dispersion is the possibility to observe faint stars; however it is difficult to correctly estimate errors due to lack of redundancy in the data; furthermore, it is easy to overweight results if observational errors (often due more to reduction and calibration than to photon statistics) are not adequately accounted for.

In the next subsections we will comment on the most important steps of abundance analysis; thorough, although somewhat dated discussion of these points can be found elsewhere (Gustafsson 1989; Gratton 1991). These authors concluded that for G-dwarfs and K-giants, uncertainties due to atmospheric modelling are $\sim 0.15 \div 0.2$ dex in the overall metal abundances, and ~ 0.1 dex in the element-to-element abundance ratios.

2.2. *Observational material*

A commonly used indicator of the line absorption is the equivalent width (EW): this is simply a measure of the area between the line profile and a reference continuum (i.e. the profile expected if the line were not present). The main advantage of EWs is that (at least in principle) they do not depend on resolution: EWs should then be independent of the instrumental apparatus used for acquiring data and results obtained by different sources can be intercompared. A further important advantage is that EWs for weak absorption lines are directly proportional to the number of absorbing atoms. However, when lines become stronger, absorption at center of lines is a substantial fraction of continuum, and saturation should be considered. The EWs for which this effect is important depend on line width (determined by the broadening mechanisms); in the limit when saturation is very strong, the EWs only depend on velocity of random motion of the absorbers. For a gaussian broadened profile (thermal motion or low scale microturbulent velocity) saturation becomes important for:

$$EW \geq 0.5w, \tag{2.14}$$

where $w \sim \lambda(v/c)$ is the full width half maximum (FWHM) of the line. For typical velocities of 2 km/s, saturation is important for lines of $EW \geq 40$ mÅ at a wavelength of 6,000 Å.

Cayrel (1988) provided a theoretical estimate of the error done when measuring the EWs of a line (with a gaussian profile) as a function of the line width w, of the detector channel width δx, and of S/N:

$$\Delta EW = 1.6(w\,\delta x)^{1/2}/(S/N) \tag{2.15}$$

Following this formula, spectra centered at $\lambda = 6,000$ Å, with $R = \lambda/w = 20,000$, and $S/N = 100$, yield $\Delta EW = 3$ mÅ.

However, Cayrel's formula does not include the error present in the definition of the local continuum, often a difficult task in line rich spectra, where few line-free regions are available. Errors ascribed to uncertainties δc in the location of the continuum are roughly:

$$\Delta EW \sim 2\,w\,\delta c \tag{2.16}$$

Not only this source may dominate the total error budget, but more dangerously it may introduce systematic errors; these are more often done at low dispersion, since in this case continuum windows may be filled by absorption of nearby lines smeared out by the instrumental profile. This underlines the importance of high resolution.

2.3. Basic Spectroscopic Data

Spectral analysis (through either fine analysis or spectral synthesis) is based on spectroscopic data for the lines of interest. For atomic lines, there are mainly five parameters of interest: wavelengths, excitation potentials, oscillator strengths gf's, damping parameters (mainly collisional damping), and hyperfine splitting. In general, **wavelengths** and **excitation potentials** are not a problem: extensive and accurate tables exists; for line analysis and spectral synthesis purposes, data listed by Kurucz (1993, 1994, 1995) are generally good enough. As to gf's, they may be often derived from theoretical computations and/or laboratory experiments. Accurate theoretical computations (like e.g. the configuration interaction method or for LS transitions, close coupling technique) are available for the simplest atoms (alkalis like Li, Na, K, Rb; alkaly-earths like Be and Mg; C, N, and O). For more complicate atoms, we should rely on laboratory estimates: the most reliable (relative) data are those obtained from absorption experiments in temperature controlled furnaces (like those obtained by the Oxford group for Fe and few Fe-peak elements). However these data require determination of the zero point, and are not available for high excitation lines or ionized species (owing to difficulties in mantaining gas at very high, constant temperatures). To these purposes, best results are obtained by combining lifetimes (obtained by measures of decays of atoms selectively excited by a suitable laser pumping) with branching ratios (derived from some type of emission experiment). Often errors as small as a few per cent are obtained in recent works (errors are much larger in older estimates). Extensive compilations of data are now available (Kurucz, 1993, 1994, 1995; the Opacity Project which databases are available at the CDS: Seaton et al. 1995).

While extremely useful, data as good as those above mentioned are unluckily available only for a minority of spectral lines (although a significant fraction of those generally used in fine analysis). Most weak lines (in particular of Fe-peak elements) lack of an accurate gf determination. The usual practice in this case is to use *solar gf*'s, i.e. values adjusted so that the EW's (or the line profiles) provide the same solar abundance determined from other lines of the same specie. Note that these gf's depend somewhat on details of the solar analysis, and in particular values used for damping constants.

While **radiative damping constant** is simply given by the sum of the inverse of the lifetimes in the upper and lower level of the transition (data generally available in the literature), much less is known about **collisional damping constants**. The impact of damping constant may be exemplified by the hot debate about the solar Fe abundance (see Holweger et al. 1995; Blackwell et al. 1995), that seems to be resolved (Anstee et al. 1997) by more accurate estimates of the damping parameters allowing good fits to wings of strong Fe lines when the meteoritic abundance is adopted together with the currently best empirical model for the solar photosphere (Holweger & Müller 1974). While we lack of a general good theory for collisional damping, quite good results have been recently obtained using the model developed by Anstee & O'Mara (1991; see Anstee & O'Mara 1995 for a convenient tabular presentation for s-p and p-s transitions). For very strong lines, empirical values can be determined from an analysis of solar lines (see e.g. Simmons & Blackwell 1982). On the whole, these data suggest that broadening given by the simple van der Waals theory (Unsöld 1955) is underestimated. For this reason, the usual practice is to increase the values obtained in this way by a constant factor for all those lines lacking better determination. While rough, this technique gives reasonably good results when damping wings are not well developed.

The effect of **hyperfine splitting** (HFS) is to desaturate strong lines (they have no effect on the EWs of weak lines). HFS should then be considered whenever is present, i.e. for transitions of atoms having a non-zero nuclear magnetic moment (all odd Z-elements) or different stable isotopes (note that there may be atoms having isotopes with different nuclear magnetic moment). However, in several cases HFS is small (this is mainly true for isotopic splitting for lines of elements of low to intermediate weight). Data on HFS are available for only a few transitions: this is an important limitation in the analysis of elements like Sc, V, Co, or the rare earths. I am not aware of a comprehensive tabulation of HFS, although useful data are in given by Kurucz (1995); most of the literature is quite old since the argument raises only a modest interest in nuclear physicists.

A lot of information on abundances in stars can be acquired by consideration of **molecular bands** (we are here mainly concerned with diatomic molecules). Individual molecular transitions are between states with different rotational quantum number (generally splitted by fine structure); they are organized in bands corresponding to the same vibrational transition; a single electronic transition - system - includes different vibrational transitions. Molecular bands are then formed by a large number of lines, where wavelengths and line strength ratios may be accurately determined from a few basic parameters (Franck-Condon and Honl-London factors) that may be rather easily calculated or measured in laboratory. An additional advantage is that isotopic splitting is large for vibrational transitions of light molecules (due to the large variation of the reduced mass). Molecules offer the best tool to determine accurate isotopic ratios, usually very difficult to obtain from atomic lines. The concentration of molecules may be determined using dissociation equations (similar to the Saha equation), where the dissociation potential plays the same role of ionization potential for atoms; however, coupling between different molecules has to be considered, so that in practice a system of equations has to be solved.

The most important molecules in stellar atmospheres are H_2 (observable in the UV), the hydrides (CH, NH, OH, MgH), the molecules including C (C_2, CN, CO), and the oxides (mainly TiO and VO) observable in M-stars. Some of these molecules are formed by two metals: in these cases, the strengths of the related bands decrease rapidly (approximately quadratically) with overall metal abundance. On the other side, hydride

TABLE 4. Data for some astrophysically important molecules

Molecule	Dissociation Potential (eV)	Electronic Transition	Δv	Wavel. (μm)
CH	3.46	$A^2\Delta - X^2\Pi$	0	0.42
		$B^2\Sigma^- - X^2\Pi$	0	0.39
CN	7.66	$B^2\Sigma^- - X^2\Sigma^+$	0	0.38
			-1	0.42
		$A^2\Delta - X^2\Sigma^+$	0	0.78
C_2	6.11	$d^2\Pi - g - a^3\Pi_u$	0	0.50
CO	11.09	vibration-rotation		2.2
NH	3.40	$A^3\Pi - X^3\Sigma^-$	0	0.34
OH	4.39	$A^2\Sigma^+ - X^2\Pi$	0	0.31
MgH	1.27	$A^2\Pi^+ - X^2\Sigma^+$	0	0.51

bands have a moderate dependence on metal abundance, since the decreased concentration of the molecule with decreasing metal abundance is in part compensated by the larger transparence of atmospheres (the continuum opacity due to H^- also decreases, due to the smaller number of free electrons, mainly provided by easily ionized metals); hence they are particularly useful in the analysis of metal-poor stars. Finally, it should be reminded that molecular bands are stronger in dwarfs than in giants due to the highest density.

Sometime basic courses in astronomy do not include a good training in the spectra of diatomic molecules. I will then give a few **references**: a good classical description of diatomic molecule spectra is Herzberg (1950); a very useful (although quite old) reference for stellar molecular spectroscopy is Schadee (1964); basic methods used to solve the dissociation equilibrium equations are given in Mihalas (1967). A quite extensive, albeit somewhat old discussion of basic data and an analysis of the solar spectrum is in Lambert (1978). More updated values (dissociation potentials; wavelengths, including bands due to different isotopes; oscillator strengths) may be found in Kurucz (1993). A few data for the most important molecules are given in Table 4.

2.4. *Model atmospheres and atmospheric parameters*

Model stellar atmospheres are defined by the methods used in their computation contained in a code, and by a set of basic parameters (effective temperature T_{eff}, surface gravity $\log g$, chemical composition [A/H], microturbulent velocity v_t).

In spite of their sophistication, theoretical (constant-flux) model atmospheres are still rough representation of the complexities of real stellar atmospheres, even if we limit ourselves to best studied range of parameters ($4000 < T_{\text{eff}} < 7000$ K). Up to a few years ago the main limitation of the available grids (Gustafsson et al. 1975; Kurucz 1979) was thought to be the inadequate consideration of the opacities due to atomic and molecular lines, mainly at short wavelengths (UV and blue wavelengths). In this respect, important improvements have been obtained in the 90's, mainly thanks to the monumental work of Bob Kurucz, who has collected an enormous database of atomic and molecular lines (in large part obtained by his own calculations). Colours computed from these model atmospheres (Kurucz 1995) agree now fairly well with observations (Bessell et al. 1998: see Figure 6). However, for some purpose, the temperature structure of these models is still not close enough to observations; this is testified by a number of small inconsistencies found when analyzing the Sun or other stars (see e.g. Castelli et al. 1997; Dalle Ore 1992; etc.). While not entirely clear, it seems that the main limitation of models is in their representation of convection; in this respect, it should be noticed that better agreement with observations is obtained when the overshooting option is switched off in the Kurucz models (Castelli et al. 1997; Bessell et al. 1998). However, a simple comparison with solar granulation shows the inadequacies of these plane parallel model atmospheres. Furthermore, these models do not predict chromospheres and coronae, which heating source is still debated. Limitation of models appear to be even more significant out of the above mentioned temperature range; in particular, at low temperatures (M-stars) even the very extensive line list by Kurucz is too limited to represent observations. While other grids of models (Plez 1992; Leggett et al., 1996; Baraffe & Allard 1998) appear to improve agreement with observations in this temperature range, we caution that care should be exerted when considering abundances derived for such cool stars.

Effective temperatures for individual stars may be derived using various different techniques:

(a) from calibrations of stellar colours (see below for the techniques used in these calibrations): in this case care must be taken of interstellar reddening, which is often poorly known. The dependence of temperature on the adopted reddening is approximately $\Delta T_{\rm eff} \sim 300\ \Delta E(B-V)$ K/mag, An uncertainty of 0.03 mag in $E(B-V)$ then translates into an error of about ± 100 K in $T_{\rm eff}$, which on turn implies an error of 0.08 dex in [Fe/H]

(b) from line excitation. In general, temperatures derived from excitation are model dependent (they depend on the adopted gf's and on the model atmosphere, see below) and are not very accurate (typical errors are ± 200 K) due to scatter of results for individual lines. However, these are often systematic, and much better results are obtained in a strictly differential approach

(c) from the wings of Balmer lines, which are sensitive mainly to temperature for $T_{\rm eff} > 5500$ K (sensitivity on temperature is lost in cooler stars). This technique may provide accurate results in a strictly differential approach (Cayrel et al. 1985); it has been recently used in the analysis of metal-poor dwarfs (Fuhrmann et al. 1994). However, since H lines are very broad, this technique requires very accurate flat-fielding, difficult to achieve in echelle spectra (having a limited free spectral range)

On the whole, temperatures derived with all these techniques finally require a calibration (temperature scale: see Figure 6). After a lively debate in the last twenty years, there is now a general consensus about the temperature scale for population I stars, thanks to the growing body of data obtained using the two best (less model dependent) set of indicators (interferometric radii, which allows determination of $T_{\rm eff}$ from its own definition: see e.g. Di Benedetto & Rabia 1987; and infrared flux method†: Blackwell and Shallis, 1977). Temperature scales for population I objects obtained by different authors now differs by only a few ten K (Blackwell & Lynas-Gray 1994; Di Benedetto & Rabbia 1987). As to metal-poor stars, a large set of temperatures using IRFM have been obtained by Alonso et al. (1996); they agree fairly well with temperatures obtained from visual and near IR colours using the Kurucz model atmosphere switched off

---

† The infrared flux method is a measure of emissivity from stellar surfaces based on ratios between bolometric and monochromatic fluxes, these last measured on the Rayleigh part of the spectrum, to be as less as possible sensitive to details of the model atmosphere. For solar type and cooler stars, monochromatic fluxes have then to be measured in the near-IR. Note that temperatures from IRFM depends on the adopted interstellar reddening, approximately as temperatures derived from colours

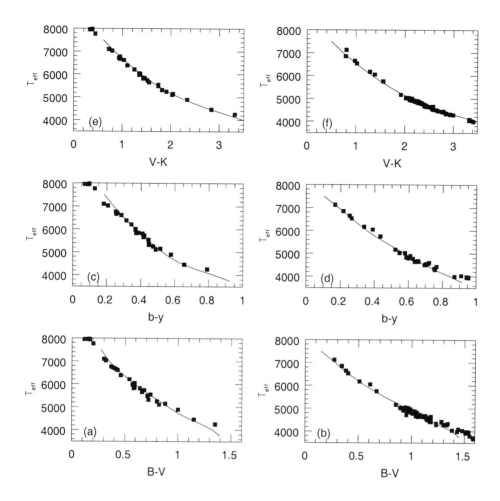

FIGURE 6. The colour-temperature relations from Kurucz models, with the overshooting options switched off (Bessel et al. 1998), compared with temperatures determined from IRFM and interferometric radii (Blackwell & Lynas-Gray 1994; Di Benedetto & Rabbia 1987) for population I stars. Panels (a), (c), and (e) show results for dwarfs compared with model atmospheres computed with $\log g = 4.5$; panels (b), (d), and (f) show results for giants compared with model atmospheres computed with $\log g = 2.5$; stars

(Bessel et al. 1998, Di Benedetto 1998). Temperatures determined from line excitation are not affected by uncertainties in the reddening estimates; however currently available atmospheric models do not well reproduce the observed difference between temperatures in the continuum and line forming regions: the effect is rather mild for dwarfs (where temperatures derived from line excitation are about 100 K higher than those derived from continuum: see e.g. Clementini et al. 1998), while it seems to be quite large in some metal-poor giants where temperatures from lines are $200 \div 300$ K lower than those derived from the continuum (Dalle Ore 1992). While there is still debate about the real interpretation of these results, it seems wiser to mantain a rather large error bar for the zero point of abundance analyses.

Surface gravities may be derived from mainly four techniques:

- from the equilibrium of ionization (in most case of Fe, which is the easiest observable element). This technique does not require very high resolution and S/N and it is then widely applied. However, derived gravities are sensitive to the adopted temperature (~ 0.003 dex/K), and may be affected by departures from LTE (although most recent estimates of departures from LTE show this effect should be small in solar type dwarfs and K-giants: see e.g. Gratton et al. 1998)

• from the equilibrium of dissociation of molecules (Bell et al. 1985); this technique is rarely applied since only in a few cases reliable abundance indicators are available from both atoms and molecular lines for a single element

• from the collisionally broadened wings of strong lines (Blackwell & Willis 1977). This technique provides accurate estimates of the surface gravity, with only moderate sensitivity to temperature and details of the model atmospheres (because damping wings form deep in the stellar atmospheres); however, it can be successfully applied only if strong, pressure broadened lines are observed (not too metal-poor dwarfs), and if accurate spectral profiles are observed (hence high resolution and high S/N are required)

• from estimates of the radii (from luminosities and temperatures) and masses. This technique is very successfull for cluster stars, where the largest uncertainty is due to possible errors in the distance modulus (usually ≤ 0.1 dex). In recent years, it has been widely used also for field metal-poor giants, on the hypothesis that they are identical to globular cluster stars of the same metallicity (an iterative procedure is then required: see e.g. Bond 1980). The main drawback is that temperatures should be previously determined (sensitivity is similar to that for ionization equilibrium gravities); in most cases this is done using colours (hence the method is affected by uncertainties in the interstellar reddening)

The **microturbulent velocity** is a parameter describing small scale non thermal motions in the stellar atmospheres, which cause desaturation of lines of intermediate strength; it is usually determined by comparing abundances obtained from lines of different strength of the same specie. Weak lines are not sensitive to microturbulent velocity, because they are not saturated; on the other side, strong, saturated lines are very sensitive to microturbulent velocity. Correct determination of abundances then requires observation of weak lines. While microturbulent velocity appears to be correlated with general properties of the atmospheres ($T_{\rm eff}$ and $\log g$), we are still not able to accurately predict its value.

Sensitivity of abundances on the adopted atmospheric parameters may change significantly from specie to specie, and from line to line. To understand these dependences, it should be remembered that line strength depends on the ratio between line and continuum opacities. In F to late K stars (the stars usually considered in the present context) continuum opacity is dominated by H^-: the dissociation equation for H^- is essentially a Saha equation (the concentration of H^- ions is proportional to the number of free-electrons, i.e. to the electron pressure, which mean value is determined by temperature and gravity, decreasing with declining values of both). From the Saha equation, abundances from neutral lines of easily ionized elements (alkalis, alkaly-earths, Fe-peak elements, rare-earths) also have a similar sensitivity to electron pressure: the two dependences cancel, so that there is practically no dependence on surface gravity; however, due to the difference in the ionization potential, there is a strong sensitivity on temperature. On the other hand, concentration of dominating specie (like singly ionized species of elements with a low ionization potential, or neutral species of elements with a high excitation potential) is nearly independent on electron pressure: hence line strength is very sensitive to gravity (due to variation in the continuum opacity), and only moderately dependent on temperature (depending on the excitation potential). While these general rules are of basic importance, prediction of sensitivity of each line is complicated by the

TABLE 5. Typical sensitivities of abundances on the adopted atmospheric parameters

Lines	$\Delta T = +100$ K		$\Delta \log g = +0.3$ dex		$\Delta [A/H] = +0.2$ dex	
	[O/H]	[O/Fe]	[O/H]	[O/Fe]	[O/H]	[O/Fe]
permitted	−0.09	−0.07	−0.11	+0.06	+0.02	+0.11
forbidden	+0.01	+0.03	−0.17	−0.01	0.00	+0.09
OH	+0.20	+0.10	−0.10	−0.08	−0.03	+0.01

dependence on excitation, on the height of formation with the atmosphere (mainly due to difference in the line strength) and finally by saturation (and hence dependence on microturbulent velocity). In practice, dependence is best determined by comparing results obtained from the same set of spectroscopic data but different values of the atmospheric parameters: such an analysis is usually included in all papers concerning abundance derivation. Table 5 contains some typical values obtained from such an analysis from a my paper; while quite representative of the sensitivities, they depend on the adopted set of lines.

2.5. *Non-LTE effects*

A common (though not essential) assumption in fine analysis is that lines form in LTE. This conditions may be relieved if reliable statistical equilibrium computations are available. A basic limitation of these studies is the poor knowledge of cross sections for collisions with neutral H atoms, which is an important thermalizing mechanism in cool stars. In most recent investigations, empirical estimates of these cross sections have been obtained using various approaches (Reetz 1998, Gratton et al. 1998). The conclusion is that non-LTE effects are small for some elements (like Fe and Mg), but are not negligible for others (e.g. for the permitted near-IR triplet of O or for the strongest Na lines)

3. The Abundance scale and element-to-element ratios in Globular Clusters

Determination of the abundance scale for globular clusters is basic both in the determination of their ages, and in a close comparison between stellar evolutionary models and observations. In this lecture I present the most recent results about the abundance scale for globular clusters, discussing uncertainties still present in their determination. Abundances of elements other than Fe are also presented.

3.1. [Fe/H] *values*

The metal abundance of globular clusters is traditionally indicated by the [Fe/H] ratio, which is usually assumed to be representative of the total metal abundance. While this last point may be questioned (see below), in the first part of this lecture we will concentrate on the estimates of [Fe/H].

The history of metal abundance derivations from high dispersion spectra of individual globular cluster stars is instructive, since it reveals the impact of high resolution and high S/N in such analysis. The very first determination of abundances from high resolution spectra was done fourty years ago by Helfer et al. (1959) on a giant in M92 and another in M13 using the Coudè spectrograph at Mt. Palomar 5 m telescope: given the low efficiency of old Coudè spectrographs and sensitivity of photographic plates, this was an heroic effort, requiring single exposures over several consecutive nights. These spectra (of a very low quality measured on modern standards) were however enough to reveal the extremely metal-poor character of these stars. Systematic observations of a large number of giants started in the late '70s, with the introduction of efficient Echelle spectrographs at KPNO and CTIO; first studies used image tubes and photographic plates (Cohen 1978; 1979, 1980, 1981; Pilachowski et al. 1980a, 1980b, 1982a, 1983c, 1984; Pilachowski & Sneden 1983; Geisler & Pilachowski 1982; Peterson 1982; Gratton, 1982a, 1982b; for a review of these early results see Pilachowski et al. 1983a). Resolution and S/N of these spectra was still rather limited (typically $\sim 20,000$ and $\sim 20 \div 30$), so that only quite strong lines could be measured, with rather large errors. While abundances for most metal-poor clusters appeared reliable, albeit noisy, systematic errors were likely present in line-rich spectra of giants in metal-rich clusters, both due to underestimates of the correct continuum level (Pilachowski et al. 1983b; Peterson 1983a; Gustafsson 1983), and to overestimates of the microturbulent velocity (Gratton 1988). This last point in particular discloses the importance of high resolution and high S/N (see Figure 7).

A decisive progress was possible when CCD detectors were utilized: the efficiency gain allowed high S/N ($\sim 50 \div 100$) spectra for somewhat fainter stars; the first spectroscopic observation of a globular cluster giant with a CCD detector was made at moderate resolution by Cohen (1983); shortly after, D'Odorico et al. (1985) used the new CASPEC spectrograph at ESO to observe a moderately warm giant in 47 Tuc: the abundance derived ([Fe/H]$= -0.8$) is close to the presently accepted value. A number of studies followed in the next years, mainly from Gratton and coworkers (who presented an extensive survey of a few giants in several clusters, aimed to the determination of the abundance scale for globular clusters: Gratton et al. 1986; Gratton 1987a, 1987b; Gratton & Ortolani 1989: collectively G8689) from the Lick-Texas group (Kraft, Sneden, and coworkers: Sneden et al. 1991, 1992, 1994, 1997; Kraft et al. 1992, 1993, 1995, 1997, 1998; Armoski et al. 1994; Langer et al. 1998; Pilachowski et al. 1997, collectively SKPL), who considered large samples of giants in a few clusters, aimed to study inhomogeneties within each single cluster; and from the Washington group (Leep et al. 1986, 1987; Wallerstein et al. 1987; Brown & Wallerstein 1989, 1993a; Brown, et al. 1990, 1991; Wallerstein & Gonzalez 1998), who addressed other points like the isotopic ratios in bright giants, the composition of stars in ω Cen (Brown et al. 1993, Brown & Wallerstein 1993b, Vanture et al. 1994, Zucker et al. 1996), abundances in evolved red giants and post-AGB stars (Gonzalez & Wallerstein 1992, 1994; Vanture & Wallerstein 1992; Whitmer et al. 1995) and more recently abundances in *young* globular clusters (Brown et al. 1997). Important contributions also came from other groups (François et al. 1988, Norris & Da Costa 1995), who studied southern globular clusters, and in particular the very interesting case of ω Cen. Very recently, an analysis of giants in the bulge cluster NGC 6553 has been presented by Barbuy and coworkers (Barbuy, 1998).

A quite updated, homogeneous determination of [Fe/H], based on high dispersion spectroscopy of 160 red giants in 24 galactic globular clusters has been presented by Carretta and Gratton (1997). The observational material used both new high resolution ($R \sim 30,000$), high S/N spectra for a few stars, as well as a reanalysis of EWs from literature, put on a homogeneous scale. Atmospheric parameters were uniformly derived using an updated temperature calibration (Gratton et al. 1997), and gravities from stellar luminosities and masses derived from their location in the c-m diagram. Oscillator strengths were derived from a careful exam of literature (mostly laboratory, but in some case solar-based) values. Last generation model atmospheres by Kurucz (1995) were adopted. Abundances were compared with those obtained for the Sun using a model atmosphere extracted from the same grid, EWs measured on high quality solar spectra,

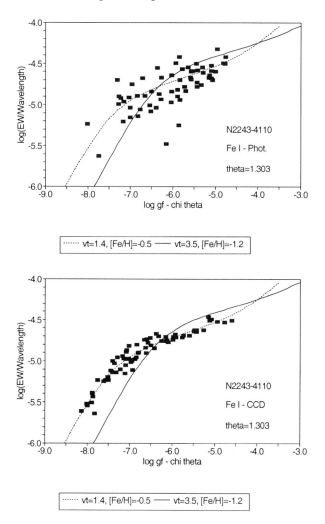

FIGURE 7. Comparison between Fe I curves of growth from photographic spectra (panel a) and CCD spectra (panel b) for star 4110 in NGC2243 (from Gratton & Contarini 1994). Note that the much better S/N of the CCD spectra allows a proper derivation of microturbulent velocity, and then of metal abundance

and the same line parameters used for the stars. Although quite strictly differential, this procedure may still hide two possible sources of errors: (i) atmospheres of giants may not be reproduced by models as well as atmospheres of dwarfs; and (ii) while line strength is not so much different in the warm, metal-rich Sun ($T_{eff} = 5770$ K) and in cool, metal-poor giants ($T_{eff} < 4700$ K, [Fe/H]< -0.8), line profiles in the spectra of giants are quite different from those in the spectra of dwarfs, where collisional damping is important. This last point in particular may be dangerous, because errors in the (not well known) collisional damping constant may introduce systematic errors in solar abundances or gf's.

Table 6 presents the abundances derived by Carretta and Gratton; Figure 8 compares the new abundances with those by Zinn & West (1984), often used for discussion of globular clusters c-m diagrams and other properties. While the correlation between the two scales is indeed good, it is not linear: abundances are similar at the extremes

TABLE 6. Mean metallicities for globular clusters compared to literature data

NGC	Messier	Stars	Mean±r.m.s.	σ	G8689	SKPL	Others	ZW
104	47 Tuc	5	-0.70±0.03	0.07	-0.82			-0.71
288		2	-1.07±0.03	0.04	-1.31			-1.40
362		1	-1.15		-1.18			-1.27
1904	M 79	2	-1.37±0.00	0.01	-1.42			-1.68
2298		3	-1.74±0.04	0.06			-1.91[a]	-1.85
3201		3	-1.23±0.05	0.09	-1.34			-1.61
4590	M 68	3	-1.99±0.06	0.10	-1.92		-2.17[b]	-2.09
4833		3	-1.58±0.01	0.01	-1.74		-1.71[b]	-1.86
5272	M 3	10	-1.34±0.02	0.06		-1.46		-1.66
5897		2	-1.59±0.03	0.05	-1.84			-1.68
5904	M 5	16	-1.11±0.03	0.11	-1.42	-1.17		-1.40
6121	M 4	3	-1.19±0.03	0.06	-1.32			-1.33
6144		1	-1.49				-1.59[b]	-1.75
6205	M 13	23	-1.39±0.01	0.07		-1.49		-1.65
6254	M 10	15	-1.41±0.02	0.10	-1.42	-1.52		-1.60
6341	M 92	9	-2.16±0.02	0.05		-2.25		-2.24
6352		3	-0.64±0.06	0.11	-0.79			-0.51
6362		2	-0.96±0.00	0.01	-1.04			-1.08
6397		10	-1.82±0.04	0.10	-1.88			-1.91
6656		3	-1.48±0.03	0.06	-1.56			-1.75
6752		12	-1.42±0.02	0.08	-1.53			-1.54
6838	M 71	13	-0.70±0.03	0.09	-0.81	-0.79		-0.58
7078	M 15	4	-2.12±0.01	0.01		-2.30	-2.23[b]	-2.17
7099	M 30	2	-1.91±0.00	0.00			-2.11[b]	-2.13

References: a = McWilliam et al.   1992, b = Minniti et al.   1993

of the scale (very metal-poor clusters like M92, or metal-rich ones like 47 Tuc), but differ by as much as 0.3 dex for clusters of intermediate metallicity (like M13, M3, or M5). It should be noted that while Carretta & Gratton calibration is based on a consistent analysis of high dispersion spectra, Zinn & West calibration uses a number of heterogeneous determinations, mainly from photometry or low dispersion indices: high dispersion abundances are used only for the most metal-poor clusters. Given the lack of a sound physical base for this calibration, the presence of a non-linear dependence on metalicity is not surprising.

A comparison may also be useful with the metallicities tabulated in the database by Harris (1996), available on the Web. The correlation is very good: the two scales differ for a nearly constant offset of 0.13 ± 0.02 dex, the present scale being higher. The r.m.s. scatter of individual values is very small (0.07 dex).

3.2. *Element-to-element abundance ratios*

While [Fe/H] provides an important estimate of overall metallicity, a full determination requires knowledge of the abundances of other elements too. Furthermore, a detailed study of element-to-element ratios may provide important clues on the formation and evolution of clusters.

In this subsection we will discuss the element-to-element abundance ratios in *normal* globular clusters. We will not discuss the lightest elements (CNO, Na, Mg, and Al),

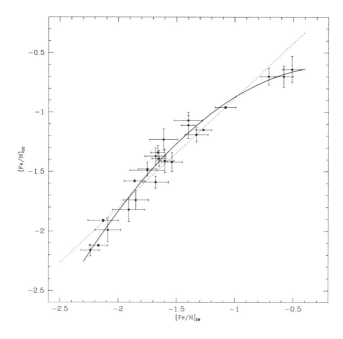

FIGURE 8. Comparison between the mean metallicities by Zinn & West (1984) and those derived from high dispersion spectra by Carretta & Gratton (1997). Note the good correlation, and the non linearity of the scale

which will be considered in the next section; there we will also discuss the case of ω Cen, that is clearly peculiar, exhibiting star-to-star variations of practically all elements.

As mentioned in Section 1, the abundance ratios of α−elements (O, Mg, Si, Ca) to Fe-peak ones provide basic constraints on the nucleosynthesis history of stars. Excesses of these elements with respect to Fe are observed in most halo as well as thick disk field stars. This is interpreted as due to the presence of two different astrophysical production sites (massive stars exploding as type II SNe, and type Ia SNe, resulting from the evolution of IMS stars in binary systems), having different evolutionary timescales (a few 10^7 yr in the first case, of the order of a Gyr in the second one). In some way this ratio is then a chronometer of star formation in the early Galaxy. However, local inhomogeneities (that may be represented as the independent evolution of individual fragments) may complicate this picture for two reasons: (i) star formation may be not simultaneous in different regions of the Galaxy; hence the same overabundance of α−elements may be observed in clusters having different ages. (ii) Evolution of *small* fragments may result in large variations of the abundance ratios at the same age, if (as expected from models) there is a large variation of the relative yields with SN progenitor mass (see Section 1). As to the second point, there are indeed evidences for large variations in the element-to-element abundance ratios amongst the most metal poor ([Fe/H]< -2) field stars (McWilliam 1997); at larger metallicities, differences are much smaller, and not much larger than observational errors.

It is then interesting to examine the abundances of these elements in an as large sample of globular clusters as possible. An earlier analysis was presented by Gratton & Sneden (1991), who derived the abundances of α− and Fe-peak elements in about 30 globular cluster stars, and compared these abundances with those obtained from a sample of well

TABLE 7. Mean abundances elements in metal-poor field stars and globular clusters according to Gratton & Sneden (1991)

Element	Field			Clusters		
	Stars	Mean	σ	Stars	Mean	σ
Si	13	+0.30	0.08	30	+0.44	0.16
Ca	13	+0.29	0.06	30	+0.33	0.15
Sc	12	−0.03	0.09	29	−0.06	0.15
Ti	11	+0.28	0.11	30	+0.21	0.16
V	12	−0.03	0.11	27	−0.01	0.12
Cr	7	−0.04	0.05	25	−0.14	0.09
Mn	17	−0.31	0.07	29	−0.44	0.12
Co	11	−0.12	0.05	26	−0.02	0.12
Ni	12	−0.04	0.08	30	−0.02	0.09

studied field stars with overall metal abundance similar to that of globular clusters. The abundance pattern found by Gratton & Sneden for the two samples are quite similar (see Table 7), and typical for the imprinting of only type II SNe. Apparently, cluster stars tend on average to have larger Si excesses and Mn deficiencies with respect to field stars: while this result might indeed be real (indicating perhaps a somewhat different mass spectrum in the progenitors of the SNe which polluted the proto-cluster medium), Gratton & Sneden were more inclined to attribute this difference to systematic errors due to the not very high resolution ($R \sim 15,000$) and S/N (~ 100) of the spectra used in the cluster analysis.

Highest quality data ($R \sim 30,000$, $S/N > 100$) has cumulated in the next years, granting a reconsideration of the issue. A new review of results about the abundances of $\alpha-$elements in globular cluster stars was presented by Carney (1996). The elements considered were Si, Ca, and Ti; additionally, Carney also included O, but only from those stars that did not exhibit signatures that their atmospheric abundances have been altered by products of the ON cycle, as signaled by large Na overabundances (see next section). Carney concluded that the [α/Fe] ratio is constant amongst the clusters surveyed; this may be an indication that age differences are rather small amongst these clusters (although Carney warned about other possible interpretations): this result appears to be similar to that obtained by Harris et al. (1997) and Stetson et al. (1996) from examinations of the relative age indicators available in the colour-magnitude diagram.

However, Carney's conclusion merits to be reconsidered from two aspects:

(a) Brown et al. (1997) have recently presented a high dispersion analysis of the two *young* clusters Pal 12 and Rup 106; these clusters are known to be significantly (a few Gyr) younger than other well studied clusters of similar [Fe/H] from the small difference between the HB and TO magnitudes (Gratton & Ortolani 1988; Stetson et al. 1989; Buonanno et al. 1990). These clusters (as well as a few others, not as well studied) have space velocity and location in space consistent with their membership to the Magellanic stream, so that they are likely to be accreted objects. Due to the large distances, stars in these clusters are quite faint, so that the spectroscopic material is only of moderate S/N (~ 50): however, Brown et al. were able to show that $\alpha-$elements are not enhanced in these clusters (see Table 8): this abundance pattern is consistent with an appreciable contribution by type Ia SNe to nucleosynthesis in the material observed in these clusters.

TABLE 8. Overabundances of α-elements in the *young* globular clusters Pal 12 and Rup 106 according to Brown et al. (1997)

Element	Pal 12	Rup 106
[Fe/H]	−0.97	−1.48
[O/Fe]		+0.02
[Mg/Fe]	+0.10	+0.02
[Si/Fe]	−0.10	+0.04
[Ca/Fe]	0.00	−0.10
[Ti/Fe]	+0.05	−0.08

They seem to have a chemical history clearly different from those of most of the other globular clusters. It is useful to remember here that there are a few halo stars which also do not exhibit excess of α-elements (King 1997; Carney et al. 1997). It must be however be noticed that while these objects have large velocities with large apocentric distances (perhaps also a signature of accretion), the vast majority of halo stars with large apocentric distances have normal excesses of α-elements (Gratton 1997).

(*b*) recent extensive studies, like those from the Lick-Texas group, confirm that most globular clusters indeed share an excess of α-elements; however this excess change slightly from cluster-to-cluster. The most striking behavior is the apparent presence of an anticorrelation of Ca and Ti (see Table 9 and Figure 9): in particular, it is interesting to compare the pair of metal rich ([Fe/H]$\sim$ −0.7) clusters M71 (Ti-rich, Ca-poor) and 47 Tuc (Ti-poor, Ca-rich). This anticorrelation (that exceed by far the star-to-star scatter in each cluster) is not easily explained by analysis errors: effects of errors of 100 K in T_{eff}, 0.25 dex in $\log g$, 0.2 dex in [A/H], and 0.5 km/s in v_t are given by the arrows shown in Figure 9. Largest errors are expected to be due to uncertainties in temperatures (a large contribution here is from possible errors in the adopted values for the interstellar reddening): however, Ca and Ti have a similar dependence on T_{eff}. Very large systematic errors in v_t might explain the observed correlation, but they appear not likely. The different abundance of Ti in M71 and 47 Tuc might help to explain the presence of severeal stars with M-type spectra in M71, and only a few in the much more massive cluster 47 Tuc . It is also interesting to note that Ca and Ti appear not equally overabundant in bulge stars (McWilliam & Rich 1994). Figure 9 suggests that (i) Ca and Ti are produced in slightly different astrophysical sites (stars of different mass?) and (ii) that in spite of the overall constancy of the overabundance of [α/Fe], the ISM was not well mixed (over a galactic scale) at the epoch of formation of globular clusters.

Armowski et al. (1994) studied the abundances of neutron capture elements in globular cluster stars (mainly in connection to the observed star-to-star variations for the proton capture elements). They concluded that the abundances of the elements considered (Ce, Nd, Ba, and Eu) were virtually indistinguishable from those determined for field stars of similar metal abundance (Gratton & Sneden 1994). This behavior is clearly different from that obtained in the case of ω Cen (see next section).

3.3. *Abundances in dwarfs*

As mentioned in Section 2.4, it is still possible that abundances derived for giants systematically differ from those derived from dwarfs, due to inadequacies of current model atmospheres. Since the high dispersion analyses we have considered insofar consider only

TABLE 9. Mean abundances of Ca, Si, and Ti in selected globular clusters

Cluster	Stars	[Si/Fe]	[Ca/Fe]	[Ti/Fe]	[α/Fe]
		Lick-Texas group			
M3	10	0.28 ± 0.05	0.22 ± 0.04	0.30 ± 0.04	0.27
M5	13	0.20 ± 0.02	0.19 ± 0.01	0.30 ± 0.04	0.23
M10	14	0.29 ± 0.03	0.29 ± 0.02	0.21 ± 0.03	0.26
M13	23	0.34 ± 0.04	0.23 ± 0.02	0.28 ± 0.03	0.28
M71	10	0.31 ± 0.04	0.14 ± 0.03	0.48 ± 0.04	0.31
NGC7006	6	0.26 ± 0.02	0.23 ± 0.03	0.22 ± 0.04	0.24
		Carretta 1995			
47 Tuc	3		0.43 ± 0.03	0.24 ± 0.03	0.34
NGC6397	3		0.23 ± 0.05	0.30 ± 0.08	0.26
NGC6752	4		0.36 ± 0.03	0.14 ± 0.04	0.25

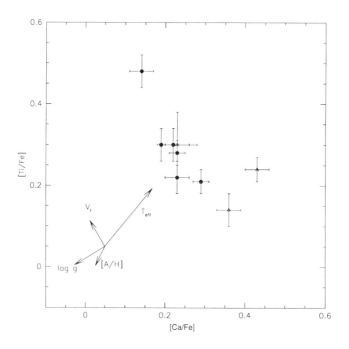

FIGURE 9. Comparison between mean [Ca/Fe] and [Ti/Fe] ratios in globular clusters observed by the Lick-Texas group, and by Carretta (1995). Each point represents a cluster. Arrows on bottom left gives the expected effects of errors in temperatures T_{eff} (+100 K), gravity $\log g$ (+0.25 dex), overall metal abundance [A/H] (+0.2 dex), and microturbulent velocity v_t (+0.5 km/s)

giants, such a difference may cause systematic errors when comparing globular clusters with e.g. local subdwarfs. As a test of the possible impact of such a systematic error on the derivation of basic data for globular clusters, Gratton (1997) considered changes in distance moduli and ages derived using the main sequence fitting method due to a systematic offset, the abundances for cluster giants being assumed to have been measured too high with respect to those obtained for subdwarfs. Gratton found that distance moduli and ages are very sensitive to such an offset ($\sim 0.4 \div 1$ mag/dex, and $\sim 6 \div 14$ Gyr/dex respectively), highest sensitivity being obtained at higher metal abundances. For typical uncertainties of ~ 0.1 dex (see Section 2.4) in the relative abundance scale for dwarfs and giants, the impact of such errors are amongst the largest source of errors in age derivation for globular clusters.

Direct determination of abundances from high dispersion spectra of dwarfs in globular clusters would be then of paramount importance. Unluckily, globular cluster dwarfs are faint. Two recent papers presented first attempts in this direction.

King et al. (1998) obtained high dispersion ($R \sim 45,000$), but quite low S/N (~ 30) spectra of three subgiants near the TO of the metal-poor cluster M92 using the HIRES spectrograph at Keck telescope. For comparison, they also observed the well known nearby subgiant HD140283, which has similar abundance and evolutionary status. The Fe abundance they obtained for this star ([Fe/H]$= -2.58$) is about 0.15 dex lower than that of other recent studies (Gratton et al. 1997), used to derive the subdwarf abundance scale; this difference is due to a lower adopted value for $T_{\rm eff}$. The metal abundance obtained by King et al. for M92 ([Fe/H]$= -2.52$) is 0.36 dex lower than obtained from studies of giants (Sneden et al. 1991; Carretta & Gratton 1997: [Fe/H]$= -2.16$). However, if the same offset found for HD140283 is considered, the Fe abundance for M92 from these subgiants would be [Fe/H]$=-2.37$, still lower than, but not totally inconsistent with the result obtained for giants. Additionally, King et al. obtained other interesting results:

(*a*) The [Ca/Fe]$=+0.33$ and [Ti/Fe]$=+0.15$ ratios agree very well with the pattern described in Figure 9

(*b*) Low Mg and high Na abundances are obtained

(*c*) while two of the subgiants surveyed share the Spites' plateau Li abundance, the Li abundance derived for the remaining one is substantially larger (this result is actually described in accompanying papers: Deliyannis et al. 1995; Boesgaard et al. 1998)

The last two points will be discussed again in the next Section.

Pasquini & Molaro (1996, 1997) exploited the presence of closer clusters (e.g. NGC6397 and 47 Tuc) in the southern emisphere to obtain high resolution ($R \sim 20,000$), albeit low S/N (~ 30) spectra of stars very near the TO using EMMI spectrograph at NTT. In spite of the moderate size of the telescope, Pasquini & Molaro were able to derive quite reliable Li abundances of $\log n({\rm Li}) = 2.37 \pm 0.08 \pm 0.07$, slightly larger than the value for the Li plateau for field stars derived using the same analysis procedure ($\log n({\rm Li}) = 2.19 \pm 0.016$). Pasquini & Molaro argues that these data suggests that no significant depletion affected these stars and it implies a mild Li galactic enrichment between the epoch of pop. II formation and the formation of 47 Tuc (at [Fe/H]$=-0.7$), in agreement with predictions of spallation scenarios for the additional production of the galactic Li.

While extremely important, these two papers must be considered as explorative. Important progresses are expected in the next few years with availability of the UVES (Dekker et al. 1992) spectrograph at VLT2. UVES will allow high resolution ($R \geq 20,000$), quite high S/N (> 50) observations of dwarfs in a quite large number of southern globular clusters.

4. Inhomogeneities in Globular Clusters

While globular clusters are usually assumed to be a simple (single age, single chemical composition) population, it is known since about 25 years that there are star-to-star variations in their surface composition. In at least one case (ω Cen) almost the abundance of all elements change from star-to-star; while in most clusters only few elements vary. In this lecture I review the large set of relevant observations, and discuss the main mechanisms proposed to explain them.

The very first discovery of significant star-to-star scatter in the composition of cluster stars is due to Osborne (1971), who detected variations in the strength of CN bands using DDO photometry. DDO photometry is an intermediate band photometric system, with four band passes in the blue part of the spectrum, particularly suitable to study CN and CH bands. In the next few years, a number of papers, based either on low dispersion spectroscopy or on DDO photometry, showed the existence of a complicate pattern of abundance inhomogeneities amongst globular cluster giants (for reviews, see Smith 1984b; Kraft 1994). It is now becoming increasingly clear that complexity is likely due to the superposition of a number of different effects: explanation of some of these effects seems to require modifications in more or less basic assumptions in models of stellar evolution, and has then a potential broad impact on a variety of astrophysical problems, ranging from the distance scale to globular clusters, to the age determination, and to the second parameter problem. Others provide insights into the mechanisms of metal-enrichment in dense environments. For these reasons, they have raised a considerable interest in the last few years.

Trying to put some order on the complex phenomenology, in the following we will consider separately three different issues (with the caveat that they may mix their effects).

4.1. *Mixing episodes in low-mass field stars*

Standard evolutionary models of small mass stars (Iben 1967) predict that surfaces abundances are modified by two episodes (dredge-up):

(*a*) the first dredge-up, occurs at base of the red giant branch, when the outer convective envelope penetrates inward in regions which have experienced some H-burning through the CN-cycle during the main sequence phase. First dredge-up causes a raise in the surface abundances of ^{13}C and ^{14}N, at the expenses of a depletion of ^{12}C. In metal-poor stars, the effect of first dredge-up is expected to be mild: the ^{12}C/^{13}C ratio is decreased to ~ 30, ^{12}C abundance decreases by $\sim 30\%$, while ^{14}N abundance raises by a factor of 2 (assuming that the original C/N ratio is a solar scaled one). The inward expansion of the outher convective envelope into regions which had experienced higher temperatures during the previous evolution also causes a dilution of the Li survived in the outer stellar regions: this is estimated to be a factor of ~ 25 for old, metal-poor stars

(*b*) the third dredge-up (Iben & Renzini 1983), occurs in the thermal-pulsing phase on the asymptotic-giant-branch: in this case products of He-burning reaction (in particular freshly produced ^{12}C) and of the $s-$process are brought up to the surface (the neutron source is in this case ^{13}C). Minimum luminosity of the thermal-pulses phase is quite large ($\log L/L_\odot \sim 3$: Marigo 1998), so that effects of third dredge-up may be observed almost only on post-AGB stars. Observation of these stars shows that large excesses of typical products of third dredge-up (CNO elements, $s-$process nuclei) are indeed observed in post-AGB stars for $\log L/L_\odot > 2.7$ (Gonzalez & Wallerstein 1994). It should be also noticed that most of CNO elements in these bright post-AGB stars appear as N and O; the primary products of the triple-α process are ^{12}C and ^{16}O. The CN-cycle must be responsible for enhancing the N. It probably converts most of the C and some of the O into N as the triple-α products pass through the H-burning shell

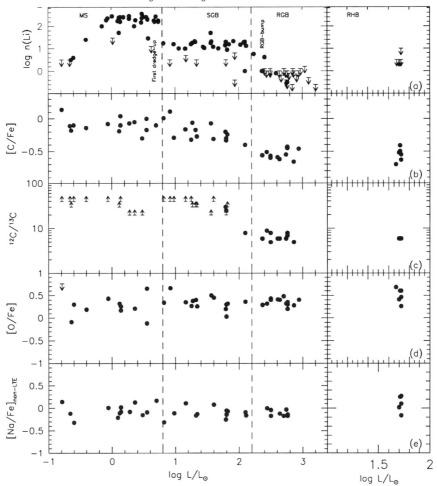

FIGURE 10. Run of the abundance of Li (panel a); of the overabundances of C (panel b), O (panel d), and Na (panel e) with respect to Fe; and of the $^{12}C/^{13}C$ isotopic ratio (panel c)as a function of luminosity in a sample of about 60 metal-poor ([Fe/H]< −1) field red giants (from Carretta et al. 1998)

Observation of field stars shows that this simple pattern is complicated by at least two factors:

• a more evolved companion (now a white dwarf) in a quite close binary system may have polluted the outer layers of a star (mass transfer mechanism): depending on the mass and degree of evolution of the companion, and on the evolutionary status of the presently observed star, this may produce a quite complicate pattern. Typical examples of such a pollution mechanism are the population I Ba stars (McClure 1980), the S stars without Tc (Van Eck et al. 1997), and the warm C-stars classified as R-stars (Dominy 1984). Pollution by once thermally-pulsing companions (which have primary production of ^{12}C and s-process elements) is much more effective in metal-poor stars, so that the effects are much more evident: various classes of stars with anomalous spectra are explained in this way (CH stars: McClure & Woodsworth 1990; Vanture 1992; C-dwarfs). Amongst field stars, pollution by a companion can be considered a good explanation for a class of stars if a high incidence of spectroscopic binaries with periods ≤ a few yrs is observed. For

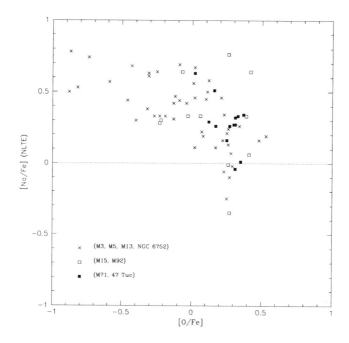

FIGURE 11. Na-O anticorrelation (from Carretta 1995)

these separations the present WD nearly filled the Roche lobe when it was on the AGB, so that accretion on the secondary might have been quite efficient. On the other side, such a binary system might have been disrupted by later dynamical interactions in the dense environments of globular cluster cores, so that this condition have to be released. It should be also noted that systems with these characteristics are not at all rare, and mild CH-stars or C-dwarfs are a sizeable fraction of all metal-poor stars

• observation of *bona fide* single metal-poor field stars reveals that a further mixing episode occurs for stars climbing up the RGB after the expanding H-burning shell reaches the chemical discontinuity left by deepest penetration of convection (RGB bump). Once this chemical discontinuity is canceled (Charbonnel 1995), mixing may occurr through e.g. rotationally driven meridional circulation (Sweigert & Mengel 1979, Zahn 1992). Observational proofs of the presence of this further mixing episode have been obtained by Pilachowski et al. (1993), and its relation with the RGB bump has been clearly showed by Carretta et al. (1998: see Figure 10). Observation for field stars indicates that this second mixing mechanism causes a further dredge-up of substantial amounts of material processed through the CN-processing: in a rather short time (a few 10^7 yr) the $^{12}C/^{13}C$ ratio reaches a value of ~ 6, close to (though actually slightly larger than) the equilibrium value. Also all Li remaining in the outer atmospheric layers is burnt. However, at least in field stars mixing do not reach regions warm enough for complete CNO cycle, and the O abundance remains constant: the same O abundance is also obtained for HB stars (both red HBs: Carretta et al. 1998; and RR Lyrae: Clementini et al. 1995)

4.2. *The Na/O anticorrelation and related issues*

While more noisy, results for some (metal-poor) globular clusters seem to conform to the relatively simple scheme outlined in the former subsection (see results for M92 by Langer et al. 1986; and NGC6397 by Bell & Dickens 1980). However, a more complicate pattern is given by other clusters, as evidenced by the extensive survey by the Lick-Texas group. The main result of this study was the presence of clear anticorrelations between abundances of pair of elements (Na and O, Al and Mg: see Figure 11) amongst red giants belonging to the same globular cluster, which else show to have the same chemical composition (in particular, they have the same abundance of Fe-group elements, and of the heavier α-elements Si, Ca, and Ti). Such anticorrelations are not seen in analogous samples of field stars. This result confirmed and extended results obtained previously using low dispersion spectra. While observations are still scarce (acquisition of high quality spectra for a significant sample of giants in several clusters requires quite a lot of precious large-telescope time), there seems to exist a link between the Na-O anticorrelation and the colour of the HB: for instance, the most extreme cases of Na-rich, O-poor stars are found in the blue horizontal branch M13, where Na-rich, O-poor stars are the vast majority of bright red giants, while very few such stars are found in M3.

The presence of such star-to-star abundance variations and anticorrelations may be explained by the contamination of the outer atmospheric layers of some globular cluster red giant by material which underwent complete CNO processing at temperatures of $T \sim 50\,10^6$ K (Langer & Hoffman 1995; Cavallo et al. 1996), with conversion of most O into N. At these high temperatures, the proton-capture reaction ^{22}Ne$+p \to ^{23}$Na may produce substantial amounts of Na (Denissenkov & Denissenkova 1990). For still slightly higher temperatures of $T \sim 65\,10^6$ K, proton capture on Mg nuclei may also produce Al. Note that abundance changes as large as those observed require that (at least in the most extreme cases) a large fraction of the outer envelope of these red giants was processed through complete CNO cycle: this on turn implies conversion of a significant fraction of H into He ($\geq 10\%$). While on the RGB, this large He abundance has not a clear observational impact. On the other side, HB stars resulting from the evolution of stars with He-enriched envelopes would be much bluer than normal HB stars: this difference might perhaps be so large to explain the difference between the HB of clusters like M13 and M3 (Sweigart 1997).

While it is quite clear now that the outer layers of some globular cluster red giants (but not field stars) have been contaminated by ON-cycle processed material, the origin of this contaminating material is still unclear; there are two possibilities:

(a) the ISM from which cluster stars formed was not well mixed (**primordial origin**). A variant consider pollution on already formed stars, possibly in temporary binaries (D'Antona et al. 1983). In this scenario, the site of ON-cycle are the interiors of massive stars, that may well reach high enough temperatures: for various reasons the most appealing candidates are stars undergoing hot bottom burning (Denissenkov et al. 1998). Within this scheme, the Na-O anticorrelation must be present also amongst less evolved cluster stars, including MS stars. While variations of the strength of CN and CH bands have been observed for stars on the MS of 47 Tuc (Bell et al. 1983; Briley et al. 1991; Cannon et al. 1998) and NGC6752 (Suntzeff & Smith 1991), it is not clear if the Na-O anticorrelation is also present (see however below). Within the primordial scheme, star-to-star variation of the He content should also be present for MS and TO stars: this causes systematic variations in the location of the main sequence; possible evidences in favour of such a possibility have been recently considered by Johnson & Bolte (1998)

(b) mixing is much more extended in cluster stars than in field ones (**deep mixing scenario**); most author favour mixing on the upper red giant branch as the most promising candidate (temperatures are too low during the MS evolution). Even so, it is not clear how mixing in red giants may extend to temperatures as high as those required to explain the Al-Mg anticorrelation. In order to explain this occurrence, low energy resonances (Langer & Hoffman 1996) and/or large differences between mono-dimensional models and real physical description of turbulent motions (Arnett 1998) are required. Deep mixing scenario (i) predicts that no deep mixing has occurred for stars fainter than the RGB bump (the star-to-star variations seen amongst 47 Tuc MS stars require then a different scenario); and (ii) requires a tuning mechanism, to explain the systematic difference observed between different clusters, and from cluster to field stars. This tuning mechanism might be rotation: this sounds feasible because meridional circulation driven by core rotation might indeed cause mixing, and because Peterson (1983b) and Peterson et al. 1995) measured high rotational velocities in some M13 BHB stars, while only upper limits were obtained for BHB stars in M3. It is not clear why stars in M13 should rotate faster than those in M3; however it is interesting to note that a spin-orbit coupling was proposed by Fusi Pecci et al. (1993) to explain different HB morphology in clusters of similar metal abundance. Fast core rotation might help to explain bluer HBs because the additional contribution to pressure allows larger core masses, the longer lifetime at bright luminosities on the RGB allowing a larger mass loss

A clarification of this point should come from deep, extensive high dispersion studies; these should be better done using multi-object fiber spectrographs, which allow simultaneous observation of a large number of giants in a single cluster. Unfortunately, such instruments are not efficient enough on 4m class telescopes, as shown by consideration of the results presented by Pilachowski et al. (1997). These authors observed Na and Mg lines in 130 red giants ($M_V > +1.2$) in M13 with Hydra at the 4m KPNO telescope: resolution was quite modest ($R \sim 11,000$); however comparison with synthetic spectra allowed reliable abundances, at least for the brightest stars (S/N was low for the fainter ones, so that in several cases no abundance was given). Pilachowski et al. found a rather large spread in Na abundances ($-0.3 <$[Na/Fe]< 0.5) at all luminosities, but the incidence of Na-rich stars was higher amongst the brightest stars ($M_V < -1.7$). They considered this as an evidence in favour of a deep mixing explanation for the Na-O anticorrelation. However, it is not entirely clear that this result implies that the Na-O anticorrelation is due to deep mixing for two reasons: (i) the lines used (the yellow doublet at 5682-88 Å; the doublet at 6154-60 Å usually adopted for stars at the tip of the RGB being too weak to be measured in fainter, warmer giants) have moderate non-LTE effects which cause lines to be stronger than predicted in LTE, and that are a function of line strength and luminosity; Na abundances are then expected to be overestimated in the brightest stars; (ii) the less luminous stars observed by Pilachowski et al. are somewhat fainter than the bump (which is at $M_V \sim +0.4$, $\log g \sim 2.3$): deep mixing is not expected in these stars, but the observed incidence of Na-rich stars seems similar to that found for more luminous stars. However, since there may be selection effects causing the Na abundance to be overestimated in these stars (only features enhanced by noise might be detected in low resolution, low S/N spectra), the results obtained by Pilachowski et al. cannot be considered as a conclusive evidence in favour of a primordial origin either.

A much stronger argument favouring a non *in situ* origin of the ON-cycle material is given by the observation of M92 subgiants by King et al. described in the previous Section. These authors found a very low [Mg/Fe] ratio of [Mg/Fe]-0.15, and a large [Na/Fe] ratio ([Na/Fe]$=+0.60$) in these stars, likely indicative of pollution by ON-cycle processed material (it should be noted here that the few M92 giants surveyed insofar did not show extensive signatures of such an O-Na anticorrelation: Sneden et al. 1991). Additionally, rather large Li abundances have been determined for all three stars (Deliyannis et al. 1995; Boesgaard et al. 1998). While the implication of these results for Big Bang nucleosynthesis is still argument of debate (see e.g. Pasquini & Molaro 1997), the simultaneous presence of large [Na/Fe], low [Mg/Fe], and surviving Li argues against an *in-situ* explanation for the O-Na anticorrelation because Li is easily destroyed at the high temperatures required by the proton capture mechanism. It is also interesting to note that the *pollution* scenario proposed by D'Antona et al. predicts a much higher incidence of Na-rich stars amongst unevolved stars, since they have outer convective layers thinner than those of giants.

Summarizing, both scenarios called for to explain the Na-O and Al-Mg anticorrelation have pro's and con's. Non-standard deep mixing is certainly present, but it is not clear if it is deep enough to modify O and Na abundances. Theoretical arguments and current observation seems to favour a *pollution* scenario; however, high S/N, high resolution observations of a significatively large sample of stars fainter the RGB bump might provide a clear answer to this issue.

4.3. *Inhomogeneities in heavy elements: the cases of ω Cen and M92*

Two particularly interesting cases of star-to-star variation in abundances are those of ω Cen and M92.

ω Cen is the most luminous and likely most massive galactic globular cluster (7×10^6 $M_\odot$: Richer et al. 1991). Cannon & Stobie (1973) first noticed its unusually broad RGB, and suggested that it may be due to a spread in the abundances of heavy elements. This was confirmed by Freeman & Rodgers (1975), who measured the strength of the Ca II K-line in a large number of RR Lyraes, and found it could be interpreted by a large ($\sim$ 2 dex spread) in the abundances of heavy elements. An IR photometric study by Persson et al. (1980) also revealed the spread of abundances: furthermore, Persson et al. found that CO strengths do not correlate with $V - K$ colors, so that independent variations of CNO and heavy elements were required.

The metallicity distribution of the stars in ω Cen was studied by Butler et al. (1978), and more recently and accurately by Suntzeff & Kraft (1996), who considered a sample of over 300 radial velocity members to the cluster; metallicities were measured using the IR Ca II triplet (see next Section). After correction for the (known) biases, they found no very metal-poor stars; a sudden rise in the metal-poor distribution to a modal [Fe/H] value of -1.70, consistent with an homogeneous, unresolved metallicity; a tail to higher metallicities; and a weak correlation between metallicity and radius such that the most metal-rich stars are concentrated to the cluster core. However, this last result might be due to dynamical evolution of the cluster.

Suntzeff & Kraft compared this metallicity distribution with predictions of a simple chemical evolution model (eq 1.13), assuming that the cluster formed from ISM with an initial abundance of $\alpha_0 = 0.0003$. The observed peak of the distribution requires very low yields; low effective yields can be obtained by introducing gas loss, based on the idea that supernovae should be related to star formation (Hartwick 1976). In this case the yield is replaced by an effective yield which is a factor $(1 + \Lambda)$ smaller than the actual yield, where Λ is the ratio of mass loss to star formation. In order to reproduce observations we have $\Lambda > 10$ for ω Cen, implying almost the whole cluster mass is lost. A cluster may survive such a severe mass loss only if this is slow-enough that the cluster can adjust adiabatically, i.e. it occurs on the order of a few crossing time (Smith 1984a). In this case the cluster will bloat out such that the cluster radius is inversely proprtional to the mass. This may be the case for a dwarf spheroidal, but it is difficult for a globular cluster (Gunn 1980). However Smith pointed out that the cluster might survive if roughly half of the cluster formed stars before chemical enrichment happened: in this case the remaining gas could self-enrich and ultimately be swept out of the cluster by supernovae, without having the cluster disrupted. A significant self-enrichment may then be related to the low central contration of ω Cen (Trager et al. 1995).

While overall metal distribution may well be studied at low dispersion, element-to-element ratios require high dispersion studies. The first high dispersion study of ω Cen was presented by Cohen (1981), followed by others (Mallia & Pagel 1981; Gratton 1982b). These results broadly confirmed the existence of a spread in metal abundances; the observed range was smaller than found by Freeman & Rodgers ($\sim 0.5 \div 1$ dex), but this could be due to the small size of the samples and biases in star selection. More recent studies used CCD detectors and allowed much more reliable determination of abundances: first such studies were done by François et al. (1988), Paltoglou & Norris (1989), Milone et al. (1992), and Brown & Wallerstein (1993b); a more extensive survey (40 stars) has been performed by Norris & Da Costa (1995).

As mentioned above, the observed large spread in heavy elements is attributed to the ability of this clusters to retain (a small) part of the ejecta of type II Sne. Given the large mass of ω Cen, we expect that it should have been produced $10^2 \div 10^3$ individual SN events. This value is so large that it is not easy to disentangle the contribution due to each of these events. The abundance pattern emerging has been summarized by Norris & DaCosta:

(*a*) the metal abundances range from $-1.8 <$[Fe/H]< -0.8, in agreement with the results by Suntzeff & Kraft

(*b*) the ratios of abundances of $\alpha-$elements (Mg, Si, Ca, and Ti) with respect to Fe is constant with [Fe/H]

(*c*) the bulk of the stars have $-0.9 <$[C/Fe]< -0.3 and [O/Fe]~ 0.3, typical of bright metal-poor giants. However, there exist a group of (CN strong) stars having [C/Fe]~ -0.7 and [O/Fe]~ -0.5; the [O/Fe] ratios are anticorrelated with [Na/Fe] and [Al/Fe]

(*d*) the abundance of heavy neutron-capture elements Y, Ba, La, and Nd rises as [Fe/H] increases, in sharp contrast with what is found in other clusters; the element-to-element abundance pattern is suggestive of $s-$processing

This abundance pattern suggests that different nucleosynthesis mechanisms are acting. The general overabundance of heavy elements agrees with expectations that the enrichment of Fe-group elements is due to contribution from type II SNe alone. This is not a surprise, since it is expected that when type Ia SNe explode, much of the ISM within the cluster has been already dispersed, so that there is no efficient breaking mechanism to keep the ejecta tied to the cluster. The observed pattern of abundances for CNO elements, as well as Na and Al, while complicate, repeat that observed in other clusters (it appears intermediate between M13 and M3, in reasonable agreement with the HB morphology). The most intriguing feature is the large overabundance of the heavier $n-$capture elements, suggestive of $s-$processing. As mentioned in Section 1, $s-$process is expected to be active in intermediate mass stars ($1 < M < 3\ M_\odot$), with lifetimes of ≥ 1 Gyr: such a long star formation phase appears surprising for a globular cluster, and need to be confirmed.

Other clusters have been proposed to have star-to-star variation in their heavy element abundances: the case for M22 has been made by Hesser et al. (1983; see also Norris & Da Costa 1995 and Anthony-Twarog et al. 1995). However, it is possible that this result is contaminated by differential reddening within the cluster (for a quite accurate, recent determination see Anthony-Twarog et al. 1995). Very recently, Langer et al. (1998) have done a careful differential study of three giants in M92, which show the existence of a small difference between the iron abundances the stars (at a level of 0.1 dex). Since O abundances are nearly the same for all the stars, the implication is that the proto-cluster ISM was not well mixed on scales of some pc. Comparison with current star forming regions, like the Orion nebula (Cunha & Lambert 1994), shows that such inhomogeneities are rather common: as a matter of fact, we think it is the homogeneity of the early ISM, more than dishomogeneity, that needs to be explained.

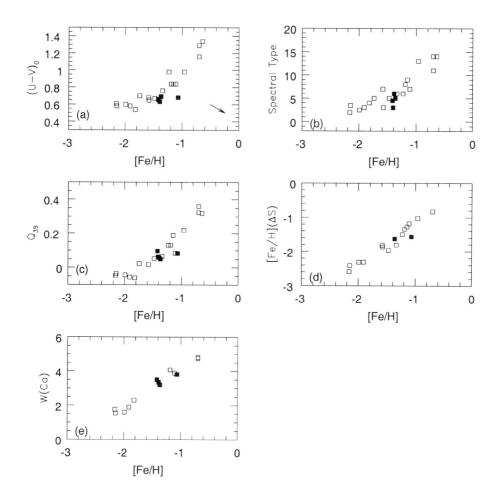

FIGURE 12. Calibration of abundance indices. Individual points for M13-like clusters (Dickens HB-type 0) are marked with filled squares, while those for the other clusters are marked with open symbols. The arrow on bottom right of panel (a) shows the effect of increasing reddening by 0.1 mag

5. Abundance Indices for Globular Clusters

In this final lecture, photometric and low-dispersion indices used for the determination of abundances in globular clusters are critically reviewed. Advantages and drawbacks of the various indices are discussed. Updated calibration based on the latest high dispersion results are presented.

Indices used to measure the abundances of globular clusters divide into different groups. They will be considered separately in the next subsections. Calibrations of all these indices will be given against the high dispersion metal abundances by Carretta & Gratton (1997). Calibration of some indices based on the colour-magnitude diagram ($(B - V)_{0,g}$, $(V - I)_{0,g}$, and $\Delta V_{1.2}$) are given by Carretta & Bragaglia (1998)

5.1. *Integrated indices*

Integrated indices are appealing, because they allow to measure the abundances in very far and faint clusters: they are the only method suitable for determination of abundances in globular clusters in galaxies out of the Local Group. However, the interpretation of data is made difficult by three facts:

• since the whole cluster population contributes to the observed features, interpretation of data is simply and direct only if we may assume that the only parameter changing from cluster-to-cluster is metal abundance. However, age may also be an important parameter, as well as morphology of the HB

• in any case, a calibration is required: this may be done either using galactic globular clusters (which however span only a limited range of parameters), or using synthesis of population (in this case, it is important to check the used methods against suitable calibrators)

• background light may contribute significantly: subtraction may be very difficult for e.g. bulge clusters

Finally, the weights of different elements influencing integrated parameters is not at all easy to be determined. The strongest spectral features in visible spectra are due to H and He in warm stars; C, Ca, Mg, and Fe in solar type stars and red giants (dominating the spectra of old stellar populations); and to TiO and C molecules in very cool stars. However, since the strength of the spectral features also depends on temperature (which is affected by overall metallicity) and continuum opacity (where the low temperature electron donors must be considered), it is not easy to tell what exactly an integrated parameter is measuring.

Integrated photometry of clusters is a fast and sensitive way to derive information about their metal abundance. Abundance sensitivity is due to the combination of a warmer red giant branch, and of a reduced line blanketing in more metal-poor clusters. Reed et al. (1988) have derived integrated $(U - V)$ colors for a number of galactic globular clusters. Photometric data must be corrected for interstellar reddening; panel a of Figure 12 shows the $(U - V)_0$ colors of Reed et al. (1988) against the present metal abundances. The correlation is quite good, although clearly not a linear one (sensitivity of $(U - V)_0$ decreases with decreasing metal abundance); if two quite discrepant clusters are eliminated (NGC288 and NGC3201) the best fit regression parabola through data is:

$$(U - V)_0 = 0.45[\text{Fe/H}]^2 + 1.72[\text{Fe/H}] + 2.22 \qquad (5.17)$$

the r.m.s. scatter along this line is 0.046 mag, corresponding to only ~ 0.09 dex at $[\text{Fe/H}] = -1.35$. While this scatter appears to be small, it is still not dominated by observational errors, since there is clear segregation effect due to the second parameter (see Figure 12a, where the M13-like clusters having Dickens HB type of 0 are marked with a different symbol). Usefulnes of integrated photometry for external galaxies is limited to explorative studies, because contamination by background galaxies cannot be checked, ages affects colours more than metal abundance, and reddening within the galaxy is not known.

The oldest parameter used to measure metal abundance of globular clusters is their integrated spectral type (Morgan 1959). A modern version of this method has been presented by Hesser & Shawl (1986). The basic reason why integrated spectral types may be used to measure abundances of globular clusters is that with declining metal abundance, red giant branches (which dominate integrated light from the clusters) become warmer; this combines with the generally lower metal abundance to yield weaker lines, which are interpreted as an earlier spectral type. Typically, metal-rich clusters are classified as G-type, and metal-poor ones are classified as F-type. The overall correlation with metal abundance [Fe/H] is quite good (see panel b of Figure 12); however, spectral types may be misclassified if there is a strong background contamination. The calibration of spectral types against [Fe/H] by Carretta & Gratton is:

$$[Fe/H] = 1.06S - 2.09 \tag{5.18}$$

where a $S = 0$ is assigned to F0, and $S = 10$ to G0 spectral types. The r.m.s. scatter along this mean line is 0.21 dex. Note that a very blue HB or a younger age may mimic a lower metal abundance. In fact, metal abundances derived from spectral type for M13-like clusters of Dickens HB type of 0 are about 0.15 dex too low.

For galactic globular clusters a quite accurate abundance indicator is the Q_{39} spectrophotometric parameter by Zinn (1980; Zinn & West 1984). Q_{39} is a measure of the strength of the Ca II H and K lines in the integrated spectra of clusters. The correlation of this parameter with metal abundances [Fe/H] derived from high dispersion spectroscopy is good (see panel c of Figure 12), although clearly not linear: the parameter changes very rapidly with metal abundance at large values of [Fe/H]. The physical meaning of Q_{39} is not very different from spectral types: ambiguous result would be obtained for a population where parameters other than metal abundances are variable from object-to-object. However it is better defined, and the authors who measured it were very careful in the consideration of background contamination. Unfortunately, as all other low dispersion indices, its definition is dependent on the instrumental apparatus used, so that it is difficult to repeat these observations; furthermore, flux at the short wavelengths of Ca II H and K lines is not large. For this reason, there are no measure of Q_{39} other than those of Zinn and coworkers; however, similar spectral indices (like the Mg b index) have been broadly used in extragalactic astronomy. The calibration of Q_{39} against the [Fe/H] of Carretta & Gratton is:

$$[Fe/H] = 5.10 \, Q_{39}^2 - 6.26 \, Q_{39} - 1.71 \tag{5.19}$$

The r.m.s. scatter around this mean line is 0.14 dex.

5.2. *Colour-magnitude diagram indices*

Detailed studies of the colour-magnitude diagram have allowed the definition of a number of metallicity sensitive indices. Most of them are based on the sensitivity of the colour of the red giant branch to metal abundance, which ultimate bases are the sensitivity to opacity of the equations of energy transfer (metal-poor stars are warmer than metal-rich ones), and the larger blanketing effect in metal-rich atmospheres (which redistributes radiation from short to long wavelengths). Hence these parameters measure an overall metal abundance: however, weights may change as a function of temperature and overall metal abundance. For instance, while weights for Ti and O are small in warm red giants, strong TiO bands in the spectra of the brightest, most metal-rich giants in globular clusters cause a strong absorption at blue and visual wavelengths, deeply affecting the c-m diagram (Ortolani et al. 1991). Further complications are due to our still poor knowledge of convection, that prevents a theoretical calibration of these indices.

Various parameters have been considered to measure the colour of the RGB.

(*a*) The colour of the red giant branch at the horizontal branch $(B-V)_{0,g}$ (Sandage & Smith 1966; Sandage 1982; Sarajedini & Layden 1997) is the most frequently used (along with their equivalent $(V-I)_{0,g}$: Sarajedini 1994). The reason is that this parameter is easily measurable, at least in most favourable cases. However, it is strongly sensitive to the assumed reddening and to uncertainties in the photometric calibrations. Furthermore, the magnitude of the HB depends on its colour, and its location is uncertain for clusters with very blue HBs. At large metallicity, some confusion is also possible between first ascending RGB stars and red HB stars. Finally, differential interstellar reddening may greatly complicate data interpretation. The calibration given by Carretta & Bra-

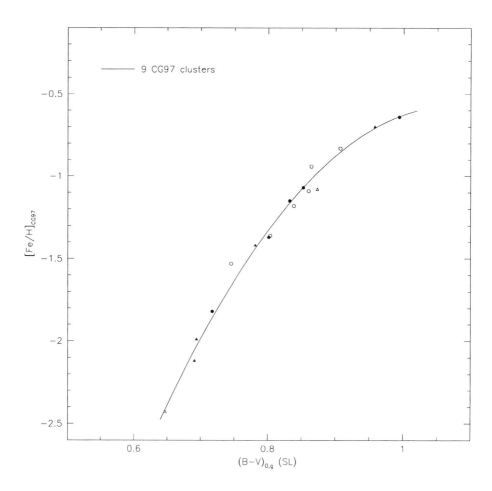

FIGURE 13. Calibration of $(B-V)_{0,g}$ according to Carretta & Bragaglia (1998) for the Sarajedini & Layden (1997: SL) clusters; the calibration is based on the [Fe/H] values of Carretta & Gratton (1997)

gaglia (1998: see Figure 13) using all the 17 clusters considered by Sarajedini & Layden (1997) is:

$$[\text{Fe/H}] = -9.253 \, (B-V)_{0,g}^2 + 20.129 \, (B-V)_{0,g} - 11.532, \qquad (5.20)$$

with a very low r.m.s. scatter of $\sigma = 0.059$ dex (fully compatible with internal errors in the metallicity calibration alone). The analogous calibration given by Carretta & Bragaglia for the $(V-I)_{0,g}$ index (Sarajedini 1994; see Figure 14) is:

$$[\text{Fe/H}] = 42.981 \, (V-I)_{0,g}^2 - 17.772 \, (V-I)_{0,g} - 26.122, \qquad (5.21)$$

which also gives an r.m.s. scatter of 0.06 dex.

(*b*) Sensitivity to reddening may be eliminated (or at least reduced) by measuring the slope of the red giant branch. However, measure of the slope is more difficult than a single measure of the colour, and it is affected by uncertainties in the colour-trasformations. Since the slope is not constant, distances should be known; this is done by reference to the HB: however, definition is difficult for clusters having blue HBs. At the brightest edge,

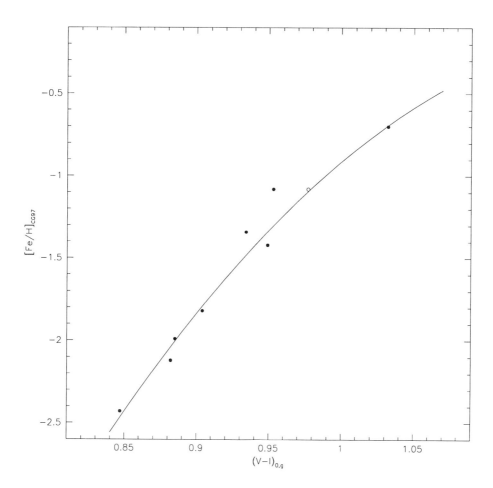

FIGURE 14. Calibration of $(V - I)_{0,g}$ according to Carretta & Bragaglia (1998) for the Sarajedini (1994) clusters; the calibration is based on the [Fe/H] values of Carretta & Gratton (1997)

confusion between first ascent RGB stars and AGB stars is possible, while at the lower edge disentangling of RHB from RGB is required. Various parameters have been devised to measure the slope of the RGB: the classical one is $\Delta V_{1.4}$ parameter that measures the difference in V magnitude between the HB and the level of the RGB at the de-reddened colour $(B - V)_0 = 1.4$ (Sandage & Wallerstein 1960). We consider the index $\Delta V_{1.2}$ of Sarajedini & Layden (1997). The calibration by Carretta & Bragaglia (see Figure 15) is:

$$[\text{Fe/H}] = -0.245 \, \Delta V_{1.2}^2 + 0.236 \, \Delta V_{1.2} - 0.627, \qquad (5.22)$$

with an r.m.s. scatter of 0.07 dex. Given the very different sensitivity to reddening of $(B - V)_{0,g}$ and $\Delta V_{1.2}$, both reddening and metallicity may be simultaneously obtained if both indices are known (Sarajedini 1994).

Other features of the c-m diagram may also be used to measure metal abundances. Recent theoretical works (Cassisi & Salaris 1997; Cassisi et al. 1997) have shown that the difference in magnitude between HB and RGB bump is expected to be very sensitive to metal abundance. There is also some sensitivity to mass (and hence age), but this

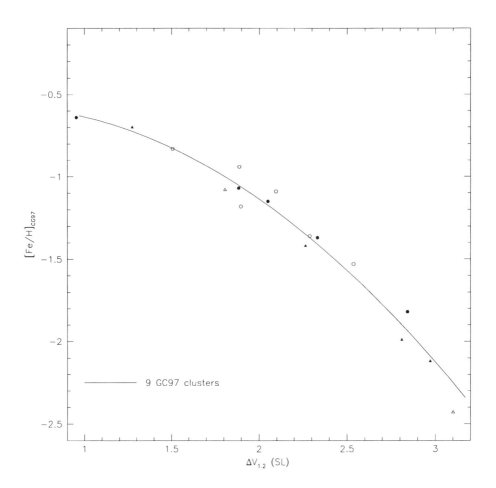

FIGURE 15. Calibration of $\Delta V_{1.2}$ according to Carretta & Bragaglia (1998) for the Sarajedini & Layden (1997: SL) clusters; the calibration is based on the [Fe/H] values of Carretta & Gratton (1997)

should have only a minor effect amongst globular clusters. However, the location of the RGB bump depends on details of convection, so that an accurate theoretical calibration is not possible at present. Extensive c-m diagrams are required to clearly show the presence of the RGB bump; however excellent observational material have been acquired in recent years so that adequate data are now available for several clusters. An empirical calibration against [Fe/H] values from Carretta & Gratton (1997) has been presented by Desidera (1998: see Figure 13), who considered separately clusters with RR Lyrae, and those with red HBs, since the definition of the HB magnitude is different in the two cases. The correlation is on the whole very good, although some uncertainty exists for clusters having very blue HBs.

5.3. *Intermediate band photometry and low-dispersion spectroscopy indices*

Intermediate band photometry of individual stars has been widely used to derive metal abundances. For field stars, accurate metal abundances have been obtained using Strömgren photometry, which uses four band passes in the UV, blue, green and yellow part of the

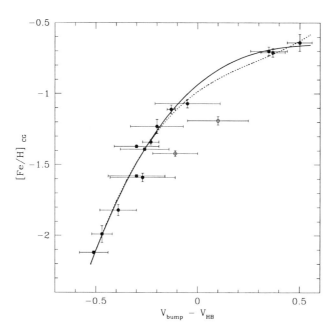

FIGURE 16. Calibration of the difference between the magnitude of the horizontal branch and that of the bump on the RGB against high dispersion metal abundance (from Desidera 1998)

spectra (u, v, b, and y; note that u band is entirely shortward of the Balmer limit); often additionally narrow band photometry with a filter centered at H_β is also added. The system was originally devised to study F-stars: it allows the definition of a temperature index $(b - y)$, similar to $B - V$, but with a much smaller dependence on metallicity and surface gravity; of a gravity index $c_1 = (u - v) - (v - b)$ which is a measure of the Balmer jump; and of a metallicity index $m_1 = (v - b) - (b - y)$. The strength of H_β is a (not monotonic, with a maximum for AO stars) temperature index: comparison with $b - y$ allows accurate determination of reddening. Extensive photometry for bright stars have allowed accurate calibrations of the Strömgren indices for both population I (Crawford 1975, 1978, 1979) and metal-poor (Schuster & Nissen 1989) dwarfs. However, quite un-expectedly Strömgren photometry revealed very useful also in the analysis of metal-poor giants (Bond 1980; Twarog & Anthony-Twarog 1994). Unluckily, until recently the low efficiency of CCD's at short wavelength have prevented accurate Strömgren photometry for globular clusters (since large and uncertain colour terms were required, and exposure times in u were prohibitively large): this situation is now changing thanks to the use of thinned CCD's with efficient UV coatings, and useful data are now being obtained (see Anthony-Twarog et al. 1995). CCD Strömgren photometry in cluster stars have two potential advantages: (i) accurate abundances can be obtained for dwarf stars, directly comparable to those obtained for nearby subdwarfs with accurate parallaxes; and (ii) gravity estimates may provide masses (this might be useful e.g. to find the evolutionary progenies of blue stragglers on the HB).

A very useful abundance indicators for clusters with RR Lyrae is the ΔS parameter (Preston 1959), which measures the difference between the spectral type for an RR Lyrae as measured from the strength of H and metal lines (practically, the Ca II K line). ΔS changes somewhat along the cycle; the value is then assumed to be at minimum phase (this is selected because colours of RR Lyrae at minimum are nearly independent of the period for both $ab-$type and $c-$type RR Lyrae). ΔS is also slightly affected by interstellar absorption lines, blended to the stellar line. The physical base of ΔS is that it measures the absorption in the wings of the collisional damping broadened K line in stars which have similar temperatures; the small temperature dependence (temperature at minimum light decreases with increasing metal abundance) is taken into account by examining the strength of the H lines. The ΔS parameter may be calibrated using field RR Lyrae; a recent calibration, revising that traditionally used (Butler 1975), has been presented by Clementini et al. (1995; we use eq. 6 of Clementini et al., because this is obtained without using globular cluster stars). Panel d of Figure 12 compares the abundances provided by ΔS (from Costar & Smith, 1988, calibrated using eq. 6 of Clementini et al.) with those given by high dispersion results. The comparison is quite good, although the slope of the best fit line is somewhat different from unity:

$$[\text{Fe}/\text{H}](\Delta S) = (1.16 \pm 0.08) \, [\text{Fe}/\text{H}](CG) - (0.05 \pm 0.14) \qquad (5.23)$$

Alternatively, the Carretta & Gratton [Fe/H] can be used to recalibrate the ΔS index. The calibration we obtain is:

$$[\text{Fe}/\text{H}] = -(0.176 \pm 0.012) \, \Delta S - (0.03 \pm 0.12) \qquad (5.24)$$

which is only marginally consistent with the calibration by Clementini et al..

    Finally, quite accurate estimates of metal abundances in clusters can be obtained from measures of the strength of the Ca II triplet in the near IR on low dispersion ($R \sim 3000$), moderate S/N ($\sim 20 \div 30$) spectra of red giants. Extensive data using this technique have been obtained in the early '90s (Olszewski et al. 1991; Armandroff & Da Costa 1991; Armandroff et al. 1992; Da Costa et al. 1992; Suntzeff et al. 1992, 1993). This method has several advantages: (i) spectra of good enough quality can now be obtained for even the furthest galactic globular clusters and dwarf spheroidal; and (ii) the physical mechanism is quite well understood. In fact, since Ca is mostly ionized in the atmosphere of red giants, these lines are sensitive to gravity and metal abundance, but only marginally to temperature. Practically, the gravity dependence can be replaced by a luminosity dependence, since luminosities are generally quite well known for globular clusters (uncertainties in the distance moduli, while very important for age derivations, are small enough not to have a broad impact in this context). This method is then nearly independent of uncertainties in reddening, and it may be applied also to sparse and heavily contaminated clusters (for which integrated spectra and photometry cannot be obtained), provided membership is proven (e.g. from radial velocities, that might also be obtained using the same spectroscopic material). The main drawbacks of this technique are that it requires to be calibrated, and that it looses sensitivity at high metal abundances. Panel e of Figure 12 compares these results (as compiled by Suntzeff et al. 1993) with those provided by high dispersion spectroscopy. The best fit regression line through data is:

$$[\text{Fe}/\text{H}] = (0.426 \pm 0.022)W_{Ca} - (2.79 \pm 0.09) \qquad (5.25)$$

where the error on the constant term represent the r.m.s. scatter of data for individual clusters along the mean line. Note that this scatter is much larger than expected from uncertainties in W_{Ca}, so that it flags errors in the analysis procedure (in particular the adopted values for interstellar reddening in the abundance analysis, and the assumption that [Ca/Fe] is constant).

    It is a pleasure to thank Paola Mazzei, Eugenio Carretta, Riccardo Claudi, and Silvano Desidera for critical reading of the text and help in its preparation

REFERENCES

ALONSO, A., ARRIBAS, S., & MARTINEZ-ROGER, C. 1996, *A&AS*, **117**, 227

ANSTEE, S.D., & O'MARA, B.J. 1991, *MNRAS*, **235**, 549

ANSTEE, S.D., & O'MARA, B.J. 1995, *MNRAS*, **276**, 859

ANSTEE, S.D., O'MARA, B.J., & ROSS, J.E. 1997, *MNRAS*, **284**, 202

ANTHONY-TWAROG, B.J., TWAROG, B.A., & CRAIG, J. 1995, *PASP*, **107**, 32

ARMANDROFF, T.E., & DA COSTA, G.S. 1991, *AJ*, **101**, 1329

ARMANDROFF, T.E., DA COSTA, G.S., & ZINN, R. 1992, *AJ*, **104**, 164

ARMOSKI, B.J., SNEDEN, C., LANGER, G.E., & KRAFT, R.P. 1994, *AJ*, **108**, 1364

ARNETT, W.D. 1973, *ARA&A*, **11**, 73

ARNETT, W.D. 1989, *ApJ*, **343**, 834

ARNETT, W.D. 1995, *ARA&A*, **33**, 115

ARNETT, W.D. 1998, in *Galaxy Evolution: Connecting the distant Universe with the Local Fossil Record*, eds. M. Spite & F. Spite, Kluwer, Dordrecht, in press

BARAFFE, I., & ALLARD, F. 1998, *Fundamental Stellar Properties: the Interaction between Observation and Theory*, IAU Symp. 189, ed. T.R. Bedding, A.J. Booth, J. Davis (Kluwer: Dordrecht), p. 227

BARBUY, B. 1998, in *Galaxy Evolution: Connecting the distant Universe with the Local Fossil Record*, eds. M. Spite & F. Spite, Kluwer, Dordrecht, in press

BELL, R.A., & DICKENS, R.J. 1980, *ApJ*, **242**, 657

BELL, R.A., & GUSTAFSSON, B. 1978, *A&AS*, **34**, 29

BELL, R.A., HESSER, J.E., & CANNON, R.D. 1983, *ApJ*, **269**, 580

BELL, R.A., EDVARDSSON, B., & GUSTAFSSON, B. 1985, *MNRAS*, **212**, 497

BESSELL, M.S., CASTELLI, F., & PLEZ, B. 1998, *A&A*, **333**, 231

BLACKWELL, D.E., & SHALLIS, M.J. 1977, *MNRAS*, **180**, 177

BLACKWELL, D.E., & WILLIS, R.B. 1977, *MNRAS*, **180**, 169

BLACKWELL, D.E., & LYNAS-GRAY, A.E. 1994, *A&A*, **282**, 899

BLACKWELL, D.E., LYNAS-GRAY, A.E., & SMITH G. 1995, *A&A*, **296**, 217

BOESGAARD, A.M., & KING, J.R. 1993, *AJ*, **106**, 2309

BOESGAARD, A.M., DELIYANNIS, C.P., STEPHENS, A., & KING, J.R. 1998, *ApJ*, **492**,

BOND, H.E. 1980, *ApJS*, **44**, 517

BOOTHROYD, A.I., & SACKMANN, I.-J. 1988, *ApJ*, **328**, 641

BRANCH, D., LIVIO, M., YUNGELSON, L.R., BOFFI, F.R., & BARON, E. 1995, *PASP*, **107**, 717

BRANDL, B. ET AL. 1996, *ApJ*, **466**, 254

BRILEY, M.M., HESSER, J.E., & BELL, R.A. 1991, *ApJ*, **373**, 482

BROWN, J.A., & WALLERSTEIN, G. 1989, *AJ*, **98**, 1643

BROWN, J.A., & WALLERSTEIN, G. 1993a, *AJ*, **106**, 133

BROWN, J.A., & WALLERSTEIN, G. 1993b, *AJ*, **104**, 1818

BROWN, J.A., WALLERSTEIN, G., & OKE, J.B. 1990, *AJ*, **100**, 1561

BROWN, J.A., WALLERSTEIN, G., & OKE, J.B. 1991, *AJ*, **101**, 1693

BROWN, J.A., WALLERSTEIN, G., CUNHA, K., & SMITH, V.V., 1993, *AJ*, **106**, 133

BROWN, J.A., WALLERSTEIN, G., & ZUCKER, D. 1997, *AJ*, **114**, 180

BROWN, J.H., BURKERT, A., & TRURAN, J.W. 1991, *ApJ*, **376**, 15

BROWN, J.H., BURKERT, A., & TRURAN, J.W. 1995, *ApJ*, **440**, 666

BROWN, J.H., BURKERT, A., & HENSLER, G. 1992, *ApJ*, **391**, 651

BUONANNO, R., BUSCEMA, G., FUSI PECCI, F., RICHER, H.B., & FAHLMAN, G.G. 1990 *AJ*, **100**, 1811

BURBIDGE, E.M., BURBIDGE, G.R., FOWLER, W.A., & HOYLE, F. 1957, *Rev. Mod. Phys.*, **29**, 547

BURKERT, A., BROWN, J.H., & TRURAN, J.W. 1993, in *The Globular Cluster- Galaxy Connection*, eds. G.H. Smith & J.P. Brodie, ASP Conf. Ser. 48, p. 656

BUTLER, D. 1975, *ApJ*, **200**, 68

BUTLER, D., DICKENS, R.J., & EPPS, E. 1978, *ApJ*, **225**, 148

CAMERON, A.G.W., & FOWLER, W.A. 1971, *ApJ*, **164**, 111

CANNON, R.D. & STOBIE, R.S. 1973, *MNRAS*, **162**, 207

CANNON, R.D., CROKE, B.F.W., BELL, R.A., HESSER, J.E., & STATHAKIS, R.A. 1998, *MNRAS*, **298**, 601

CARNEY, B.W. 1996, *PASP*, **108**, 900

CARNEY, B.W., WRIGHT, J.S., SNEDEN, C., LAIRD, J.B., & AGUILAR, L.A. 1997, *AJ*, **114**, 363

CARRETTA, E. 1995, Ph. D. thesis, Un. Padova

CARRETTA, E., & BRAGAGLIA, A. 1998, *A&A*, **329**, 937

CARRETTA, E., & GRATTON, R.G. 1997, *A&AS*, **121**, 95

CARRETTA, E., GRATTON, R.G., SNEDEN, C., & BRAGAGLIA, A. 1998, in *Galaxy Evolution: Connecting the distant Universe with the Local Fossil Record*, eds. M. Spite & F. Spite, Kluwer, Dordrecht, in press

CASSISI, S., & SALARIS, M. 1997, *MNRAS*, **285**, 593

CASSISI, S., SALARIS, M., & DEGL'INNOCENTI, S. 1997, *MNRAS*, **290**, 515

CASTELLI, F., GRATTON, R.G., & KURUCZ, R.L. 1997, *A&A*, **318**, 841

CAVALLO, R.M., SWEIGART, A.V., & BELL, R.A. 1996, *ApJ*, **464**, L79

CAYREL, R. 1988, in *The Impact of Very High S/N Spectroscopy on Stellar Physics*, ed. G. Cayrel de Strobel & M. Spite, Kluwer, Dordrecht, p.345

CAYREL, R., CAYREL DE STROBEL, G., & CAMPBELL, B. 1985, *A&A*, **146**, 249

CHABOYER, B., DELIYANNIS, C.P., DEMARQUE, P., PINSONNEAULT, & M.H., SARAJEDINI, A. 1992, *ApJ*, **388**, 372

CHARBONNEL, C. 1995, *ApJ*, **453**, 41

CHIAPPINI, A., MATTEUCCI, F., & GRATTON, R.G. 1997, *ApJ*, **477**, 765

CLEMENTINI, G., CARRETTA, E., GRATTON, R.G., MERIGHI, R., MOULD, J.R., & MCCARTHY, J.K. 1995, *ApJ*, **110**, 2319

CLEMENTINI, G., CARRETTA, E., GRATTON, R.G., & SNEDEN, C. 1998, *MNRAS*, in press

COHEN, J.G. 1978, *ApJ*, **223**, 487

COHEN, J.G. 1979, *ApJ*, **231**, 751

COHEN, J.G. 1980, *ApJ*, **242**, 981

COHEN, J.G. 1981, *ApJ*, **247**, 869

COHEN, J.G. 1983, *ApJ*, **270**, 654

COSTAR, D., & SMITH, H.A. 1988, *AJ*, **96** 1925

CRAWFORD, D.L 1975, *AJ*, **80**, 955

CRAWFORD, D.L 1978, *AJ*, **83**, 48

CRAWFORD, D.L 1979, *AJ*, **84**, 1858

CUNHA, K., & LAMBERT, D.L. 1994, *ApJ*, **426**, 170

DA COSTA, G.S., ARMANDROFF, T.E., & NORRIS, J.E. 1992, *AJ*, **104**, 154

DALLE ORE, C. 1992, Ph. D. Thesis, Un. California at S. Cruz

D'ANTONA, F., GRATTON, R., & CHIEFFI, A. 1983, *MSAIt*, **54**, 173

DEKKER, H., DELABRE, B., HESS, G., & KOTZLOWSKI, H 1992, in *Progresses in Telescopes and Instrumentation Technology*, ed. M.-H. Ulrich, ESO, Garching, p. 581

DELIYANNIS, C.P., BOESGAARD, A.M., & KING, J.R. 1995, *ApJ*, **452**, L13

DENISSENKOV, P.A. & DENISSENKOVA, S.N. 1990, *SVAL*, **16**, 275

DENISSENKOV, P.A., DA COSTA, G.S., NORRIS, J.E., & WEISS, A. 1998, *A&A*, **333**, 926

DESIDERA, S. 1998, private communication

DRAINE, B.T. 1980, *ApJ* **241**, 1021

DRAINE, B.T., ROBERGIE, W.G., & DALGARNO, A. 1983, *ApJ* **264**, 485

DI BENEDETTO, G.P. 1998, *A&A*, **339**, 858

DI BENEDETTO, G.P., & RABBIA, Y. 1987, *A&A*, **188**, 114

D'ODORICO, S., GRATTON, R.G., & PONZ, D. 1985, *A&A*, **142**, 232

DOMINY, J.F. 1984, *ApJS*, **55**, 27

DUNCAN, D.K., PRIMAS, F., REBULL, L.M., BOESGAARD, A.M., DELIYANNIS, C.P., HOBBS, L.M., KING, J.R., & RYAN, S.G. 1997, *ApJ*, **488**, 338

EDVARDSSON, B., ANDERSEN, J., GUSTAFSSON, B., LAMBERT, D.L., NISSEN, P.E., & TOMKIN, J. 1993, *A&A*, **275**, 101

EGGEN, O.J., LYNDEN-BELL, D., & SANDAGE, A.R. 1963, *ApJ*, **136**, 748

ELMEGREN, B.G., 1983, *MNRAS*, **203**, 1011

ELMEGREN, B.G., 1985, in *Birth and Infancy of Stars*, eds. R. Lucas, A. Omont, & R. Stora, North Holland, Amsterdam, p. 257

FRANÇOIS, P., SPITE, M., & SPITE, F. 1988, *A&A*, **191**, 267

FREEMAN, K.C., & RODGERS, A.W. 1975, *ApJ*, **201**, L71

FROGEL, J.A., PERSSON, S.E., & COHEN, J.G. 1983, *ApJ*, **275**, 773

FROGEL, J.A., MOULD, J., & BLANCO, V.M. 1990, *ApJ*, **352**, 96

FUHRMANN, K., 1998, *A&A*, **338**, 161

FUHRMANN, K., AXER, M., & GEHREN, T. 1994, *A&A*, **285**, 585

FUSI PECCI, F., FERRARO, F.R., BELLAZZINI, M., DJORGOWSKI, S., PIOTTO, G., & BUONANNO, R. 1993, *AJ*, **105**, 1145

GEISLER, D., & PILACHOWSKI, C.A. 1982, in *Astrophysical Parameters for Globular Clusters*, eds. A.G. Davis Philip, D.S. Hayes, Davis Press, Schenectady, p. 91

GONZALEZ, G., & WALLERSTEIN, G. 1992, *MNRAS*, **254**, 343

GONZALEZ, G., & WALLERSTEIN, G. 1994, *AJ*, **108**, 1325

GRATTON, R.G. 1982a, *A&A*, **115**, 171

GRATTON, R.G. 1982b, *A&A*, **115**, 336

GRATTON, R.G. 1987a, *A&A*, **177**, 177

GRATTON, R.G. 1987b, *A&A*, **179**, 181

GRATTON, R.G. 1988, in *The Impact of Very High S/N Spectroscopy on Stellar Physics*, ed. G. Cayrel de Strobel & M. Spite, Kluwer, Dordrecht, p.525

GRATTON, R.G. 1991, *MSAIt*, **61**, 647

GRATTON, R.G. 1997, in Views on Distance Indicators, F. Caputo eds., in press

GRATTON, R.G. 1998, in *Galaxy Evolution: Connecting the distant Universe with the Local Fossil Record*, eds. M. Spite & F. Spite, Kluwer, Dordrecht, in press

GRATTON, R.G., & CONTARINI, G. 1994, *A&A*, **283**, 911

GRATTON, R.G., & ORTOLANI, S. 1988, *A&AS*, **73**, 137

GRATTON, R.G., & ORTOLANI, S. 1989, *A&A*, **211**, 41

GRATTON, R.G., & SNEDEN, C. 1990, *A&A*, **234**, 366

GRATTON, R.G., & SNEDEN, C. 1991, *A&A*, **241**, 501

GRATTON, R.G., & SNEDEN, C. 1994, *A&A*, **287**, 927

GRATTON, R.G., QUARTA, M.L., & ORTOLANI, S. 1986, *A&A*, **169**, 208

GRATTON, R.G., CARRETTA, E., MATTEUCCI, F., & SNEDEN, C., 1996, in *Formation of the Galactic Halo... Inside and Out*, eds. Morrison, H., Sarajedini, A., ASP Conf. ser. 92, 307

GRATTON, R.G., CARRETTA, E., & CASTELLI, F. 1997, *A&A*, **314** 191

GRATTON, R.G., CARRETTA, E., GUSTAFSSON, B., & ERIKSSON, K., 1998, submitted to *A&A*

GUNN, J.E. 1980, in *Globular Clusters*, ed. D. Hanes & B. Madore (Cambridge University Press, Cambridge), p. 301

GUSTAFSSON, B. 1983, *Highlights in Astronomy*, **6**, 101

GUSTAFSSON, B. 1989, *ARA&A*, **27**, 701

GUSTAFSSON, B., BELL, R.A., ERIKSSON, K., & NORDLUND, A. 1975, *A&A*, **42**, 407

HARRIS, W.E. 1991, *ARA&A*, **29**, 543

HARRIS, W.E. 1996, *AJ*, **112**, 1487

HARRIS, W.E., ET AL. 1997, *AJ*, **114**, 1030

HARTQUIST, T.W., & DYSON, J.E. 1997 in *Herbig-Haro Flows and the Birth of Low Mass Stars*, IAU Symp. 182, eds. B. Reipurth & C. Bertout, Kluwer, Dordrecht, p. 537

HARTWICK, F.D.A. 1976, *ApJ*, **209**, 418

HAYWOOD, M., PALASI, J., GÓMEZ, A., & MEILLON, L., in Hipparcos Venice '98, eds. M.A.C. Perryman & P.L. Bernacca, ESA, Noordwijk, p. 489

HELFER, H.L., WALLERSTEIN, G., & GREENSTEIN, J.L. 1959, *ApJ*, **129**, 700

HERZBERG, G. 1950, *Molecular spectra and molecular structure. Vol. 1: Spectra of diatomic molecules*, Van Nostrand, New York

HESSER, J.E., & SHAWL, S.J. 1986, *PASP*, **97**, 465

HESSER, J.E., HARTWICK, F.D.A., & McCLURE, R.D. 1983, *ApJ*, **207**, L113

HOLWEGER, H., & MÜLLER, E.A. 1974, *Solar Phys.*, **39** 19

HOLWEGER, H., KOCK, M., & BARD, A. 1995, *A&A*, **296**, 233

HUNTER, D.A., SHAYA, E.J., HOLTZMAN, J.A., LIGHT, R.M., O'NEIL, E.J., & LYNDS, R. 1995, *ApJ*, **448**, 179

HUNTER, O'NEIL, E.J., LYNDS, R. D.A., SHAYA, E.J., GROTH, & E.J., HOLTZMAN, J.A., 1996, *ApJ*, **459**, L27

IBEN, I.JR. 1967, *ARA&A*, **5**, 571

IBEN, I.JR., & RENZINI, A. 1983, *ARA&A*, **21**, 27

IBEN, I.JR., & TUTUKOV, A.V. 1984, *ApJS*, **54**, 335

ILLINGWORTH, G. 1976, *ApJ*, **204**, 73

ISRAELIAN, G., GARCÍA LOPEZ, R.J., & REBOLO, R. 1998, *ApJ*, **507**, 805

IWAMOTO ET AL. 1998, *Nature*, in press

JEHIN, E., MAGAIN, P., NEUFORGE, C., NOELS, A., & THOUL, A.A. 1998, *A&A*, **330**, L33

JOHNSON, J.E. & BOLTE, M. 1998, *AJ*, **115**, 693

KING, J.R. 1997, *AJ*, **113**, 2302

KING, J.R., DELIYANNIS, C.P., & BOESGAARD, A.M. 1996, *AJ*, **112**, 2839

KING, J.R., STEPHENS, A., BOESGAARD, A.M., & DELIYANNIS, C.P. 1998, *AJ*, **115**, 666

KOBAYASHI, C., TSUHIMOTO, T., NOMOTO, K., HACHISU, I., & KATO, M. 1998, *ApJL*, , astro-ph 9806335

KRAFT, R.P. 1994, *PASP*, **106**, 553

KRAFT, R.P., SNEDEN, C., PROSSER, C.F., & LANGER, G.E. 1992, *AJ*, **104**, 645

KRAFT, R.P., SNEDEN, C., LANGER, G.E., & SHETRONE, M.D. 1993, *AJ*, **106**, 1490

KRAFT, R.P., SNEDEN, C., LANGER, G.E., SHETRONE, M.D., & BOLTE, M. 1995 *AJ*, **109**, 2586

KRAFT, R.P., SNEDEN, C., SMITH, G.H., SHETRONE, M.D., LANGER, G.E., & PILACHOWSKI,

C.A. 1997, *AJ*, **113**, 279

KRAFT, R.P., SNEDEN, C., SMITH, G.H., SHETRONE, M.D., & FULBRIGHT, J. 1998, *AJ*, **115**, 1500

KURUCZ, R.L. 1979, *ApJS*, **40**, 1

KURUCZ, R.L. 1993, *CD-ROM*, **15**

KURUCZ, R.L. 1994, *CD-ROM*, **20, 21, 22**

KURUCZ, R.L. 1995, *CD-ROM*, **23**

LAMBERT, D.L. 1978, *MNRAS*, **182**, 249

LANGER G.E. & HOFFMAN, R. 1995, *PASP*, **107**, 1177

LANGER G.E., KRAFT, R.P., CARBON, D.,F., FRIEL, E., & OKE, J.B. 1986, *PASP*, **98**, 473

LANGER G.E. ET AL. 1998, *AJ*, **115**, 685

LARSON, R.B. 1974, *MNRAS*, **166**, 585

LARSON, R.B. 1991, in *Fragmentation of Molecular Clouds and Star Formation*, IAU Symp. 147, eds. E. Falgarone, F. Boulanger, & G. Duvert, Kluwer, Dordrecht, p. 261

LATTANZIO, J.C., & BOOTHROYD, A.I. 1997 in *Astrophysical Implications of the Laboratory Study of the Presolar Materials*, eds. Bernatowitz, T., Zinner, E., AIP: Sunnyside NY

LEEP, E.M., WALLERSTEIN, G., & OKE, J.B. 1986, *AJ*, **91**, 1117

LEEP, E.M., WALLERSTEIN, G., & OKE, J.B. 1987, *AJ*, **93**, 338

LEGGETT, S.K., ALLARD, F., BERRIMAN, G., DAHN, C.C., & HAUSCHILDT, P.H. 1996, *ApJS*, **104**, 117

LEMOINE, M., VANGIONI-FLAM, E., & CASSE, M. 1998, *ApJ*, **499**, 735

MALLIA, E.A., & PAGEL, B. 1981, *MNRAS*, **194**, 421

MARIGO, P. 1998, Ph. D. Thesis, Un. Padova

MASSEY, P., & HUNTER, D.A. 1998, *ApJ*, **493**, 180

MATTEUCCI, F. 1983, *MSAIt*, **54**, 289

MATTEUCCI, F., & FRANÇOIS, P. 1992, *A&A*, **262**, L1

MCCLURE, R.D., & WOODSWORTH, A.W. 1990, *ApJ*, **352**, 709

MCCLURE, R.D., FLETCHER, J.M., & NEMEC, J.M., 1980, *ApJL*, **238**, L35

MCCRAY, R., & KAFATOS, M.C. 1987, *ApJ*, **317** 190

MCCRAY, R., & SNOW, T.P.JR 1979, *ARAA* **17**, 213

MCWILLIAM, A. 1997, *ARA&A*, **35**, 503

MCWILLIAM, A., & RICH, M.J. 1994, *ApJS*, **91**, 749

MCWILLIAM, A., GEISLER, D., & RICH, M. 1992, *PASP*, **104**, 1193

MENEGUZZI, M., AUDOUZE, J., & REEVES, H. 1971, *A&A*, **15**, 337

MIHALAS, D. 1967, *Methods in Comp. Phys.*, **7**, 1

MILLER, G.E., & SCALO, J.M. 1979, *ApJS*, **41**, 513

MILONE, A., BARBUY, B., SPITE, M., & SPITE F. 1992, *A&A*, **261**, 551

MINNITI, D., GEISLER, D., PETERSON, R.C., & CLARIA, J.J. 1993, *ApJ*, **413**, 548

MORGAN, W.W. 1959, *AJ*, **64**, 432

NISSEN, P.E. & SCHUSTER, W.J. 1997, *A&A*, **326**, 751

NOMOTO, K. 1998, in *Galaxy Evolution: Connecting the distant Universe with the Local Fossil Record*, eds. M. Spite & F. Spite, Kluwer, Dordrecht, in press

NOMOTO, K., THIELEMANN, F.-K., & YOKOI, Y. 1984, *ApJ*, **286**, 644

NOMOTO, K., IWAMOTO, K., & KISHIMOTO, N. 1997, *Science*, **276**, 1378

NORRIS, J.E. 1994, *ApJ*, **431**, 645

NORRIS, J.P. & DA COSTA, G.S. 1995, *ApJ* **447**, 680

OLSZEWSKI, E.W., SCHOMMER, R.A., SUNTZEFF, N.B., & HARRIS, H.C. 1991, *AJ*, **101**, 515

ORTOLANI, S., BARBUY, B., & BICA, E. 1991, *A&A*, **249**, 31

OSBORNE, W. 1971, *Observatory*, **91**, 223

PAGEL, B.E.J. 1992, in *The Stellar Populations of Galaxies*, IAU Symp. 149, ed. B. Barbuy & A. Renzini (Kluwer, Dordrecht), p. 133

PALTOGLOU, G., & NORRIS, J. 1989, *ApJ*, **336**, 185

PASQUINI, L., & MOLARO, P. 1996, *A&A*, **307**, 761

PASQUINI, L., & MOLARO, P. 1997, *A&A*, **322**, 109

PERSSON, S.E., FROGEL, J.A., COHEN, J.G., AARONSON, M., & MATTHEWS, K. 1980, *ApJ*, **235**, 452

PETERSON, R.C. 1982, in *Astrophysical Parameters for Globular Clusters*, eds. A.G. Davis Philip, D.S. Hayes, Davis Press, Schenectady, p. 121

PETERSON, R.C. 1983a, *PASP*, **95**, 98

PETERSON, R.C. 1983b, *ApJ*, **275**, 737

PETERSON, R.C., ROOD, R.T., & CROCKER, D.A. 1995, *ApJ*, **453**, 214

PILACHOWSKI, C.A., & SNEDEN, C. 1983, *PASP*, **95**, 229

PILACHOWSKI, C.A., CANTERNA, R., & WALLERSTEIN, G. 1980a, *ApJL*, **235**, L21

PILACHOWSKI, C.A., WALLERSTEIN, G., & LEEP, M. 1980b, *ApJ*, **236**, 508

PILACHOWSKI, C.A., WALLERSTEIN, G., LEEP, M., & PETERSON, R.C. 1982a, *ApJ*, **263**, 187

PILACHOWSKI, C.A., SNEDEN, C., & WALLERSTEIN, G. 1983a, *ApJS*, **52**, 241

PILACHOWSKI, C.A., OLSZEWSKI, E.W., & ODELL, A. 1983b, *PASP*, **95**, 713

PILACHOWSKI, C.A., BOTHUN, G.D., OLSZEWSKI, E.W., & ODELL, A. 1983c, *ApJ*, **273**, 187

PILACHOWSKI, C.A., SNEDEN, C., & GREEN, E.M. 1984, *PASP*, **96**, 932

PILACHOWSKI, C.A., SNEDEN, C., & BOOTH, J. 1993, *ApJ*, **407**, 699

PILACHOWSKI, C.A., SNEDEN, C., & KRAFT, R.P. 1997, *AJ*, **112**, 545

PLEZ, B., 1992, *A&AS*, **94**, 527

PRESTON, G.W. 1959, *ApJ*, **130**, 507

RAITERI, C.M., VILLATA, M., & NAVARRO, J.F. 1997, *A&A* **315**, 105

RAYMOND, J.C., COX, D.P., & SMITH, S.W., 1976, *ApJ* **204**, 290

REED, B.C., HESSER, J.E., & SHAWL, S.J. 1988, *PASP*, **100**, 545

REETZ, J. 1998, in *Galaxy Evolution: Connecting the distant Universe with the Local Fossil Record*, eds. M. Spite & F. Spite, Kluwer, Dordrecht, in press

RICHER, H.B., FAHLMAN, G.G., BUONANNO, R., FUSI PECCI, F., SEARLE, L., & THOMPSON, I.B. 1991, *ApJ*, **381**, 147

RICHER, H.B., & FAHLMAN, G.G. 1996, Astro-ph 9611193

RUBIO, M., BARBÁ, R.H., WALBORN, N.R., PROBST, R.G., GARCÍA, J., & ROTH, M.R. 1998, *AJ*, **116**, 1708

SALPETER, E.E. 1955, *ApJ*, **121**, 161

SAMLAND, M., HENSLER, G., & THEIS, CH. 1997, *ApJ*, **476**, 544

SANDAGE, A.R. 1982, *ApJ*, **252**, 553

SANDAGE, A.R., & SMITH, H.A. 1966, *ApJ*, **144**, 886

SANDAGE, A.R., & WALLERSTEIN, G. 1960, *ApJ*, **131**, 598

SARAJEDINI, A., 1994, *AJ*, **107**, 618

SARAJEDINI, A. & LAYDEN, A. 1997, *AJ*, **113**, 264

SCHADEE, A. 1964, *BAN*, **17**, 311

SCHMIDT, M. 1959, *ApJ*, **129**, 243

SCHMIDT, M. 1963, *ApJ*, **137**, 758

SCHUSTER, W.J., & NISSEN, P.E. 1989, *A&A*, **222**, 69

SCOVILLE N.Z., & HERSH, K. 1979, *ApJ* **229**, 578

SEARLE, L., & ZINN, R. 1978, *ApJ*, **225**, 357

SEATON, M.J. 1995, *The Opacity Project*, (IP Publ., Bristol)

SHU, F.H., ADAMS, F.C., & LIZANO, S. 1986, *ARA&A*, **25**, 23

SIMMONS, G.J. & BLACKWELL, D.E. 1982, *A&A*, **112**, 209

SMITH, G.E. 1984, *AJ*, **89**, 801

SMITH, H.A. 1984, *PASP*, **99**, 67

SMITH, V.V., & LAMBERT, D.L. 1990, *ApJ*, **361**, L69

SNEDEN, C., KRAFT, R.P., PROSSER, C.F., & LANGER, G.E. 1991, *AJ*, **102**, 2001

SNEDEN, C., KRAFT, R.P., PROSSER, C.F., & LANGER, G.E. 1992, *AJ*, **104**, 2121

SNEDEN, C., KRAFT, R.P., PROSSER, C.F., & SHETRONE, M.D. 1994, *AJ*, **107**, 1773

SNEDEN, C., KRAFT, R.P., SHETRONE, M.D., SMITH, G.H., LANGER, G.E., & PROSSER, C.F. 1997, *AJ*, **114** 1964

SPITE, F., & SPITE, M. 1982, *A&A*, **115**, 357

STETSON, P.B., VANDENBERG, D.A., BOLTE, M., HESSER, J.E., & SMITH, G.H. 1989 *AJ*, **97**, 1360

STETSON, P., VANDENBERG, D.A., & BOLTE, M. 1996, *PASP*, **108**, 560

SUNTZEFF, N.B., & SMITH, V.V. 1991, *ApJ*, **381**, 160

SUNTZEFF, N.B., & KRAFT, R.P. 1996, *AJ*, **111** 1913

SUNTZEFF, N.B., SCHOMMER, R.A., OLSZEWSKI, E.W., & WALKER, A.R. 1992, *AJ*, **104**, 1743

SUNTZEFF, N.B., MATEO, M., TENDRUP, D.M., OLSZEWSKI, E.W., GEISLER, D., & WELLER, W. 1993, *ApJ*, **418**, 208

SWEIGART, A.V. 1997, *ApJ*, **474**, L23

SWEIGART, A.V., & MENGEL, J.G. 1979, *ApJ*, **229**, 624

THIELEMANN, F.-K., NOMOTO, K., & HASHIMOTO, M. 1996, *ApJ*, **460**, 408

TIMMES, F.X., WOOSLEY, S.E. & WEAVER, T.A. 1995, *ApJS*, **98**, 617

TOOMRE, A. 1977, in *The Evolution of Galaxies and Stellar Populations*, ed. B.M. Tinsley & R.B. Larson (New Haven, Yale Univ. Press), 401

TRAGER, S.C., KING, I.R., & DJORGOVSKI, S. 1995, *AJ*, **109**, 218

TRURAN, J.W. 1973, *Space Sci. Rev.*, **15**, 23

TRURAN, J.W., & CAMERON, A.G.W. 1971, *Ap. Space Sci.*, **14**, 179

TURATTO, M. ET AL. 1998, *ApJ*, **498**, L129

TWAROG, B.A., & ANTHONY-TWAROG, B.J. 1994, *AJ*, **107**, 1371

UNSÖLD, A. 1955, *Physik der Sternatmosphären*, Springer, Berlin

VAN DEN BERGH, S., MORBEY, C., & PAZDER, J. 1991, *ApJ*, **375**, 594

VAN ECK, S., JORISSEN, A., UDRY, S., MAYOR, M., & PERNIER, B. 1997, in *Hipparcos Venice '97*, eds. M.A.C. Perryman & P.L. Bernacca, ESA, Nordwijk, p. 327

VANGIONI-FLAM, E., RAMATY, R., OLIVE, K.A., & CASSE, M. 1998, *A&A*, **337**, 714

VANTURE, A.D. 1992, *AJ*, **104** 1997

VANTURE, A.D. & WALLERSTEIN, G. 1992 *PASP*, **104**, 888

VANTURE, A.D., WALLERSTEIN, G., & BROWN, J.A. 1994 *PASP*, **106**, 835

WALLERSTEIN, G., & CONTI, P.S. 1969, *ARA&A*, **7**, 99

WALLERSTEIN, G., & GONZALEZ, G. 1998, *A&AS*, **192**, 7903

WALLERSTEIN, G., LEEP, E.M., & OKE, J.B. 1987, *AJ*, **93**, 1137

WALLERSTEIN, G., ET AL. 1997, *Rev. Mod. Phys.*, **69**, 995

WHEELER, J.G., SNEDEN, C., & TRURAN, J.W. 1989, *ARA&A*, **27**, 279

WAGENHUBER, J., & GROENEWEGEN, M. 1998, *A&A*, **340** 183

WEBBINK, R.F. 1984, *ApJ*, **277**, 355

WHELAN, I, & IBEN, I.JR. 1973, *ApJ*, **186**, 1007

WHITMER, J.C., BECK-WINCHATZ, B., BROWN, J.A., & WALLERSTEIN, G. 1995, *PASP*, **107**, 127

WYSE, R.F.G., & GILMORE, G. 1995, *AJ* **110**, 2771

WOOSLEY, E.E., & WEAVER, T.A. 1995, *ApJS*, **101**, 181

WOOSLEY, E.E., HARTMANN, D.H., HOFFMAN, R.D., & HAXTON, W.C. 1990, *ApJ*, **356**, 272

ZAHN, J.-P. 1992, *A&A*, **265**, 115

ZINN, R. 1980, *ApJS*, **42** 19

ZINN, R., & WEST, M.J. 1984, *ApJS*, **55**, 45

ZUCKER, D., WALLERSTEIN, G., & BROWN, J.A. 1996 *PASP*, **108**, 911

REBECCA A. W. ELSON was born in Montreal (Canada) on 2 January 1960.

She now lives in the United Kingdom but retains double Canadian and American nationality.

After graduating in Astronomy at Smith College (USA), she obtained a master's degree at the University of British Columbia (Canada) and concluded her work on dynamical friction models.

In 1986 she completed her doctorate at the Institute of Astronomy (Cambridge), the subject of her thesis being the structure and evolution of rich stellar clusters in the Large Magellanic Cloud, a topic on which she continued to work during her stay as a postdoc at the Institute of Advanced Studies, Princeton, New Jersey (USA).

From 1989 onwards, her work has centred more on the formation and early evolution stellar clusters. During 1989-90 she gave a course at Harvard University on "Science and Ethics".

Since 1991, she has been an Associate Postdoctoral Researcher at the Institute of Astronomy, Cambridge (United Kingdom), working mainly with Hubble Space Telescope data in various projects related to galactic and extragalactic stellar populations, globular clusters and globular cluster systems.

Stellar Dynamics in Globular Clusters

By R E B E C C A A. W. E L S O N ,

Institute of Astronomy, Madingley Rd., Cambridge CB3 0HA, United Kingdom

Globular clusters provide ideal laboratories for studying the dynamical behaviour of N-body systems. If one includes in one's definition of 'globular cluster' the young and intermediate age rich star clusters in the Magellanic Clouds, then one has at hand a set of objects that can serve as a testing ground for theories that describe self-gravitating systems of point masses at any stage in their evolution. These different stages include violent relaxation, a gradual approach to quasi-static equilibrium through two-body relaxation, the dramatic collapse, probably followed by oscillations, of the cluster core, and ultimately dissolution of the cluster as it contributes its stars to the parent/host galaxy's field (usually halo) population. Understanding the mechanisms that hasten the dissolution of a cluster can help us reconstruct the original population of clusters in a given galaxy. This in turn can guide theories of globular cluster formation, and, to the extent that globular clusters trace the early stages of galaxy evolution, the formation of galaxies themselves. This chapter provides an overview of the life of a globular cluster (Section 1), derives the time scales relevant to various stages of cluster evolution (Section 2), and discusses the main observable qualities of clusters relevant to their dynamical evolution: their surface brightness profiles (Section 3) and their internal velocity dispersions (Section 4). In Section 5 some recent results from a large HST project to study the formation and evolution of rich star clusters in the Large Magellanic Cloud are described.

1. Overview: Why Study the Dynamics of Globular Clusters?

The globular clusters in the halo of our Galaxy initially attracted the attention of stellar dynamicists because of the homogeneity of their appearance. They are perhaps more similar to one another, both in structure and range of absolute magnitudes, than any other kind of stellar system. This uniformity demanded a simple physical explanation, and prompted quantitative descriptions as early as the 1910s, when structural data were first becoming available (eg. Plummer 1915; Jeans 1916). Since then, observations have revealed many interesting trends and complexities which theorists have been tackling. This, then, is perhaps the foremost reason for studying the structure and dynamics of globular clusters: they provide a unique laboratory for testing our understanding of the physics of dissipationless self-gravitating systems.

The first physical models to be developed matched in their simplicity the apparent structure of the clusters themselves. These "King models" (King 1966a) are based on the so-called "lowered Maxwellian" velocity distribution as a solution of the Fokker-Planck equation: a distribution of velocity decreasing exponentially up to a high velocity cutoff, beyond which stars are no longer gravitationally bound to the cluster. The velocity distribution was assumed to be isotropic, and the stars all of the same mass. Neither of these assumptions is strictly valid, but the models, a single parameter family with central gravitational potential as the only variable, nevertheless gave a very satisfactory representation of the observed surface brightness profiles of clusters, and continue to form the basis of dynamical studies of clusters to this day.

In the mid-1970s velocity dispersions in the nearest clusters began to become accessible to observation, adding a new dimension which theories could incorporate. The original King models were extended, still analytically, to include velocity anisotropies, and computationally, to include a more realistic range of stellar masses.

Meanwhile, out of the apparently homogenous set of clusters in the Milky Way, one, M15 (NGC 7808), had been singled out in early studies as having an anomoulous surface brightness profile (cf. Newell & O'Neil 1978). Rather than flattening out towards the centre, like the classic isothermal cores seen in other clusters, the surface brightness continued to rise to the limit of resolution. Three things became imperative: to probe the core of M15 with increased resolution, both in velocity space and in surface brightness; to search for other clusters in which observations obtained with the technology of the 1980s might reveal anomolous cores where that of the 1960s had failed to do so; and to explain the so-called 'cusps' theoretically. New observations did indeed reveal anomolous cusps in perhaps a fifth of the entire globular cluster population (Djorgovski & King 1986), and various explanations were advanced, ranging from dynamical processes to aggregations of dark matter, perhaps massive black holes, at the cluster centres. The theory of runaway collapse of a cluster core, the "gravothermal catastrophe", first discussed by Lynden-Bell & Wood (1968), began to be explored more thoroughly using computer simulations.

Nowadays, with the availability of ultra-fast computers designed specifically to solve the N-body problem (cf. Aarseth 1998), the ambition has become to create real-size models of globular clusters with realistic distributions of stellar masses, and including all the relevant physical processes, from cumulative distant encounters, to close encounters and collisions, to the formation and disruption of binary systems, the effect of external tidal fields (static or varying), and the effect of stellar evolution. These sophisticated models, in turn, demand detailed observations of the structure and stellar content of rich clusters, which can both guide and constrain theories of their formation and evolution.

Beyond their role as stellar dynamics laboratories, globular clusters provide us with clues to help us reconstruct the early stages of the evolution of our Galaxy, and ultimately of other galaxies as well. To interpret these clues, an understanding of the dynamical evolution of the clusters is essential. The apparent homogeneity of globular cluster systems from one galaxy to the next led to their much publicized use as distance indicators, the assumption being that the mean magnitude of globular clusters is the same in every galaxy. While globular cluster systems certainly provide crude distance indicators, their reliability as such has been somewhat undermined by the realization that environmental factors will inevitably alter the distribution of masses and therefore absolute magnitudes of globular clusters in any given galaxy, in a way that must depend on the galaxy's morphology. Understanding these environmental effects requires a good knowledge of the dynamical evolution of clusters both from a theoretical and an observational point of view.

Every globular cluster that could in principle have formed in our Galaxy would not have survived to the present. Both internal and external forces will have acted over the lifetime of the Galaxy to disrupt some fraction of the original population, which may have included clusters with a much wider range of masses and densities than we see today. These processes of selective disruption include: evaporation of stars due to internal relaxation; tidal shocking through encounters with other clusters or with the Galactic bulge or disk; and disruption of the most massive objects through dynamical friction. Figure 1 illustrates this idea. Here, Fall & Rees (1977) explore in simple physical terms the relative effects of tidal shocking and internal evaporation on a hypothetical set of clusters with a large range of size and mass. They start out with the assumption that density fluctuations in the proto-Galaxy would follow a power-law: $\delta\rho/<\rho>\propto m^{-\alpha}$. They then derive simple expressions for the various disruption processes as a function of time, and of the structure of the clusters and the Galaxy. The lines in Fig. 1 indicate their predicted range of masses and sizes for Galactic globular clusters which would have survived to the present. The points represent the observed globular clusters. The fact

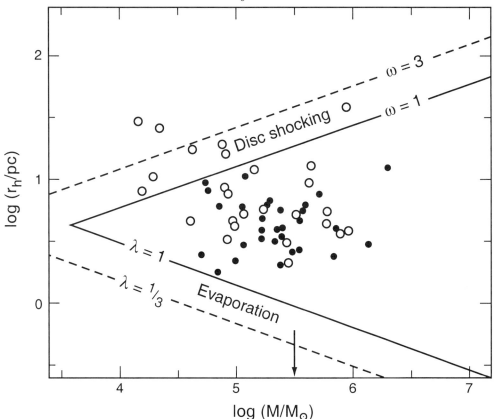

FIGURE 1. A 'survival triangle' for the globular clusters in the Milky Way, predicted from simple physical assumptions about disruption of clusters through evaporation and tidal shocking. Filled circles are clusters with Galactocentric distance < 10 kpc and open circles are clusters at > 10 kpc. λ and ω are parameters that depend on the structure of the clusters and of the Galactic disk. The figure is from Fall & Rees (1977).

that most of the points lie within the 'survival triangle' suggests that disk shocking and evaporation are indeed instrumental in whittling down what may have initially been a much larger population to produce the one we now observe. A recent study of this same problem by Gnedin & Ostriker (1998) using numerical modelling produces very similar results.

Through studies like these, analytical theories and numerical modelling can thus help us reconstruct what the original population of clusters, and hence the very young Galaxy (and the early stages of evolution in other galaxies), might have looked like. It is particularly interesting to consider such studies in the context of HST discoveries of what appear to be young globular clusters forming in galaxies which have undergone recent mergers, such as NGC 1275 (cf. Zepf *et al.* 1995), the Antennae galaxies (Whitmore & Schweizer 1995) and NGC 7252 (Schweizer & Seitzer 1998).

In exploring the origin of globular clusters it is also interesting to consider them in relation to other pure population II systems. This can be done, for example, by comparing their surface brightnesses and absolute magnitudes with those of giant and dwarf elliptical galaxies, as in the plot by Binggeli (1994) reproduced in Fig. 2. What is the relation between the most massive globular clusters in the halo of our Galaxy, and the smallest dwarf elliptical satellites whose luminosities are very similar to those of the brightest

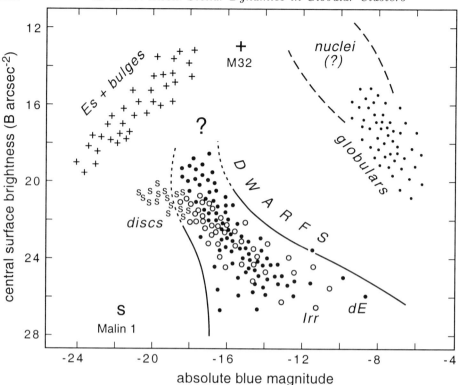

FIGURE 2. Central surface brightness plotted against absolute magnitude for globular clusters, dwarf galaxies, spiral galaxies, and elliptical galaxies. This figure is from Binggeli (1994).

globulars? Here, studies of velocity dispersions can help, by probing the non-luminous content of both globular clusters and dwarf galaxies. With mass-to-light ratios in the range $M/L_V \sim 1 - 4$ (in solar units), globular clusters appear to contain no dark matter, while dE galaxies, with $M/L_V \gtrsim 100$, do. It has been suggested that the presence of dark matter may be correlated with the length scale of the system. However, the smallest dE galaxies in the Local Group, Carina, Draco, and Ursa Minor, have tidal radii only a factor $\sim 2 - 3$ greater than the largest globular clusters: $\sim 500 - 600$ pc (cf. Irwin & Hatzidimitriou 1995) compared to $\sim 200 - 250$ pc (NGC 2419, NGC 5466; Harris 1996). Their absolute magnitudes are similar: $M_V \sim -8.5 \pm 0.5$ for the three dwarfs and $M_V = -9.53$ and -7.06 for the two globular clusters. Yet mass-to-light ratios for Carina, Draco, and Ursa Minor are $59 \pm 47, 245 \pm 155$ and 95 ± 43 respectively, compared to only 1.2 and 2.3 for the two globular clusters.

The aim of this chapter is to outline the basic theoretical and observational foundations needed to address questions about the dynamical evolution of globular clusters. Here a point of semantics must be clarified. Traditionally "globular cluster" has referred to the ancient clusters in the halo of our Galaxy. Open clusters (or more confusingly, "galactic" clusters) are the much smaller, looser associations that inhabit the disk. But these categories, invented for the Milky Way, are not necessarily appropriate for other galaxies. Indeed, the Large Magellanic Cloud (LMC) has clusters as rich and "globular" as some of our halo objects, but which are only a few million years old. And "young" globular clusters have recently been identified in galaxies where violent mergers appear to be underway. I will therefore use the term "globular cluster" to refer to any rich, centrally concentrated star cluster with mass upwards of $\sim 10^4 M_\odot$. The focus of this chapter will thus include dynamical processes in young globular clusters as well as old.

The life of a typical star cluster is illustrated in Fig. 3. At a million years the newly born cluster is still mostly hidden in its progenitor cocoon of gas. Winds from stars of $\sim 100 M_\odot$ blow bubbles in the gas. The first supernova explosions may blow all the remaining gas from the cluster, thus halting star formation altogether. By ten million years, little or no gas remains in the cluster and all stars more massive than about $10 M_\odot$ have evolved to remnants. At this point the cluster still perserves traces of its initial state. It has not yet had time to relax to a quasi-static equilibrium through the process of "two-body relaxation" dominant in older clusters. It may, however, have gone through a period of so-called "violent relaxation" as the stars react to the background potential of the cluster which is fluctuating dramatically. Expansion due to mass loss through stellar evolution may, depending on the slope of the initial mass function (IMF), cause the cluster to overflow its eventual tidal radius, creating a halo of unbound stars. Up until $\sim 10^9$ years the cluster core, which starts out fairly compact, may continue to expand, while the unbound halo is gradually stripped away, and the surface brightness profile takes on the characteristic tidal radius turndown in the outer parts. The cluster continues to fade as the more massive stars leave the main-sequence. After a few $\times 10^9$ years the core may undergo a runaway collapse as more and more energy is transferred out of the core to the outer parts of the cluster. This collapse may be halted and even reversed by the formation of binaries in the core, and a series of collapses and rebounds may follow. This is the state that many of the globular clusters in the Milky Way may currently be in. The highest energy stars will continue to escape from the cluster and eventually, after many billions of years, the cluster will evaporate altogether. Tidal shocking through encounters with a disk or bulge will accelerate this process.

The clusters available for studying these processes are not just those in the halo of our Galaxy, but also those in our neighbour Small and Large Magellanic Clouds. Peterson (1993) has measured velocity dispersions for globular clusters in M31, and with the Hubble Space Telescope (HST) it is now possible to obtain colour-magnitude diagrams for clusters as far away as M31 (cf. Fusi Pecci *et al.* 1996), and to measure structural parameters in globular clusters as far away as the Fornax cluster (~ 18 Mpc) (Elson & Schade 1994). The Magellanic Cloud clusters are particularly interesting because, spanning the full range of ages from $\sim 10^6$ to $\sim 10^{10}$ years, they provide "snapshots" of globular clusters at all stages of evolution.

An outline of this chapter is as follows: Section 2 derives expressions for the basic time scales relevant to the different stages of cluster evolution. Sections 3 and 4 discuss observations of surface brightness profiles and velocity dispersions in clusters in the Milky Way and the LMC. Section 5 describes a large HST project to study the formation and evolution of rich clusters in the LMC, and presents some recent results. Topics including the IMF and the population of binary stars in clusters, while of great interest for the dynamical evolution of clusters, are discussed only briefly; they are explored in more depth elsewhere in this volume (see the chapter by King).

There are many excellent reviews as well as whole text books on the subject of stellar dynamics in globular clusters. Among these is a recent ~ 140 page review by Meylan & Heggie (1997) which covers all aspects of the dynamical evolution of globular clusters, and includes an extensive set of references to the large body of publications, both observational and theoretical, research and review, in this field.

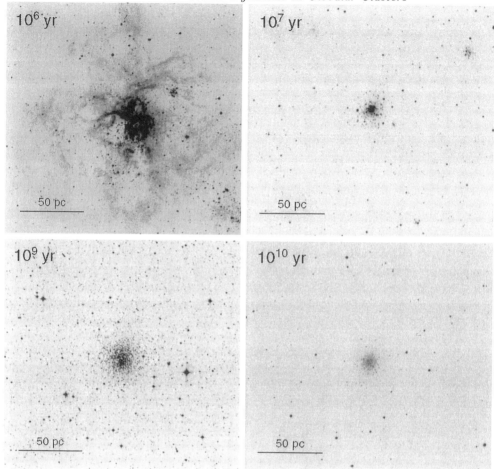

FIGURE 3. Images from the Digital Sky Survey of LMC clusters of increasing age, illustrating the qualitative changes in a cluster's appearance as it evolves. (See text for explanation.)

2. Evolutionary Time scales in Globular Clusters

2.1. *Introduction*

This section provides an overview and derivation of the basic time scales relevant to the evolution of rich star clusters. Estimates of these time scales for a particular cluster tell you which physical processes are likely to be the most important in determining its dynamical state (and which processes can, for the sake of simplicity, be ignored). The expressions for each time scale should always be applied keeping in mind whether the assumptions that went into it are valid in a given context, and what the uncertainties are in the various quantities involved. The time scales include the following:

(a) crossing time (dynamical time)
(b) time scale for violent relaxation
(c) time scale for close stellar encounters
(d) two-body relaxation time
(e) disk shocking time
(f) core collapse time
(g) evaporation time.

Of these, the two-body relaxation time is most relevant to the evolutionary processes taking place in old globular clusters, while the crossing time and violent relaxation time are most relevant in young globular clusters such as the $\sim 10^7$ year old rich star clusters in the LMC. The time scales are not independent. For instance, external processes such as disk shocking may accelerate internal processes such as two-body relaxation and core collapse. And in a cluster undergoing core collapse, the two-body relaxation time will decrease dramatically in the core as densities soar. Caution must be used in applying only these expressions, without N-body simulations, to assess what processes are underway in a cluster. Nevertheless, they provide a useful guide. In the following subsections, each of the above time scales is discussed in detail, and examples are given for globular clusters in our Galaxy and in the LMC.

2.2. Crossing Time

The crossing time is the time it takes for a star in a cluster travelling with the average velocity, to traverse the cluster. This will depend on the star's orbit and the size of the cluster. It is the time scale on which orbits in the cluster mix, and is often referred to as the 'dynamical time'. There are various definitions of the crossing time in the literature. It may be defined as the the characteristic radius of the cluster divided by the velocity dispersion. If the radius is in pc and the velocity in km/s, then the crossing time is

$$t_{cross} \sim 10^6 \left(\frac{r_h}{v_m} \right) \ years. \qquad (2.1)$$

Here, r_h is the half-mass radius (the radius enclosing half the cluster's total mass), and v_m is the mean velocity of stars in the cluster. If one assumes that the virial theorem holds (perhaps not a good assumption in young clusters, but safe in old ones), then

$$v_m^2 \approx \frac{1}{2} \frac{GM}{r_h}$$

and the crossing time can be written in terms of the cluster size and mass, M, as

$$t_{cross} \approx 2.1 \times 10^7 \left(\frac{r_h^3}{M} \right)^{1/2} \ years. \qquad (2.2)$$

Here again, the radius is in pc, and the mass is in $M_\odot$.

A typical mass for a Galactic globular cluster is $10^5 \, M_\odot$, and a typical half-mass radius is 3 pc (Harris 1996). Equation (2.2) thus gives a typical crossing time of $\sim 3.5 \times 10^5$ yr, which is very much smaller than the age of these clusters.

In young LMC clusters, a typical mean velocity might be $\sim 2 - 3$ km/s (Lupton *et al.* 1989). The clusters may not be tidally truncated, and the half-mass radius may therefore be ill-defined (see Section 3.4). The profiles may be traced to radii of $\sim 15 - 20$ pc. For these radii, equation (2.1) gives crossing times of $\sim 10^7$ yr. While uncertain, this number suggests that the ages of the young LMC clusters are comparable to their crossing times. An interesting question is whether age spreads in these clusters (the time scale on which star formation was completed) are large or small compared to the crossing times. This has implications for the mechanism which triggered the protocluster to fragment into stars.

2.3. Time Scale for Violent Relaxation

The term 'violent relaxation' was coined by Lynden-Bell (1967). The concept was inspired by the observation that the profiles of elliptical galaxies are exceedingly smooth and regular even though these systems have not had time to relax through the normal process that produces smooth profiles in old globular clusters (see Section 2.5). What, therefore, can account for their relaxed appearance?

Briefly, during violent relaxation stars can be thought of as relaxing due to the effect of a violently fluctuating background gravitational potential. This potential fluctuates due to dramatic mass loss as high mass stars evolve to remnants and most of their mass is expelled from the system. (The velocity of stellar winds is much greater than the escape velocity from a globular cluster.) During violent relaxation the energies of individual stars are not conserved. The energy, E, changes in relation to the potential, Φ, as

$$\frac{dE}{dt} = -m\frac{d\Phi}{dt}. \tag{2.3}$$

The violent relaxation time scale, t_{vr}, depends on how fast the background potential is changing. The whole system 'vibrates', as kinetic energy is converted to potential energy, which is converted back to kinetic energy, and so on. Arguments based on the time-dependent virial theorem,

$$\frac{1}{2}\frac{d^2 I}{dt^2} = 2T + W \tag{2.4}$$

(where $I = \Sigma mr^2$, T is the kinetic energy, and W is the potential energy), give a violent relaxation time scale

$$t_{vr} \approx 3/[4(2\pi G\rho_m)^{1/2}]$$

or, with the mean density ρ_m in units of $M_\odot$ pc^{-3},

$$t_{vr} \approx 4.5 \times 10^6 \rho_m^{-1/2} \quad years. \tag{2.5}$$

Thus, from equation (2.2) we can write

$$t_{vr} \approx 0.2 t_{cross}. \tag{2.6}$$

Thus, even the $\sim 10^7$ year old LMC clusters will probably have already undergone violent relaxation, while very young objects like the central cluster in the 30 Doradus nebula in the LMC may be currently undergoing it. Note that violent relaxation is independent of stellar mass, so will not produce any mass segregation. Thus, any mass segregation observed in a very young cluster must be primordial, and would contain clues to the cluster formation and star formation processes. Violent relaxation will produce a surface brightness profile with a slope which depends on the initial ratio of kinetic to potential energy, and the profiles of the young LMC clusters may preserve clues to this initial ratio; they are discussed further in Section 3.2.1.

2.4. Time Scale for Close Stellar Encounters

A time scale relevant in particular to the formation and destruction of binary stars in a cluster is the time one must wait for a stellar encounter to cause a 90 degree deflection in the direction a star is travelling. This is referred to as the close encounter time, t_{ce}.

The impact paramter p_0 for such an encounter is

$$p_0 = G(m_1 + m_2)/v_{rel}^2. \tag{2.7}$$

Here, m_1 and m_2 are the masses of the two stars involved in the encounter, and v_{rel} is their relative velocity. The cross-section, α, for such an encounter, is

$$\alpha = \pi p_0^2. \tag{2.8}$$

The time scale for a close encounter is then given by

$$t_{ce} = \frac{1}{\alpha n v_{rel}} \tag{2.9}$$

where n is the local number density of stars. Substituting equations (2.7) and (2.8) into (2.9) we have

$$t_{ce} = \frac{v_{rel}^3}{\pi G^2 (m_1 + m_2)^2 n}. \tag{2.10}$$

Setting $v_{rel}^2 = 2 v_m^2$, this becomes

$$t_{ce} = 4.8 \times 10^{10} \frac{v_m^3}{(m_1 + m_2)^2 n} \quad years \tag{2.11}$$

where the mean velocity v_m is in km/s, m_1 and m_2 are in $M_\odot$, and n is has units pc^{-3}.

In a globular cluster we may assume $m_1 \approx m_2 \approx 0.3 M_\odot$. This median mass comes from adopting a power-law IMF with slope equal to the Salpeter value 1.35, between a lower mass of $0.2 M_\odot$ and an upper mass of $0.8 M_\odot$. The value of the median mass is very insensitive to the upper mass cutoff: for an upper cutoff of $10 M_\odot$ the median mass changes by only $0.03 M_\odot$. It is sensitive to the lower mass cutoff, and adopting $0.1 M_\odot$ instead of 0.2 gives a median value of $0.16 M_\odot$, introducing an uncertainty in t_{ce} of a factor of four. However, recent evidence suggests that in Galactic globular clusters a Salpeter IMF extending as low as $0.1 M_\odot$ is not realistic (see King, this volume), and the real uncertainty in the median mass is probably not this large. Note also that these estimates do not take into account any population of white dwarfs or neutron stars.

Taking $v_m \approx 5$ km/s, and $n \approx 10^3$ pc^{-3}, the time scale for close encounters is $\sim 1.7 \times 10^{10}$ yrs. In the core of a young LMC cluster, with $v_m \approx 3$ km/s, and $n \approx 10^2$ pc^{-3}, (and for the same mean mass) the encounter time is $\sim 3.6 \times 10^{10}$ yr. Both these time scales are long compared with the cluster ages, and we would only expect stellar encounters to be important in the densest cluster cores. Since close encounters are essential for the formation of (non-primordial) binaries, and for binary destruction, these processes will be accelerated in dense cores, for example in clusters undergoing a runaway core collapse (see Section 2.7). For a more detailed discussion of formation and destruction rates of binaries in clusters, see for example Hut, Mcmillan & Romani (1992).

2.5. *Two-body Relaxation Time Scale*

The most important time scale for the evolution of a globular cluster is the two-body relaxtaion time. This is the time scale on which mass segregation develops, as the heavier stars sink to the cluster centre due to equipartition of energy. It is the time required for a star to experience a net deflection of 90 degrees due, not to a single close encounter, but to the cumulative effect of many distant encounters. Equivalently, it is the time required for a star to experience a cumulative velocity change comparable to its original velocity. This time scale is often referred to as the two-body time, t_{2body}. It depends on the stellar density and is therefore a function of radius in the cluster. A convenient definition is the 'reference relaxation time' t_{rh} (Spitzer & Hart 1971; Spitzer 1975). This is the two-body relaxation time calculated for the mean density of the cluster inside the half-mass radius.

Lightman & Shapiro (1978) provide a clear derivation of the two-body relaxtaion time, on which the following discussion is based.

The velocity change due to each distant encounter is $|\Delta v| = v_{rel}\theta$, where θ is the angle of deflection caused by the encounter. The cross-section, α, for encounters is the Rutherford cross-section:

$$\alpha(\theta) = \frac{4p_0^2}{\theta^4}. \tag{2.12}$$

The mean square deflection per unit time is

$$\frac{d<|\Delta v|^2>}{dt} = nv_{rel} \int |\Delta v|^2 \alpha(\theta) d\Omega$$

$$= 8\pi p_0^2 nv_{rel}^3 \int_0^{p_{max}} dp/p \tag{2.13}$$

where n is the number of stars per unit volume. The two-body relaxation time is the time such that

$$t_{2body} \frac{d<|\Delta v|^2>}{dt} = v_{rel}^2 \tag{2.14}$$

so

$$t_{2body} = \frac{1}{8\pi p_0^2 nv_{rel} \int_0^{p_{max}} \frac{1}{p}dp}. \tag{2.15}$$

The integral may be evaluated by setting $p_{max} = r_h$, applying the virial theorem, and assuming that all stars have the same mass. Of course this is not strictly true, although in an old cluster the range of stellar masses still on the main-sequence is small ($\sim 0.2 - 0.8M_\odot$), and the mass of white dwarfs, which are the remnants of stars up to $\sim 8M_\odot$ which have already evolved off the main-sequence, is in the same range ($\sim 0.6M_\odot$). This gives for the integral the expression $\ln(0.5N)$ where N is the total number of stars in the cluster. Substituting p_0 from equation (2.7) above gives

$$t_{2body} = \frac{v_{rel}^3}{8\pi G^2 (m_1 + m_2)^2 n \ln(0.5N)}. \tag{2.16}$$

If we set $v_{rel}^2 = 2v_m^2$, then t_{2body} measures the time for a self-gravitating N-body system to reach quasi-Maxwellian equilibrium.

The two-body time scale is a local quantity. It is inversely proportional to the local density, so that the inner regions of a cluster will relax most quickly. It is inversely proportional to the square of the stellar mass, so more massive stars will relax more quickly. It is important to bear this in mind particularly in the context of young clusters, where, although the mean relaxation time may be much longer than the age of the cluster, the two-body relaxation time for the more massive stars in or near the core may be such that sufficient time has elapsed for dynamical mass segregation to occur. Mass segregation observed in a young cluster is not necessarily primordial, and realistic simulations are necessary to determine its origin.

A convenient version of the two-body time is the reference relaxation time. This is the two-body time calculated for the mean density inside the half-mass radius r_h, with the stars assumed to have the rms velocity of the cluster as a whole. The reference relaxation time, t_{rh}, can be written:

$$t_{rh} = \frac{0.6 M^{1/2} r_h^{3/2}}{m_* G^{1/2} \log(0.4N)} \qquad (2.17)$$

where m_* is the mean stellar mass and M is the total mass.

In a typical globular cluster the two-body relaxation time in the core is $\sim 10^8$ yrs, while at the half-mass radius it is a few $\times 10^9$ yrs (cf. Harris 1996). Thus, globular clusters are relaxed at all but the outermost radii. For young LMC clusters, relaxation times may be long compared to their ages, and they may thus retain traces of their initial states. However, as already mentioned, caution must be exercised in applying equation (2.17) to young clusters where the assumption of virial equilibrium and of a single stellar mass are not necessarily appropriate. The presence of a realistic spectrum of masses can speed up two-body relaxation considerably. In general, equations (2.16) and (2.17) should be used as rough guides, while simulations are required to accurately assess the dynamical state of any cluster.

In comparison to the relaxation times for globular clusters, relaxation times for dwarf spheroidal galaxies are a few $\times 10^{13}$ yrs, and for elliptical galaxies, a few $\times 10^{18}$ yrs. Thus, both dwarf spheroidals and ellipticals must owe their relaxed appearance to a dynamical process such as violent relaxation and not two-body relaxation.

2.6. *Disk Shocking Time Scale*

Tidal shocking by, for example, passage through a disk, heats a cluster by compressing it in the z direction (perpendicular to the disk). This heating is most effective in the outer parts of the cluster, and will thus cause the cluster halo to puff up. For the case of the Milky Way globular cluster system, the orbital velocities of stars in a cluster are much slower than the typical speed of passage of the cluster through the disk, so in calculating the effect of disk shocking, the impulse approximation may be adopted.

Tidal shocking has been explored analytically (cf. Ostriker, Spitzer & Chevalier 1972), as well as numerically (cf. Chernoff & Weinberg 1990). The following discussion is based on that given by Lightman & Shapiro (1978). We define z as the height of a cluster star above the Galactic plane, z_c as the height of the cluster center, and $Z \equiv z - z_c$. Then

$$\frac{dv_z}{dt} = g(z) - g(z_c) \approx Z \frac{dg}{dz}(z_c). \qquad (2.18)$$

where $g(z)$ is the gravitational acceleration at height z. If we set

$$V_{zc} \equiv \frac{dz_c}{dt}$$

then

$$\Delta v_z = \frac{2 Z g_m(z_m)}{V_{zc}}$$

Here, z_m is the value of Z above which $g(z)$ is roughly constant, and equal to g_m. The heating rate of the cluster is given by the time derivative of the energy change per unit mass:

$$\frac{dE}{dt} = \frac{0.5(\Delta v_z)^2}{0.5 P_c} = \frac{4 g_m^2 Z^2}{P_c V_{zc}^2} \qquad (2.19)$$

Note that the orbital period P_c includes two passes through the Galactic plane. The disk shocking time scale t_{sh} is defined as

$$t_{sh} = \frac{-E_h}{dE/dt}$$

Here E_h is the mean energy of a cluster with mass M and size r_h:

$$E_h = -0.2\frac{GM}{r_h},$$

and dE/dt is evaluated at r_h. Then

$$t_{sh} = \frac{3GMP_cV_{zc}^2}{20r_h^3g_m^2}. \tag{2.20}$$

For the Galaxy we can take $z_m = 250$ pc and $g_m = 4.7 \times 10^{-9}$ cm s^{-2}. Then

$$t_{sh} = 9 \times 10^{12} \left(\frac{M}{10^5\,\mathrm{M_\odot}}\right)\left(\frac{r_h}{5pc}\right)^{-3}\left(\frac{P_c}{4 \times 10^8 yr}\right)\left(\frac{V_{zc}}{300 km/s}\right)^2 \quad years. \tag{2.21}$$

Disk shocking is most effective in low density clusters, and in clusters with short orbital periods, and slow orbital velocities. If a cluster has undergone equipartition of energy, then the lower mass stars will have preferentially higher velocities. The impulse imparted to them by disk shocking may cause their velocities to exceed the escape velocity from the cluster. They will be particularly vulnerable if, due to mass segregation, they are preferentially located in the outer parts of the cluster. Recent evidence for the effectiveness of disk shocking in stripping low mass stars from globular clusters comes from deep HST luminosity functions which show turnovers at faint magnitudes which are more dramatic in clusters closer to the Galactic plane (cf. Section 5). A full investigation of this effect requires computer simulations which incorporate a knowledge of each cluster's orbit (cf. Dinescu, Girard & van Altena 1998).

2.7. *Core Collapse Time Scale*

Core collapse is the phase of evolution which sets in after a number of two-body relaxation times have elapsed, and is especially effective in clusters which started out with relatively high central concentrations. The basic mechanism is that thermal energy, in the form of escaping stars, flows from the core to the halo of the cluster. The core pressure decreases, and a collapse ensues. Potential energy in the core is converted to kinetic energy, and because of the negative heat capacity of self-gravitating systems, if the core is sufficiently compact, a runaway collapse follows.

The time scale for core collapse is best estimated numerically. It will depend on the IMF, the degree of anisotropy, and the initial density. Estimates of the time scale range from $\sim 16t_{2body0}$ (the central two-body relaxation time) for models with stars all of the same mass, to $\sim 0.9t_{2body0}$ for models which include a realistic mass function. In addition to the above factors, tidal shocking may accelerate core collapse and mass loss from stellar evolution may delay it. Simulations suggest that, following the initial collapse, a series of rebounds and re-collapses may ensue, although during the re-expansion the core never reaches its pre-collapse size. This is illustrated in Fig. 4 with recent numerical simulations of core collapse from Grabhorn *et al.* (1992). The effect of core collapse on the surface brightness profile is discussed in Section 3.

2.8. *Time Scale for Evaporation*

Models of post-core-collapse evolution suggest that the time scale for a cluster to evaporate altogether is $\sim 10 - 20t_{rh}$. This depends on many factors including the environment

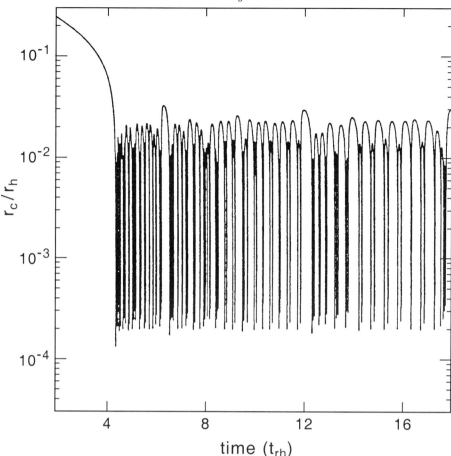

FIGURE 4. Simulations of a cluster undergoing gravothermal oscillations, from Grabhorn *et al.* (1992). The ratio of core radius to half-mass radius is plotted as a function of the half-mass relaxation time. Note that the core never re-expands to its initial size.

in which the cluster finds itself (presence of a disk, bar, bulge, giant molecular clouds, etc.), and is best determined numerically. Hut & Djorgovski (1992) estimate from simple arguments using measured core and half-mass relaxation times for a Galactic globulars, that the current death rate of clusters is ~ 5 per 10^9 year. If it has been similar in the past, then at least ~ 60 clusters will have already disintegrated, contributing their stars to the Galactic halo.

3. Surface Brightness Profiles of Globular Clusters

The property of a star cluster most easily observed, apart from its integrated magnitude and colours, is its surface brightness profile. Surface brightness profiles for the brightest globular clusters were derived using star counts from photographic plates as early as the beginning of the 1900s. The 1950s brought the first experiments with electronographic observations. Using this technique, Gascoigne & Burr (1956) determined the surface brightness profiles of 47 Tuc and ωCen, at radii 0.1 to 40 arcmin $(0.1 - 50$ pc$)$ and compared the results with the $r^{1/4}$ law of deVaucouleurs, known to represent well the surface brightness profiles of elliptical galaxies. The two clusters showed similar but not identical profiles, diverging in the inner regions both from each other and from the $r^{1/4}$ law.

Important advances were made in the 1960s with the work of King (1962, 1966a,b), who for the first time, approached the clusters simultaneously from the observational and theoretical points of view. He determined profiles first in two and later in seven more bright globular clusters, using the best technology then available, to push the observations further into the inner regions. To quantify the surface brightness profiles, he explored first a set of empirical models, and a few years later, a set of analytical models based on the internal stellar dynamics of globular clusters. These models, widely referred to as King models, form the basis of all studies of the dynamical evolution of rich star clusters.

In general the surface brightness profiles of globular clusters display a core with roughly constant surface brightness. The core radius is taken to be the radius where the surface brightness has decreased to $\sim$ half its central value (exactly half in the case of empirical King models). Outside the core the surface brightness falls off roughly as a power law until the outermost regions, where a sharp turndown indicates the tidal limit of the cluster. A useful scale is the half-mass (or light) radius, which is the radius containing half the total mass (or light) of the cluster.

Empirical and analytical King models are discussed in Section 3.1. Observations of surface brightness profiles are discussed in Section 3.2, including interesting deviations from the general trend. Section 3.3 describes an intriguing trend of core radius with age among LMC clusters, and Section 3.4 includes a comment on numerical modelling.

3.1. Models of Globular Cluster Surface Brightness Profiles

3.1.1. Empirical King models

The first parameterization of surface brightness as a function of radius tailor-made for globular clusters, were the empirical models of King (1962). At the time, a frequently cited law describing the density of star clusters was that of Jeans (1916), which fell off as r^{-3} (in projection) in the outer parts. However, early starcounts (cf. Bailey 1915) appeared to fall off less steeply than this, and the need for a better parameterization was evident. Using the data of Gascoigne & Burr for 47 Tuc and ωCen, together with profiles for two other clusters (M13 and M15) which he derived using starcounts on plates taken with the 48-inch telescope at Mt. Palomar, King (1962) found that in the outer parts of the clusters the surface density, f, fell with radius according to

$$f = f_1 \left[\frac{1}{r} - \frac{1}{r_t} \right]^2 .   \tag{3.1}$$

Here f_1 is a constant, and r_t is the radius at which $f = 0$, in other words, the tidal radius. In the inner parts of the clusters he found a good match to the data with a function of the form

$$f = \frac{f_0}{1 + (r/r_c)^2} .   \tag{3.2}$$

Here f_0 is the central surface density, and the scale factor, r_c, is the radius at which the surface density drops to half its central value, in other words, the core radius. A single expression that shares the characteristics of both equations (3.1) and (3.2) is

$$f = k \left[\frac{1}{(1 + (r/r_c)^2)^{1/2}} - \frac{1}{(1 + (r_t/r_c)^2)^{1/2}} \right]^2 .   \tag{3.3}$$

In a typical cluster, $r_t/r_c \approx 30$ so the second term has a value ≈ 0.03. So for $r \ll r_t$ we recover equation (3.2). On the other hand, if $r \gg r_c$ then equation (3.3) becomes

$$f = k \left[\frac{r_c}{r} - \frac{r_c}{r_t} \right]^2. \tag{3.4}$$

Comparing with equation (3.1), we see that equation (3.4) indeed reduces to equation (3.1) in the outer parts of the cluster for $f_1 = kr_c^2$. Equation (3.3) thus represents a family of curves which provide good representations of observed cluster profiles, with two free parameters (see King 1962, Fig. 5). These are the core radius, r_c, and the concentration parameter, $c = \log(r_t/r_c)$. The concentration parameter gives the curvature of the surface brightness profile at intermediate radii, which is much more easily observed than the actual tidal radius. Equation (3.3) provides a convenient expression for quickly estimating the core radius and concentration parameter for a cluster, or for generating artificial images of globular clusters as they might appear around distant galaxies, to assess, for example, the feasibility of using them to probe that galaxy's mass distribution.

3.1.2. Dynamical King models

A few years after publishing his empirical models, King (1966a) developed a set of dynamical models of the surface brightness profiles of globular clusters based on the physics of spherically symmetric, isotropic, self-gravitating systems made of stars with a single mass. The basis of any model of a cluster is a distribution function $f(r, v, m, t)$ such that $f(d^3 r d^3 v dm)$ is the mean number of stars with positions in an interval dr, velocities in an interval dv, and masses in an interval dm, as a function of time. A star cluster can be thought of as a gas in which a kind of diffusion is taking place, with many small changes to the particles' (=stars') motions due to distant encounters. The distribution function must satisfy the Boltzmann equation:

$$\frac{df}{dt} + v \frac{df}{dr} - \frac{d\Phi}{dr} \frac{dr}{dv} = \left[\frac{df}{dt} \right]_{enc}. \tag{3.5}$$

Here Φ is the smoothed gravitational potential per unit mass, and the right hand term is the evaluation of the Boltzmann collision integral over interactions between a test star and other stars. Once orbits are mixed, which happens on the scale of a crossing time, the phase space density no longer changes significantly, so $df/dt \approx 0$. The cumulative effect of distant encounters will change the distribution function very slowly on the time scale of the two-body relaxation time. This is generally very long, so equation (3.5) may be reduced to the collisionless Boltzmann equation:

$$v \frac{df}{dr} - \frac{d\Phi}{dr} \frac{df}{dv} = 0. \tag{3.6}$$

According to Jeans' theorem, the distribution function f must be a function of the integrals of the equations of motion of a star. These are:

$$E = \frac{1}{2} v^2 + \Phi(r) \tag{3.7}$$

$$J = r v_t \tag{3.8}$$

where v_t is the transverse velocity. The distribution function f must correspond to the mean density $\rho_m(r)$ which gives the cluster potential Φ. Poisson's equation gives

$$\nabla^2 \Phi(r) = 4\pi \rho_m(r)$$

$$= 4\pi G \int m f d^3 r d^3 v dm. \qquad (3.9)$$

There are many possible solutions. One is the isothermal sphere, in which the space density falls off as r^{-2} (and projected density as r^{-1}). This gives a total mass which diverges when integrated over radius and is not, therefore, a good model for a real cluster. A second solution is the Plummer model, which is often used by theorists (see Section 3.4). A third is the family of analytical King models, which give more realistic representations of the actual surface density profiles of globular clusters.

King derived his models using the Fokker-Planck equation, which is the Boltzmann equation (3.5) with the Liouville terms set to zero, and the term $(df/dt)_{enc}$ written out in full. This approach follows on from the work of Spitzer & Härm (1958), and comes from making the approximation that the cumulative effect of distant encounters is very much more important that the effect of close encounters. This will be true in all but the densest regions of globular clusters. This gives

$$\frac{df}{dt} = -\frac{1}{v^2}\frac{d}{dv}\left[a(v)\left(\frac{df}{dv} + 2vf\right)\right]. \qquad (3.10)$$

He then separates variables: $f(v,t) = g(v)h(t)$. The equation for the change with velocity is

$$0 = \frac{d}{dv}\left[a(v)\left(\frac{dg}{dv} + 2vg\right)\right] + \lambda v^2 g(v) \qquad (3.11)$$

The eigenvalue λ is determined by imposing the condition that $g(v) = 0$ at the escape velocity v_{esc}. Solving (3.11) gives the velocity distribution known as a "lowered Maxwellian", which has the form

$$f(E) = K(e^{-\beta E} - e^{-\beta E_0}) \qquad (E < E_0) \qquad (3.12)$$

$$f(E) = 0 \qquad (E > E_0)$$

Thus, one first solves the steady-state Fokker-Planck equation (3.10), and then converts the resulting velocity distribution into a density distribution. The result is a family of models, shown in Fig. 5, characterized by one free parameter, the central potential, denoted W_0.

3.1.3. Refinements

Many assumptions have gone into the above derivation of dynamical King models: that all the stars have the same mass; that there is no mass loss due to stellar evolution (the stars do not evolve); that the system is spherically symmetric; and that the velocity distribution is isotropic. Despite these simplifying assumptions, the King models represent the observed profiles of globular clusters remarkably well, and allow general studies of relations between their structure, Galactocentric radius, height above the disk, etc. (cf. Djorgovski & Meylan 1994).

However, some clusters show interesting deviations from these models. Explaining these requires extensions of the basic theory to include a range of stellar masses and velocity anisotropies. Deviations from King models in the outermost regions of old globular

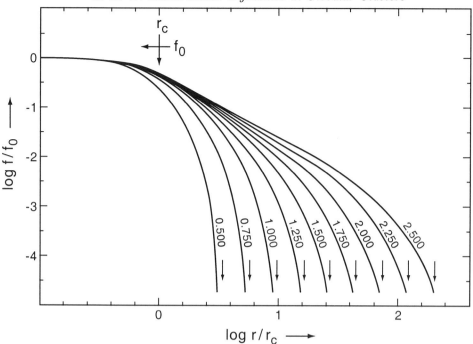

FIGURE 5. Surface density as a function of radius for a family of analytical King models from King (1966a). The number associated with each curve is the concentration parameter.

clusters can be attributed to the presence of radial anisotropies. The stellar orbits in a dynamically old cluster (ie. one which is many times older than its two-body relaxation time), are expected to be isothermal, no matter how they started out. Typically, globular clusters are much older than their two-body times out to at least the half-mass radius. However, in the outer parts, where denisties are low and stellar encounters are relatively rare, they may not be fully relaxed, and traces of their early state (after their initial violent relaxation) may persist in the form of predominantly radial orbits. The degree of radial anisotropy in an unrelaxed cluster may be indicative of whether the initial collpase was from relatively warm initial conditions (high ratio of kinetic to potential energy), or cool initial conditions (low ratio of kinetic to potential energy). Estimating the degree of radial anisotropy from the surface brightness profiles of clusters has been explored, for example, by Gunn & Griffin (1979) (M3); Lupton (1985) (M13, M92); Lupton *et al.* (1987) (M13); and Meylan & Mayor (1991) (NGC 6397).

Deviations due to other effects, including core collapse, are discussed in Section 3.2.1 and 3.2.2 below.

3.2. *Observations of Surface Brightness Profiles*

The surface brightness of a typical globular cluster may vary by five orders of magnitude from the core to the outer regions and near the tidal radius it may drop below the background surface brightness. This is illustrated in the surface brightness profile of 47 Tuc from Illingworth & Illingworth (1976) reproduced in Fig. 6. This large variation in density means that different methods must be used to determine the structure in different regions.

With a typical tidal radius of 30 pc, a cluster at a typcial distance of 10 kpc covers an area on the sky of ~ 300 arcmin^2. In the outer regions, star counts on photographic

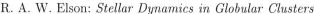

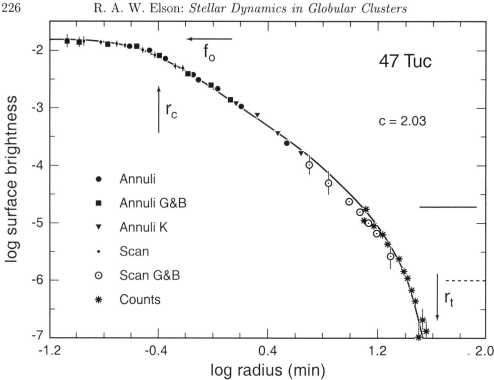

FIGURE 6. Surface brightness profile of 47 Tuc from Illingworth & Illingworth (1976). 'Annuli' are centered apertures measurements, 'scan' refers to photoelectric scans, and 'counts' refers to star counts on photographic plates. The parameters f_0, r_c and r_t are the central surface brightness and core and tidal radii from King's (1966a) models. c is the concentration parameter. The horizontal lines indicate the background level for the different observations. Surface brightness units are $V = 10.0$ mag arcsec^{-2}.

plates generally provide the spatial coverage needed, and are not affected much by errors due to the crowding of stellar images. Photographic plates also generally provide a large area beyond the tidal limit of the cluster in which the background stellar density may be determined accurately. The exact value of the tidal cutoff in a globular cluster is very sensitive to background subtraction since the outer parts may have densities below that of the background. Ideally such counts could be performed from observations made with a mosaic of CCD images rather than photographic plates. Large format CCDs mounted on small telescopes (cf. Fischer *et al.* 1993) will eventually replace meausrements made from photographic plates, but for the moment these remain, for many clusters, the best available data in the outer regions.

At intermediate radii, the surface density of stars in a typical globular cluster is sufficiently high that the surface brightness profile can be determined accurately from observations of the integrated light. This can be done either through spot measurements with a small aperture, through concentric apertures with appropriate diameters centered on the cluster, or with driftscans. This technique can be improved by observing in the U-band, where giants have roughly the same magnitude as stars near the main-sequence turnoff, rather than in the more commonly used V-band.

The structure of the central regions of a cluster is best determined using CCD observations. A typical core radius for a globular cluster is ~ 1 pc which, at a typical distance of 10 kpc, corresponds to 20 arcsec. Observations made with HST's Faint Object Camera (FOC) in the core of 47 Tuc reveal $\sim$ 440 stars down to $\sim$ 3 magnitudes below the main-sequence turnoff (Paresce, De Marchi & Jedrzejewski 1995). With a field size 7 x 7 arcsec, this corresponds to a surface density of $\sim 3 \times 10^4$ stars per arcmin^2. This is well beyond reach of star counts on photographic plates, and requires either measurements of integrated light, or star counts on CCD images. Here the limiting factor for ground based observations is seeing, typically imposing an inner limit to the radius at which structure can be resolved of $\sim$ 1 arcsec. HST's Wide Field and Planetary Camera 2 (WFPC2) and FOC cameras have proved ideal instruments for studies of compact cores in globular clusters.

For many brighter clusters, the range of radii covered by these three observational approaches is adequate for charcterising the surface brightness profile for the purpose of exploring the cluster's internal dynamics. Examples of such profiles may be seen in King (1966b) and Illingworth & Illingworth (1976). A catalogue of surface brightness profiles for 125 Galactic globular clusters has been published by Trager, King & Djorgovski (1995).

Other clusters, however, show interesting deviations from the standard models, both in the outer and inner parts. This is illustrated, for example, in the profile of M15 (NGC 7078) reproduced in Fig. 7. The deviations in the outer parts may be due to the presence of a range of stellar masses which have undergone mass segregation such that the more massive stars form a more centrally concentrated subsystem, and the less massive stars, a more extended one. Such a deviation can also indicate the presence of stellar orbits which are predominantly radial rather than isothermal. Such velocity anisotropies are important because they may represent rare traces of the early phases in the clusters' evolution. In the case of young globular clusters (in the LMC) deviations from King models at large radii are pronounced, and suggest the presence of halos of unbound stars not yet stripped by the ambient tidal field.

At intermediate radii, surface brightness profiles of young LMC clusters show sharp cores, kinks, and bumps, which may also contain information about the clusters' initial evolution through violent relaxation (Section 3.2.1). Finally, some clusters such as M15 deviate from King models in their innermost regions. Rather than flattening out, the surface brightness profile continues to rise to the limit of observation. Such "cusps" were once thought to be caused by the presence of central concentrations of dark matter, perhaps masive black holes. Recently, a more widely accepted explanation is that such features are present in clusters which have undergone the runaway process of core collapse (Section 3.2.2).

An interesting question which cannot be explored in the context of Galactic globular clusters is What determines the initial core radius of a cluster? and How does it evolve with time? Section 3.3 describes an interesting trend in the core radii of rich LMC clusters, in which the upper envelope of the distribution of core radii increases with age. The youngest clusters all have very small cores, while at intermediate and old ages, there is a large range of core radii.

3.2.1. *Surface brightness profiles in young LMC clusters*

The effect of violent relaxation on the surface brightness profile of a globular cluster can be explored observationally in the context of the young globular clusters in the LMC. These young clusters show profiles which differ markedly from King models. The turndown at the core radius may be sharper, the surface brightness profile at intermediate radii may show kinks, and there may be no turndown at all where the tidal radius should be. Surface brightness profiles for a sample of young LMC clusters are presented by

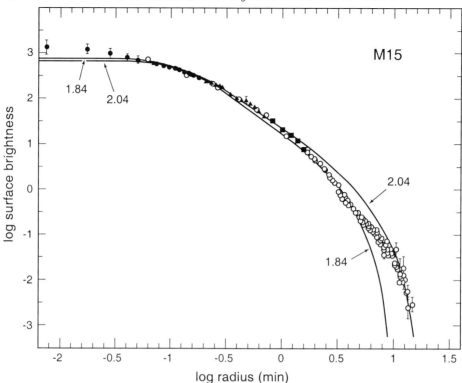

FIGURE 7. *B*-band surface brightness profile of M15 from Newell & O'Neil (1978). The logarithm of surface brightness is on an arbitrary scale. Different symbols correspond to different measurement techniques. Filled circles are electronographic measurements and open circles are star counts. King (1966a) models for two different concentrations are shown. Both models under-predict the surface brightness in the central ~ 2 arcsec.

Elson *et al.* (1987) and Elson (1991, 1992). An example of one of these profiles is shown in Fig. 8 for NGC 1818, a $\sim 3 \times 10^7$ year old cluster in the LMC.

The profiles of the young LMC clusters may be fit by a function of the form

$$\mu = \mu_0/(1 + (r/\alpha)^2)^{\gamma/2}. \tag{3.13}$$

For $\gamma = 2$ this is essentially a King (1962) model without the turndown at the tidal radius, and for $\gamma = 4$ it has the same radial dependence at large radii as the Plummer model. The parameter α is related to the King (1962) core radius according to $r_c = \alpha(2^{2/\gamma} - 1)^{1/2}$. For the 15 clusters from Elson *et al.* (1987) and Elson (1992), the mean value of γ, the slope of the power law portion of the profile, is 2.7 ± 0.5. Elson *et al.* (1987) discuss the implications of this slope for the initial ratio of potential to kinetic energy in these clusters. Simulations of dissipationless collapse by McGlynn (1984) explore the relation between the initial clumpiness and temperature of a system, and the core radius and slope of the surface brightness profile of the system after it has gone through violent relaxation. His results suggest a relation between the power-law slope of the profile at large radii, and the initial ratio of kinetic to potential energy. For the range of values of γ measured for the LMC clusters, the models imply relatively cool initial conditions, with $0.2 \lesssim 2T_0/W_0 \lesssim 0.5$.

The lack of tidal turndowns in the profiles suggests that these clusters have expanded, due to mass loss from stellar evolution, beyond the initial tidal radius of their progenitor

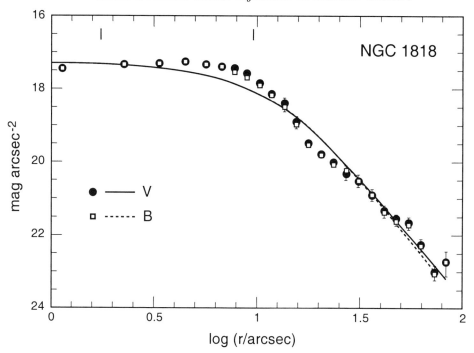

FIGURE 8. Surface brightness profile for NGC 1818, a young cluster in the Large Magellanic Cloud, from Elson (1991). The profile is derived from multi-aperture photometry on CCD images. The age of the cluster is $\sim 3 \times 10^7$ yr.

clouds, and that the unbound stars have not yet been stripped from the vicinity of the clusters. Indeed, models of the process of tidal stripping suggest that a cluster has to travel many times around the galaxy before the unbound stars will be lost (cf. Seitzer 1985). This will depend, of course, on the tidal field of the galaxy, the orbit of the cluster, and the orbits of stars within the cluster. (In the case of the LMC, the orbital period for a cluster at a galactocentric distance of 3 kpc, moving with the average disk rotation speed, is 2.6×10^8 yr.)

3.2.2. *Collapsed cores*

As an elderly cluster evolves, equipartition of energy among the stars, produced by two-body relaxation, causes the lower mass stars to speed up, while the higher mass stars slow down and sink towards the cluster centre. This process happens fastest in the core, where stellar densities are highest. The net effect is the escape of stars from the core to the halo of the cluster. The stars carry away kinetic energy or "heat", and the core "cools down" and contracts. In the denser core, two-body relaxation proceeds faster, and the whole process accelerates, producing a runaway collapse.

The process of core collapse was first explored by Hénon (1961) and by Lynden-Bell & Wood (1968), who dubbed it the "gravothermal catastrophe": for a self-gravitating system, there is no final equilibrium state. Eventually, as the core contracts further and further, a new physical process must become dominant. This is analogous to the ignition of helium in the core of a contracting star. In the case of the cluster, what is "burned" is binary stars. Through close encounters with other stars, binary systems are formed, and existing binary systems become more tightly bound. This "heat source" stops the runaway collapse, and lets the core begin to expand again. The collapse and re-expansion may be repeated many times over an extended period in what are termed "gravothermal oscillations". These oscillations were 'discovered' in computer simulations of model clusters (cf. Bettwieser & Sugimoto 1984). They have never, of course, been observed in real clusters, but circumstantial evidence for their existence is the fact that a substantial fraction of clusters appear to have 'collapsed' cores (see below). Since the primary collapse takes place so quickly (cf. Fig. 4), it would be unlikely that we would see so many clusters with collapsed cores unless each one were undergoing repeated collapses.

The first cluster in which a central enhancement in brightness was noted was M15 (cf. Newell and O'Neil 1978; Fig. 7). Ever since this discovery, efforts have been made to observe the core of M15 with increasing resolution, and with the launch of HST, it became an obvious target. Sosin & King (1997) show a surface brightness profile for the core of M15, derived from FOC images (Fig. 9). Theory predicts that a collapsed core will have a power-law profile with slope ~ -1.2 (in projection), depending on the stellar mass range observed (cf. Cohn 1985). A central black hole, on the other hand, will create a region around it with power-law slope ~ -0.75, more or less independent of stellar mass (cf. Bahcall & Wolf 1977). Stars near the main-sequence turnoff in M15 appear to follow a power law slope $\sim -0.70 \pm 0.05$, while lower mass stars, near $0.7 \, M_\odot$, have a slightly shallower slope of -0.56 ± 0.03. Sosin & King conclude that these slopes are not consistent with the predictions of either simple core collapse or black hole models. More sophisticated models are needed to decide between these two possible causes of M15's central cusp. Velocity dispersion data may also be useful in attempting to distinguish between the two possibilities. This is discussed further in Section 4.

Meanwhile, an obvious question was whether there were other clusters with similarly unresolved cores. A study by Djorgovski & King (1984) of nine clusters which were considered likely core collapse candidates revealed three clusters with central luminosity cusps. Evidently such objects were not rare. The next step was a systematic survey of clusters to find out what fraction of the clusters in our Galaxy had (according to the most popular interpretation) undergone core collapse. Out of 123 clusters, Djorgovski & King (1986) found 21 with cores which are not well fit even with high-concentration King models, and which appear to have power-law surface brightness profiles at small radii. These are identified as clusters which may have undergone core collapse, and they constitute nearly a fifth of the entire population.

3.3. *Core radius vs age in LMC clusters*

We have just seen what happens to the core of a cluster late in its life, but how does the core evolve before it reaches the collapse stage? Indeed, what factors establish the size of the core in a newborn cluster? These questions have not been explored extensively either theoretically or observationally. The objects best adapted for such a study are the LMC clusters, and a study of the structure of a sample of these clusters shows a trend of increasing core radius with age (cf. Elson 1992). This is illustrated in Fig. 15 (bottom right panel). Here, core radius is plotted against age for ~ 30 rich LMC clusters. The youngest all have small cores, with radius ≈ 8 arcsec (≈ 2 pc). Towards older ages, the upper envelope of the distribution increases.

Why should the youngest clusters all have small cores? What can this tell us about the processes through which they formed? The core of a cluster will expand due to mass loss from stellar evolution at a rate which depends on its IMF slope. Why should two intermediate age clusters with the same age have dramatically different core radii? Is this evidence for variations in IMF slope from cluster to cluster? What does the core radius vs age distribution tell us about the internal dynamics of the clusters? Have the intermediate age clusters with small cores already gone through core collapse, or did they

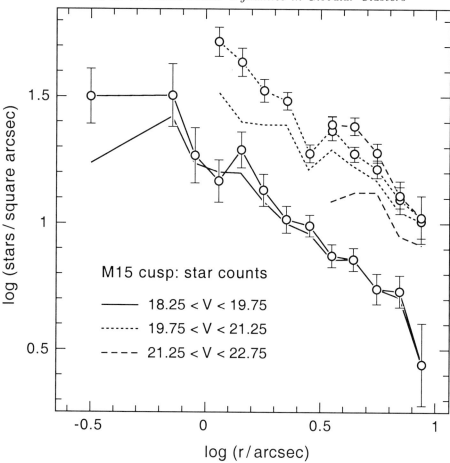

FIGURE 9. Surface brightness profile of the central regions of M15 from Sosin & King (1997). Circles are data corrected for incompleteness. Lines without points are counts prior to correction. The observations were obtained with HST's FOC.

never expand much to begin with? Some of these questions are addressed in Section 5, with some new observational results from a large HST study of rich star clusters in the LMC.

3.4. A note on numerical models

A discussion of numerical modelling techniques for globular clusters is beyond the scope of this chapter. A recent description of state-of-the-art techniques for N-body simulations of real size globular clusters, including full stellar evolution, binary interactions, etc., is provided by Aarseth (1998). One comment relative to input parameters is made here. Normally N-body simulations adopt as a starting point for the spatial distribution of stars, a King (1966a) model (for its physical basis), or a Plummer sphere (for its mathematical convenience). A Plummer sphere has surface denisty which falls off with radius according to

$$\mu(r) \propto \frac{1}{\left[1 + (r/R)^2\right]^2} \tag{3.14}$$

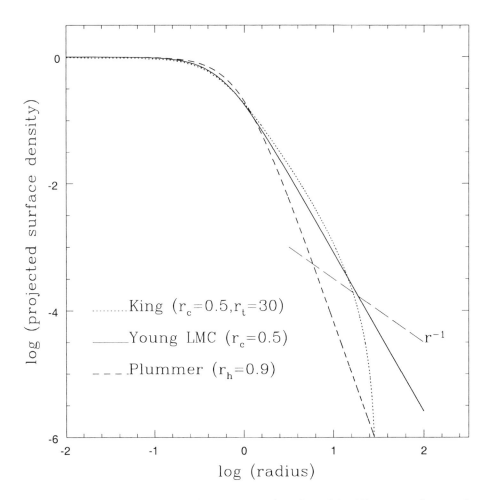

FIGURE 10. Surface brightness profiles for a King (1962) model, a Plummer sphere, and a young LMC cluster (equation (3.13) with $\gamma = 2.5$). The projected surface density at large radii for an isothermal sphere ($\propto r^{-1}$) is also indicated.

where the half mass radius (in projection) is related to the scale radius R according to $r_h = 0.996R$ (cf. Spitzer 1987). Figure 10 shows both a Plummer model and a King (1962) model for the same core radius. Also plotted is a profile for a young LMC cluster from equation (3.13) with $\gamma = 2.5$ and α such that the King (1962) core radius is the same as for the King model. The young LMC model and King model differ mainly near the tidal radius. The Plummer sphere falls off considerably more steeply than the young LMC cluster model. These differences may become less pronounced as the simulated cluster evolves, however if one wishes to make comparisons between a simulation and an observed surface-brightness profile for a young or intermediate age cluster, then it is important to start from realistic initial conditions. Also shown is the slope of an isothermal sphere at large radii, to emphasize the difference between its projected density as a function of radius and those of the other three models.

4. Velocity Dispersions in Globular Clusters

Dynamical models of globular clusters can only be partly constrained from observed surface brightness profiles. A full understanding requires observations of the velocity dispersion as a function of radius as well (and also of the IMF). The velocity dispersion can be measured either from the integrated light, or by measuring radial velocities of individual stars. Measurements of integrated light primarily provide central velocity dispersions, while measurements of individual stars can provide information about the velocity dispersion as a function of radius, and about rotation. Combined with physical models, velocity dispersions can be converted to dynamical masses and hence to mass-to-light ratios, and can be used to explore the subluminous population in a cluster (white dwarfs, neutron stars, very low mass stars, black holes). Velocity dispersions used to be the only available probe of the low mass end of the IMF, although with recent HST observations, the faintest stellar populations, including white dwarfs and main-sequence stars down to $\sim 0.1 M_\odot$, are accessible in the nearest globular clusters from direct observations of CMDs (see King, this volume).

Examples of the kinds of questions which can be addressed with velocity information are: By how much does the mass-to-light ratio vary from cluster to cluster? Is there any dynamical evidence for dark matter in globular clusters (cf. Peebles 1984)? Can we distinguish between central massive black holes and core collapse as explanations for the central luminosity cusps observed in some globular clusters? Which clusters rotate (and why)? Are velocity dispersions in all clusters isothermal all the way to the tidal limit? If not, what can the degree of anisotropy tell us about the cluster's history? What is the velocity dispersion as a function of radius like in young LMC clusters, and what can this tell us about their formation? Are there stars present in clusters with velocities greater than the escape velocity, which might be indentified with unbound halos in the process of being tidally stripped?

Velocity dispersions in globular clusters are small: typically 5 km/s in Galactic globulars, and perhaps $2 - 3$ km/s in young LMC clusters. Observations therefore require good resolution, and plenty of spectral lines. For this reason, stars with spectral types later than F are needed. This can be problematic in young LMC clusters, where the light is dominated by A and B type stars. The two methods for determining velocity dispersions are discussed in Sections 4.1 and 4.2 below. Our current knowledge of velocity dispersions in globular clusters is summarized in Section 4.3. Section 4.4 contains a brief note on the central velocity dispersion of M15.

4.1. *Velocity dispersions from integrated light*

The first determinations of velocity dispersions in globular clusters were derived from observations of their integrated light, using a method pioneered by Illingworth (1976). This is the only method available for clusters which are too distant to obtain spectra of individual stars. The basic idea is to compare a spectrum of the integrated light in the cluster centre with a template spectrum of a single star of similar spectral type to that of the stars dominating the integrated light. The lines in the cluster spectrum will appear broadened compared to the the lines in the spectrum of the individual star because of the random radial motions of the stars contributing to the integrated light.

The most straightforward method for comparing the spectra is by artificially broadening the template by convolving it with a Gaussian function with width corresponding to a given velocity dispersion. Illingworth estimated by eye the best match between the cluster and the broadened template. This automatically gives much more weight to the strongest lines, while essentially ignoring information in the weaker lines. Alternatively, the cluster and template spectra may be Fourier transformed and the slopes of their

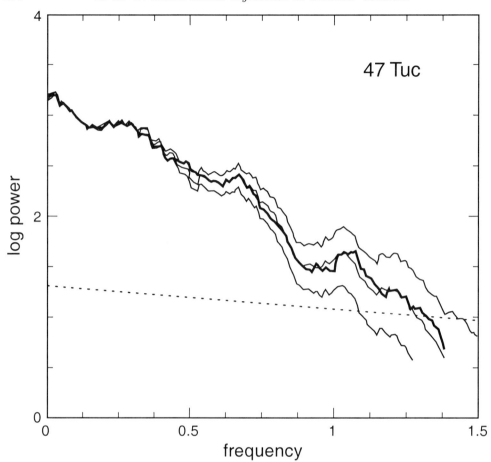

FIGURE 11. Fourier transformed spectrum of the integrated light of 47 Tuc superposed on Fourier transformed spectra of a template star convolved with Gaussian functions with width corresponding to 8, 10.5 and 13 km/s. The dotted line indicates the mean noise level.

power spectra compared. A mathematical description of this process is given by Illingworth (1976). It is illustrated in Fig. 11 for the case of 47 Tuc. The broader the lines are in the original cluster spectrum, the steeper the power spectrum will be. This second technique has the advantage of calling into service all the information in the spectrum and not just the most prominent lines. Illingworth (1976) applied both these techniques to ten bright southern globular clusters. The results given by the two techniques are in good agreement, with the Fourier method giving relatively smaller errors (±1.0 km/s compared to ±2.5 km/s for the direct comparison). A direct comparison between the cluster and template spectra may also be performed using cross-correlation techniques to determine which velocity dispersion matches best the observed spectrum. This is the method employed, for example, by Dubath *et al.* (1990) to measure the core velocity dispersion for the old LMC cluster NGC 1835.

A drawback to using integrated light is that the integrated light of a cluster, particularly in the centre, may be dominated by a small number of bright giants, and the observed velocity dispersion may not, therefore, be representative of the true velocity dispersion. This is discussed in detail by Zaggia *et al.* (1992b). Other things that can affect velocity dispersions derived this way are the presence of binary stars within the aperture, and the broadening of spectral lines due to turbulent motions in the atmospheres of stars.

The method can be refined by producing a composite template more representative of the cluster integrated light (for example, containing some component corresponding to a contribution by horizontal branch stars). It remains the only method for determining velocity dispersions in clusters too distant to obtain spectra of individual stars with sufficient resolution.

4.2. *Star-by-star velocity dispersions*

If a cluster is sufficiently nearby and uncrowded then a velocity dispersion may be derived by measuring radial velocities of individual stars. This is done by cross-correlating their spectra with a template sepctrum, and calculating the dispersion in the sample around the mean. The requirements for this are that the stars are bright enough to obtain spectra with good signal-to-noise ratio, that the resolution and stellar density are such that a spectrum is not contaminated by the light from neighbouring stars, and that the cluster is old enough that the spectral type of the stars accessible to observations is late enough to contain enough lines for successful cross-correlation.

Pioneering work in determining star-by-star velocity dispersions was carried out by Gunn & Griffin (1979), who determined radial velocities for 111 stars in M3 (NGC 5272). A crucial ingredient of their work was a special purpose spectrometer designed to do online cross-correlations and thus to determine radial velocities directly at the telescope, dramatically increasing the speed with which velocities for large numbers of stars could be measured. With this instrument they obtained an accuracy of 1 km/s.

Much work on deriving velocity dispersions from individual stars was done in the second half of the 1980s and first half of the 1990s by Meylan, Mayor, Dubath & their collaborators, using an instrument called CORAVEL. A technical description of CORAVEL is given by Barane, Mayor & Poncet (1979). It is a Cassegrain spectrometer for determining stellar radial velocities by on-line cross-correlation. It discards all information in the spectra except the Doppler shift, so can determine radial velocities very efficiently with errors better than 0.5 km/s. A second instrument which may be of interest for determining stellar radial velocities in globular clusters is the multi-object fiber-fed spectrograph Hydra, due to be available at CTIO in the near future. Its field size of 40 arcmin is well matched to the angular size of many Galactic globular clusters.

Recently, Fabry-Perot interfermometry has been used to measure velocities for, in some cases, over a thousand stars in a small sample of globular clusters. These include 47 Tuc, NGC 362 and NGC 3201 (Gebhardt & Fischer 1995) and M15 (Gebhardt & Fischer 1997).

A drawback to the star-by-star method, as with the method using spectra of the integrated light, is the systematic uncertainty introduced in the case where too few stars are used. Again, this is discussed by Zaggia *et al.* (1992a). Further uncertainties come from crowding (more than one star contributing to the light entering the slit), from the bias towards measuring velocities of the brighter (more massive) stars, and from the need to make a somewhat arbitrary judgement as to cluster membership based on radial velocity.

4.3. *Velocity dispersions in globular clusters*

A compilation of velocity dispersion data for 56 Galactic globular clusters is given by Pryor & Meylan (1993). (Velocity dispersions have also been measured for a few LMC clusters: NGC 1835 (Dubath *et al.* 1990); NGC 1866, NGC 2164, NGC 2214 (Lupton *et al.* 1989); NGC 1978 (Fischer *et al.* 1992a); NGC 1866 ((Fischer *et al.* 1992b). For 19

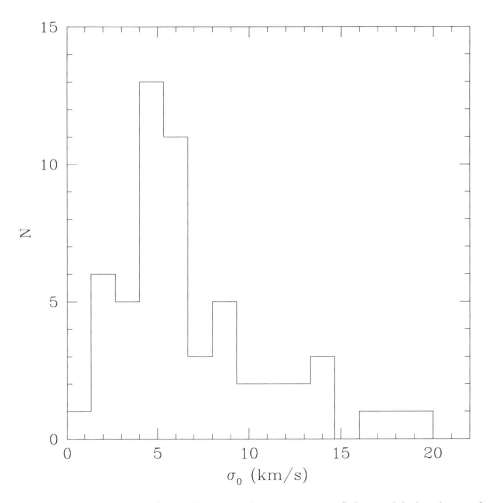

FIGURE 12. Histogram of central velocity dispersions in 56 Galactic globular clusters from Pryor & Meylan (1993), Table II.

of the Galactic globular clusters listed by Pryor & Meylan the velocity dispersions are derived only from integrated light, while for 37, star-by-star measurements are available. A histogram of the central velocity dispersions (from Table II of Pryor & Meylan) is shown in Fig. 12. The distribution peaks at a velocity dispersion of ~ 5 km/s, with a tail extending to ~ 20 km/s. The clusters with the largest velocity dispersions are NGC 6388 and NGC 6441 (both determined from integrated light). The smallest velocity dispersions measured are ~ 1.5 km/s, for NGC 5053 and NGC 6366 (from individual stars).

A simple conversion from a velocity dispersion to a mass assumes a lowered Maxwellian velocity distribution (equation 3.12) and Poisson's equation, and defines a dimensionless mass μ

$$\mu = \int_0^{R_t} \frac{\rho}{\rho_0} 4\pi R^2 dR \qquad (4.1)$$

where $R = r/r_c$ and $R_t = r_t/r_c$. The total mass can then be written as

$$M = 167 r_c \mu < v_r^2 >_0$$

(see King 1966a). Here, the total mass M has units of $M_\odot$, r_c is in pc, and v_r is in km/s. The parameter μ depends on the concentration of the cluster and ranges from ~ 4 to ~ 100 for values of $0.6 < \log c < 2.2$. It is tabulated in Table II of King (1966a). A typical cluster with $v_r = 5$ km/s, $r_c = 1$ pc, and $\log c = 1.8$ will therefore have mass $\sim 1.7 \times 10^5 M_\odot$. A better mass estimate must employ a multi-mass King model, and hence an assumed IMF, and must also take into account the degree of mass segregation present in the cluster. For deriving their total masses, Pryor & Meylan use a multi-mass King model with five mass bins, and a two-segment power-law IMF with slope 1.35 at $0.32 < m < 8 M_\odot$ and slope 0.0 at $0.16 < m < 0.32 M_\odot$.

The IMF slope, particularly at the low mass end, is the factor that affects the total mass of a cluster the most, while the metallicity can have a significant effect on the total luminosity. To explore the combined effect of these two factors on the mass-to-light ratio, it is instructive to consider simple population synthesis models. This can also provide an indication of the degree to which the mass-to-light ratio might serve as a probe of the subluminous component of a star cluster. Here I consider a single-burst population with age 12 Gyr, and a single power-law IMF with Salpeter slope (1.35) and upper mass cutoff $60 M_\odot$. The models are the same as those used to predict colours and spectra for galaxies at high redshift (Kodama 1998, private communication). They include the giant branch, horizontal branch and AGB as well as the main-sequence, and include remnant white dwarfs and neutron stars. They were run for a range of metllicities, and for three different values of the lower mass cutoff, m_l, 0.1, 0.2, and $0.3 M_\odot$. They are plotted in Fig. 13. Metallicity alone can change the mass-to-light ratio by a factor of two for the range of metallicities observed among Galactic globular clusters. A lower mass cutoff varying from 0.1 to $0.3 M_\odot$ can also change the mass-to-light ratio by a factor of two at a given metallicity. (Again, it must be stressed that recent observations of the low mass end of the stellar IMF in nearby clusters indicate that at least in these clusters the power-law slope of the IMF is much flatter than the Salpeter value at $m \lesssim 0.2 M_\odot$.)

For comparison, the measured global mass-to-light ratios for the globular clusters listed in Pryor & Meylan (1993) are also plotted in Fig. 13. This comparison is not strictly valid since Pryor & Meylan assumed a different IMF, truncated at $0.16 M_\odot$ and flat below $0.32 M_\odot$ in deriving mass-to-light ratios from velocity dispersions. However, it serves as a crude comparison of the range of observed mass-to-light ratios and of mass-to-light ratios derived from simple population synthesis models, and to give a sense of the extent to which velocity dispersions can provide useful probes of the faint end of the IMF in clusters too distant to observe it directly.

Pryor & Meylan note that "An understanding of the variations in globular cluster mass-to-light ratios is an important goal of cluster studies in the $5 - 10$ years." Five years from their writing this, the question still remains, although deep observations with HST of the IMF in nearby clusters, extending even to the hydrogen burning limit (cf. King, this volume) will undoubtedly help provide an answer.

4.4. *A note on the velocity dispersion in M15*

In Section 3.2.2 surface brightness profiles in the innermost regions of M15 were discussed, and the conclusion (from the work of Sosin & King 1997) was that structural data were unable to distinguish between the competing interpretations of a collapsed core and a central black hole to explain the central luminosity cusp. What can velocity dispersion data contribute towards answering this question?

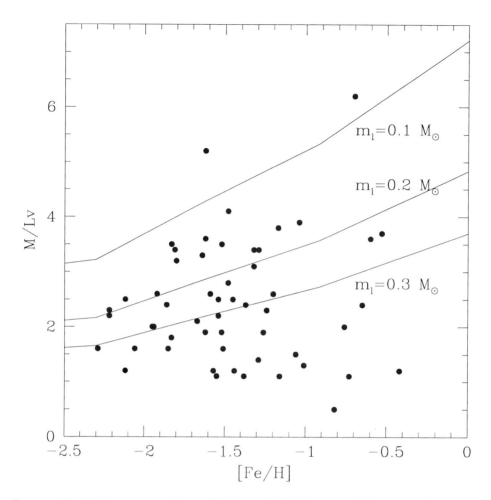

FIGURE 13. Points: measured metallicities and global mass-to-light ratios for the globular clusters listed in Pryor & Meylan (1993). Lines: values of M/L_v from single-burst population synthesis models for a power-law IMF with Salpeter slope (1.35) and three different values of the lower mass cutoff.

Dubath, Meylan, & Mayor (1994) derive a velocity dispersion from integrated light in the central 6×6 arcsec of M15. They derive a value $\sigma = 15^{+6}_{-4}$ km/s. The large errors are due to uncertainties from the inevitable sampling of small numbers of stars in a small area. These results appear at odds with those of Peterson, Seitzer & Cudworth (1989) who derived a value $\sigma = 25$ km/s from integrated light in the central 1×1 arcsec. However, Dubath *et al.* argue that, given sampling uncertainties in Peterson *et al.*'s much smaller area, the results are probably not inconsistent. Dynamical models of M15 imply central velocity dispersions consistent with the Dubath *et al.* value, and they conclude that, while a central black hole cannot be ruled, it is not required by the observations.

Gebhardt *et al.* (1994) derive a velocity dispersion for the core of M15 using Fabry-Perot measurements for 216 stars less than 2.6 arcmin from the centre. Their innermost point is at a radius of 6 arcsec. Between 36 and 6 arcsec the velocity dispersion appears to be constant. In a more recent study using Fabry-Perot interferometry, Gebhardt & Fischer (1997) measure velocities for ~ 1500 stars in M15. Over 100 are within 10 arcsec

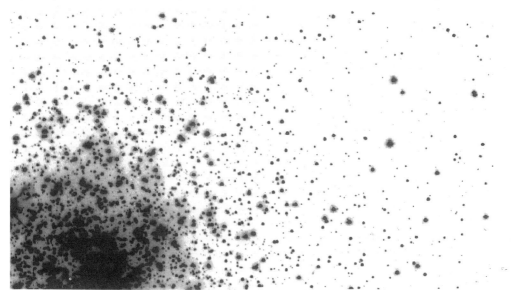

FIGURE 14. A STIS image of NGC 1868, an intermediate age star cluster in the LMC. The exposure time is $\sim 15,000$ seconds. The field size is $\sim 30 \times 50$ arcsec. The image was obtained with the long pass filter, which roughly spans V, R and I.

of the centre, and 12 are within 2 arcsec. They find a velocity dispersion profile which is flat to 1.2 arcsec, the limit of their observations. Again, their data neither support, nor rule out, the possibility of a central black hole.

5. Formation and evolution of globular clusters in the LMC

Our Galaxy contains two distinct populations of clusters: the ancient globular clusters in the halo, and the much smaller young and intermediate age open clusters in the disk. The LMC on the other hand, contains a large population of star clusters with masses similar to those of our globular clusters, $\sim 10^4 - 10^5 \, M_\odot$, and with ages spanning the full range from $\sim 10^6$ to $\sim 10^{10}$ years. It is not clear why the LMC is still forming massive clusters, while our Galaxy ceased to do so $\sim 10^{10}$ years ago; it may be due to the relatively weaker tidal field in the LMC. In any case, the LMC clusters provide an ideal laboratory for exploring the formation of rich clusters, the evolution of N-body systems, and for investigating the stellar IMF. With their span of ages, they provide "snapshots" of clusters at all evolutionary stages.

From a practical point of view, the clusters are ideally suited for such studies. They are all at a distance which can be determined accurately by many independent means. The LMC is at Galactic latitude -30 degrees, well out of the plane of the disk. Contamination from foreground disk stars is negligible, in marked contrast to the situation for Galactic open clusters. The angular size of the clusters is well suited to the instruments on board HST: a typical cluster has an apparent diameter about the size of the WFPC2 field of view (~ 3 arcmin), and core dimensions similar to those of the PC ($\sim 36 \times 36$ arcsec). At the distance of the LMC, the range of magnitudes accessible to HST's instruments in reasonable exposure times ($V \sim 15 - 28$) covers most of the range of stellar masses ($\sim 0.3 - 10 \, M_\odot$) potentially present in the clusters with ages $\gtrsim 10^7 \, M_\odot$.

An allotment of 95 orbits of HST time in Cycle 7 (~ 1998) has been devoted to a study of the formation and evolution of rich star clusters in the LMC. This project

(GO-7307) is unique in its approach of using a small, carefully selected sample of LMC clusters to address major questions concerning cluster formation and evolution. A close collaboration with a team performing state-of-the-art N-body simulations of clusters on a special purpose, ultra-fast HARP computer (cf. Aarseth 1998) ensures an optimum interplay between observation and theory. This section provides a general description of the strategy and targets, and summarizes the early results.

The primary goals of the project are as follows, and are illustrated in Fig. 15. In the youngest clusters the fraction of primordial binaries, the IMF, and the degree of primordial mass segregation will teach us about the process of star formation in a protocluster. Age spreads and a search for pre-main-sequence stars will reveal both the time scale for star formation, which has important implications for its trigger, and whether the low or high mass stars form first. In the intermediate and old clusters we will trace the development of mass segregation and the binary fraction in the core and at the half-mass radius; binaries play a crucial role in the structural evolution of a cluster, and in particular, affect the process of core collapse. Finally, we will derive deep IMFs in all the clusters. A knowledge of the form of the IMF (and its variability from cluster to cluster) will provide important constraints for models of star formation in a dense environment, and will contain clues about the state of dynamical evolution of the clusters.

Our sample includes eight clusters which are among the richest in the LMC, and have masses $\sim 10^4 - 10^5 M_\odot$. Crossing times at the half-mass radius are $\sim 10^7$ years, and two-body relaxation times are $\sim 10^8$ years at the centre and $\sim 10^9$ years at the half-mass radius (Elson, Fall & Freeman 1989). Our clusters are chosen with ages spanning this range, and should thus manifest characteristics reflecting the evolutionary processes that operate on all time scales. They are paired in age to help discriminate between trends and coincidences (eg. in the IMFs), and each pair is at a similar distance (and if possible in a similar direction) from the centre of the LMC, to minimize any differential effects of the tidal field of the LMC on the cluster's evolution. They are also at the greatest possible distance from the centre, where the effect of the tidal field is smaller, and stellar backgrounds are sparser. We also chose pairs which are particularly interesting in having members with disparate properties such as metallicity or core radius.

The youngest clusters in our sample are free of gas, and are therefore pure N-body systems. They are older than their crossing times, and have probably gone through violent relaxation, but are not yet relaxed through two-body encounters (cf. Elson *et al.* 1987, 1989). Their structure, IMF, degree, if any, of mass segregation and of clumpiness in the stellar distribution, and the fraction of binaries in the core and further out, should therefore reflect their primordial state.

The IMF slope is particularly important in understanding the influences operating early in the life of a cluster. A flat IMF implies large numbers of massive stars. Stellar winds and supernovae would therefore have been important in removing gas from the protocluster, and in possible self-enrichment, or even destruction of the cluster (cf. Elson *et al.* (1989) for a discussion of these influences in the context of the young LMC clusters.) We will determine accurate IMFs in the youngest clusters over the range $\sim 0.3 - 10 M_\odot$.

These young systems will also provide clues about the time scale and sequence of star formation in the protocluster. For example, if star formation occurred on a time scale substantially shorter than the crossing time, we may infer that a strong perturbation such as a cloud-cloud collision was required to initiate it. In the absence of such a perturbation star formation would require a time scale comparable to the crossing time (cf. Elson *et al.* 1987). The star formation time scale in a young cluster may be determined from age spreads, manifested in the broadening of the main-sequence due to post-main-sequence evolution of massive stars, and pre-main-sequence contraction of low mass stars.

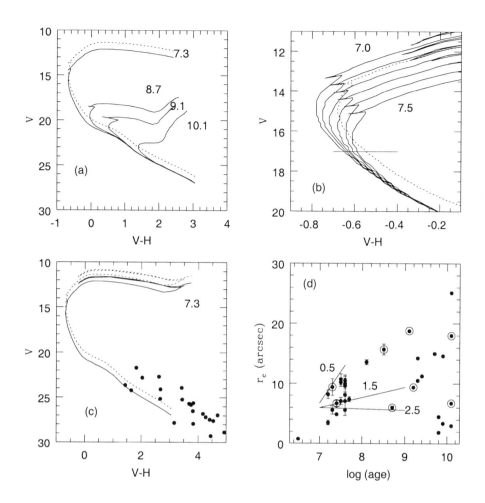

FIGURE 15. Schematic illustration of the goals of HST project GO-7307 to study the formation and evolution of rich star clusters in the LMC. Top left: isochrones for four different ages corresponding to our cluster pairs. The dashed curve indicates the position of a sequence of equal mass binaries. Top right: isochrones with ages in the range $1 - 3 \times 10^7$ yr, illustrating the scale of age spreads that would be detectable in a young cluster. Bottom left: points are pre-main-sequence stars from the Chameleon star forming region, as they would appear if viewed at the distance of the LMC. Bottom right: core radius vs age for LMC clusters. Models of core expansion for three different IMF slopes are shown (the Salpeter IMF has slope 1.35). The target clusters are circled.

Age spreads have been estimated this way for open clusters in the Milky Way (Stahler 1985; Adams, Strom & Strom 1983), but the results are limited by the relatively small number of stars, by the presence of variable reddening, and by the problem of determining membership. Attempts to estimate age spreads from the upper main-sequence in a few young LMC clusters have been limited by photometric accuracy and crowding (Robertson 1974).

Indirect evidence of variations in the IMFs of LMC clusters is given by the relation between core radius and age shown in Fig. 15 (bottom right). The clusters in our sample are circled. Dynamical models predict that the core radius will increase with age at a rate that depends on the IMF slope. Models for three IMF slopes are shown. According to this hypothesis, within each older pair of clusters we should find substantially different IMF slopes.

Our observing strategy takes advantage of the wide field of WFPC2, the near IR capability of NICMOS, and the sensitivity and dynamic range of STIS, used in imaging mode with the LP filter (which effectively spans the V, R, and I passbands). A STIS image of one of the clusters is shown in Fig. 14. STIS data will be used to derive deep luminosity functions. NICMOS2 with the F160W ($\sim H$) filter will be combined with observations with the WFPC2 F606W ($\sim V$) filter to construct $V - H$ colour-magnitude diagrams. Age spreads and the separation of any binary sequence is 2-3 times greater in a CMD constructed in $V - H$ than in one in $B - I$.

Results from a pilot study using archive data for the young cluster NGC 1818 have revealed a striking sequence of near equal mass binaries (Fig. 16) (Elson *et al.* 1998a). NGC 1805, the other young cluster in the sample, appears to have a similar binary population. The overall fraction of binaries is about 30%. These binaries have not had time to form through capture processes that can be effective in dense cluster cores, and must therefore be primordial. This is the only large sample of stars in the mass range $\sim 2 - 5 M_\odot$ for which the binary fraction has been measured. (For a discussion of binaries in old globular clusters, see King, this volume.)

It is important to investigate numerically the evolution of a more realistic cluster with a higher primordial binary fraction, as binaries can have a strong influence on cluster evolution. In the core, the formation of binaries is thought to be important in halting core collapse. Also, if a large fraction of the stars are in binary systems, then the effective mean mass in the cluster is higher, two-body relaxation times are effectively shorter, and mass segregation will proceed more rapidly. Indeed, we find a higher fraction of binaries in the core of NGC 1818 than in the outer parts. Modelling is needed to determine whether this effect is also primordial, or whether dynamical mass segregation has had time to occur.

As mentioned above, Fig. 15 (bottom right) raises an intriguing question: why should two clusters of approximately the same mass and age, in the same environment, have radically different core radii? We now have deep luminosity functions for two such clusters, NGC 1831 and NGC 1868. If very different IMFs provoked different rates of expansion, then we would expect NGC 1868 to have a much steeper IMF, in other words, proportionally many more low mass stars. Figure 17 shows preliminary luminosity functions from STIS data for these two clusters. The most striking difference between the two luminosity functions is that NGC 1868 appears to be missing its low mass stars. The luminosity function turns over at about 0.5 $M_\odot$ while that of NGC 1831 keeps rising. This is exactly the opposite of what was expected on the basis of the core-expansion hypothesis. It also raises the possibility of real IMF variations, unless we can find a dynamical explanation for the difference.

A comparison of luminosity functions in three radial ranges in NGC 1868 reveals the presence of mass segregation. Modelling is required to determine whether this is primordial, or whether the local two-body relaxation time in this $\sim 6 \times 10^8$ year old cluster is such that dynamical mass segregation could have occurred. But in any case, the mass segregation is in the normal sense: there are more massive stars in the centre, and more low mass ones in the outer parts. The luminosity function in Fig. 17 was constructed from stars in the outer parts, and so if it is affected at all by mass segregation, should contain relatively more, not fewer, low mass stars.

A possibility which remains is that the low mass star have been tidally stripped from the cluster. This effect can be seen in the luminosity functions of Galactic globular clusters. Figure 18 shows luminosity functions for eight globulars. These are from:

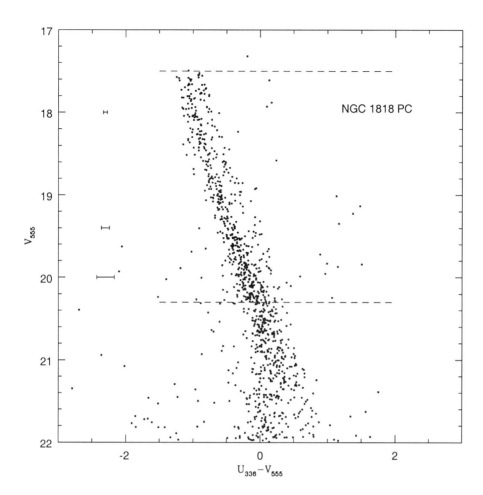

FIGURE 16. Colour-magnitude diagram for objects detected in the PC image (the central $\sim 2r_c$) of the young LMC cluster NGC 1818. A binary sequence stands out clearly from the main sequence. At $V_{555} \lesssim 17.5$ the stars are saturated. The dashed lines indicate the limits of the sample we used to analyse the binary population of this cluster, and correspond to single star masses $\sim 2 - 5.5 M_\odot$. Poisson error bars are shown. (See Elson *et al.* 1998a.)

Piotto, Cool & King (1997) (NGC 6397, NGC 6341, NGC 7078 & NGC 7099); Marconi *et al.* (1998) (NGC 5272); Elson *et al.* (1995) (ω Cen); and STIS images from HST project GO-7307 (NGC 6553), and from another Cycle 8 programme (47 Tuc). The luminosity functions have been normalized to match at $M_{814} \approx 4.8$. They all turn over at different rates at the faint end. These different rates of turnover may represent different rates of tidal stripping of the low mass stars. What could be responsible for stripping the stars at different rates? A likely explanation is tidal shocking. Tidal shocking is an effective way of heating a cluster, puffing up its outer parts, and making its low mass population particularly vulnerable to being torn from the cluster.

If this is the case, then one would expect the severity of the turnover in the luminosity function to correlate with the number of passages of the cluster through the Galactic disk. As a crude approximation, we can take the cluster's height above the Galactic disk as an indication of how much disk shocking it might have been subjected to. Figure 19

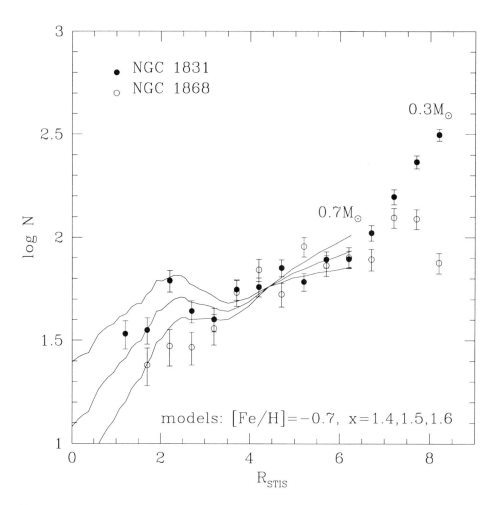

FIGURE 17. Luminosity functions from STIS for NGC 1831 (filled circles) and NGC 1868 (open circles). NGC 1868 has a much smaller core radius than NGC 1831. The solid curves are power-law IMFs with the slopes indicated (Salpeter = 1.35).

shows the relative differences in $\log N$, measured up from the horizontal dashed line in Fig. 18, plotted against the height Z above the disk. There is a striking correlation, suggesting that tidal shocking may indeed be influential in shaping the observed luminosity functions. A proper analysis of the effect of tidal stripping on the low mass end of the IMF would require detailed modelling, including a knowledge of the cluster oribts (cf. Dinescu *et al.* 1998). Also, a direct comparison of the luminosity functions is not strictly valid since the mass-luminosity relation is a function of metallicity, and metallicity varies from cluster to cluster in Fig. 18. An accurate conversion from luminosity functions to mass functions is required.

Returning to the case of NGC 1868 and NGC 1831, we might now speculate that NGC 1868 has been subjected to tidal shocking, while NGC 1831 has not. This might also provide an explanation for the difference in core radii: tidal shocking may accelerate the rate of internal evolution, and precipitate the cluster towards a premature core collapse. The only problem with this hypothesis is identifying the cause of tidal shocking. The intermediate age clusters in the LMC appear to participate in the general rotation of the

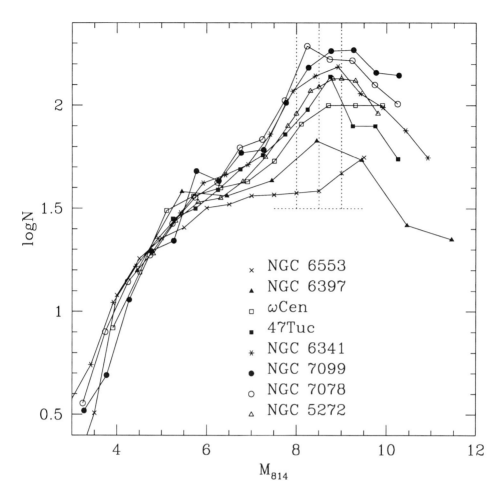

FIGURE 18. Luminosity functions for 8 Galactic globulars clusters. Normalization is arbitrary. (See Elson *et al.* 1998b.)

disk, and are not on orbits which carry them through the disk. NGC 1868 is at a large enough galactocentric radius that it should not have encountered the LMC bar. One would also need to explain why NGC 1831, which is at the same galactocentric radius and has a very similar positon angle and rotational velocity to NGC 1868, has escaped such tidal shocking. These are matters for N-body modelling to explore.

6. Acknowledgements

I would like to thank Richard Sword for his invaluable help in reproducing many of the figures in this chapter. I would also like to thank Ivan King for his many helpful comments during the Winter School, and for his careful reading of a draft of this manuscript, and many suggestions for improvements.

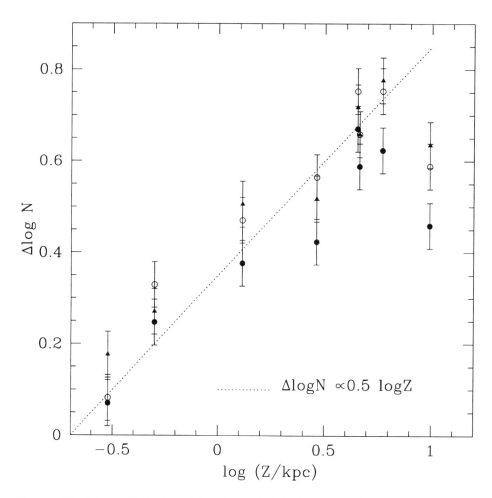

FIGURE 19. Relative flattening of the faint end of the luminosity functions in Fig. 18, measured at $M_{814} = 8.0, 8.5, 9.0$ (filled circles, open circles, triangles) plotted against distance above the Galactic plane.

REFERENCES

Aarseth, S. 1998 in IAU Colloq. 172, *Impact of Modern Dynamics in Astronomy*, eds. J. Henrard & S. Ferraz-Mello

Adams, M., Strom, K. & Strom, S. 1983 *Ap. J. Suppl.* 53, 893

Bahcall, J. N. & Wolf, R. A. 1977 *Ap. J.* 216, 883

Bailey, S. I. 1915 *Ann. Harvard Coll. Obs.* 76, No. 4

Barane, A., Mayor, M. & Poncet, J. 1979 *Vistas in Astron.* 23, 279

Bettwieser, E. & Sugimoto, D. 1984 *M.N.R.A.S.* 208, 439

Binggeli, B. 1994, in ESO/OHP Workshop on Dwarf Galaxies, ed. G. Meylan & P. Prugniel (ESO:Garching) p. 13

Chernoff, D. & Weinberg, M. 1990 *Ap. J.* 351, 121

Cohn, H. 1985 in IAU Symp. 113 *Dynamics of Star Clusters* eds. J. Goodman & P. Hut, p. 161

Djorgovski, S. & King, I. 1984 *Ap. J. (Letters)* 277, L49

Djorgovski, S. & King, I. 1986 *Ap. J. (Letters)* 305, L61

Djorgovski, S. & Meylan, G. 1994 *A. J.* 108, 1292

Dinescu, D., Girard, T. & van Altena, W. 1998 *preprint*

Dubath, P., Meylan, G., Mayor, M. & Magain, P. 1990 *Astr. Ap.* 239, 142

Dubath, P., Meylan, G. & Mayor, M. 1994 *Ap. J.* 426, 192

Elson, R. 1991, *Ap. J. Suppl.* 76 185

Elson, R. 1992, *M.N.R.A.S.* 256 515

Elson, R., Fall, S. M. & Freeman, K. 1987 *Ap. J.* 323 54

Elson, R., Fall, S. M. & Freeman, K. 1989 *Ap. J.* 336 734

Elson, R., Gilmore, G., Santiago, B. & Casertano, S. 1995 *A. J.* 110, 682

Elson, R., *et al.* 1998a, *M.N.R.A.S.* 300, 857

Elson, R., *et al.* 1998b, in IAU Symp. 190 *New Views of the Magellanic Clouds*

Elson, R. & Schade, D. 1994 *Ap. J.* 437, 625

Fall, S. M. & Rees, M. 1977 *M.N.R.A.S.* 181, 37P

Fischer, P., Welch, D. & Mateo, M. 1992a *A. J.* 104, 1086

Fischer, P., Welch, D., Côté, P., Mateo, M. & Madore, B. 1992b *A. J.* 103, 857

Fischer, P., Welch, D., Mateo, M. & Côté, P. 1993 *A. J.* 106, 1508

Fusi Pecci, F. *et al.* 1996 *A. J.* 112, 1461

Gascoigne, B. & Burr, E. J. 1956 *M.N.R.A.S.* 116, 570

Gebhardt, K., Pryor, C., Williams, T. & Hesser, J. 1994 *A. J.* 107, 2067

Gnedin, O. Y. & Ostriker, J. P. 1998 in ASP 136 *Galactic Halos*, ed. D. Zaritsky, p. 56

Grabhorn, R., Cohn, H., Lugger, P. & Murphy, B. 1992 *Ap. J.* 392 86

Gunn, J. & Griffin, R. 1979 *A. J.* 84, 752

Harris, W. 1996 *A. J.* 112, 1487

Hénon, M. 1961 *Annales d'Ast.* 24, 369

Hut, P. & Djorgovski, G. 1992 *Nature* 359, 806

Hut, P., Mcmillan, S. & Romani, R. 1992 *Ap. J.* 389, 527

Illingworth, G. 1976 *Ap. J.* 204, 73

Illingworth, G. & Illingworth W. 1976 *Ap. J. Suppl.* 30, 227

Irwin, M. & Hatzidimitriou, D. 1995 *M.N.R.A.S.* 277, 1354

Jeans, J. 1916 *M.N.R.A.S.* 76, 567

King, I. R. 1962 *A. J.* 67, 471

King, I. R. 1966a *A. J.* 71, 64

King, I. R. 1966b *A. J.* 71, 276

Lightman, A. & Shapiro, S. 1978 *Rev. Mod. Phys.* 50, 437

Lupton, R. 1985, IAU Symp. 113, *Dynamics of Star Clusters* eds. J. Goodman & P. Hut, p. 19

Lupton R., Fall, S. M., Freeman, K. C., & Elson, R. 1989 *Ap. J.* 347, 201

Lupton R., Gunn, J. & Griffin, R. 1987 *A. J.* 93, 1114

Lynden-Bell, D. 1967 *M.N.R.A.S.* 136, 101

Lynden-Bell, D. & Wood 1968 *M.N.R.A.S.* 138, 495

Marconi, G. *et al.* 1998 *M.N.R.A.S.* 293, 479

McGlynn, T. 1984, *Ap. J.* 281, 13

Meylan, G. & Heggie, D. 1997 *Astron. Rev.* 8, 1

Meylan, G. & Mayor, M. 1991 *Astr. Ap.* 250, 113

Newell, B. & O'Neil, E. 1978 *Ap. J. Suppl.* 37, 27

Ostriker, J., Spitzer, L. & Chevalier, R. 1972 *Ap. J. (Letters)* 176, L51

Paresce, F. De Marchi, G. & Jedrzejewski, R. 1995 *Ap. J. (Letters)* 442, L57

Peebles, J. 1984 *Ap. J.* 277, 470

Peterson, R. 1993 in *The Globular Cluster - Galaxy Connection*, eds. G. Smith & J. Brodie (ASP) p. 469

Peterson, R., Seitzer, P. & Cudworth K. M. 1989 *Ap. J.* 347, 251

Piotto, G., Cool, A. & King, I. 1997 *A. J.* 113, 1345

Plummer, H. C. 1915 *M.N.R.A.S.* 76, 107

Pryor, C. & Meylan, G. 1993 in *Structure and Dynamics of Globular Clusters*, (ASP Conf. Series. No. 50) eds. Djorgovski, S. G. & Meylan, G. (San Fransisco:ASP) p. 357

Robertson, J. W. 1974 *Astr. Ap. Suppl.* 15, 261

Schweizer, F. & Seitzer, P. 1998 *A. J.* 116, 2206

Seitzer, P. 1985 in IAU Symp. 113 *Dynamics of Star Clusters*, eds. J. Goodman & P. Hut (Dordrecht:Reidel) p. 343

Sosin, C. & King, I. 1997 *A. J.* 113, 1328

Spitzer, L. 1987, *Dynamical Evolution of Globular Clusters* (Princeton Universtiy Press: Princeton)

Spitzer, L. & Härm, R. 1958 *Ap. J.* 127, 544

Stahler, S. W. 1985 *Ap. J.* 293 207

Trager, S., King, I. & Djorgovski, S. 1995 *A. J.* 109, 218

Whitmore, B. & Schweizer, F. 1995 *A. J.* 109, 960

Zaggia, S., Capaccioli, M. & Piotto, G. 1992a *Astr. Ap.* 265, 43

Zaggia, S., Capaccioli, M. & Piotto, G. 1992b *Mem. Soc. Astron. Ital.* 63, 211

Zepf, S., Carter, D., Sharples, R., & Ashman, K. 1995 *Ap. J. (Letters)* 445, L19

MICHAEL W. FEAST was born in England and obtained a PhD in physics from Imperial College (London). He then spent two years as a post-doctoral fellow working with G. Herzberg at the National Research Council of Canada in Ottawa.

All his research work up to that time had been in laboratory studies of molecular spectra.

In 1952 he moved to the Radcliffe Observatory Pretoria, South Africa. This Observatory had at that time the largest telescope in the southern hemisphere (1.9m). The work there was mainly spectroscopy and centred on studies of the structure of our Galaxy and the Magellanic Clouds as well as astrophysical studies of objects of particular interest including the globular cluster 47 Tuc which was quickly shown to be unusual and is now the prototype of metal-rich globular clusters.

In 1974 Feast moved to Cape Town where the old Royal Observatory became the headquarters of the South African Astronomical Observatory (SAAO) and the 1.9m reflector was moved to the SAAO station at Sutherland. He was director of the SAAO from 1976 to 1992 and from 1983 an honorary professor in the University of Cape Town where he continues his research.

Much of his recent work has been on the use of variable stars to establish distance scales and in studies of galactic structure, the Magellanic Clouds and stellar evolution.

Pulsating Stars in Globular Clusters and Their Use

By MICHAEL W. FEAST

Astronomy Department, University of Cape Town,
Rondebosch, 7700, SOUTH AFRICA

1. About Pulsating Stars

1.1. *Introduction*

The discussions in the following sections are limited to pulsating variables and thus omit such objects as eclipsing and cataclysmic variables. Rather than try to cover every conceivable aspect of the subject an attempt is made to discuss in detail a few problems of current interest. This has meant that some types of pulsating variable are not dealt with at all, e.g. Type II Cepheids (including RV Tau stars), SX Phe variables and the recently discovered pulsating K giants in globular clusters (Edmonds and Gilliland 1996). In several of the areas covered strongly divergent views are held by different workers. In such cases an attempt is made to summarize the arguments of the various groups whilst at the same time indicating what in the present writer's opinion seems most likely to be the correct interpretation.

Pulsating stars are of importance for a variety of reasons. First, a study of their light, colour, and radial velocity, changes through the pulsation cycle tell us a great deal about the stars themselves - about their structure - which we cannot easily learn in other ways. Secondly, because pulsating variables are rather easily classified into groups with homogeneous properties it is possible to use them, provided their absolute magnitudes can be calibrated, to derive distances. Pulsating stars are at the basis of the galactic and extragalactic distance scales and are important in determining the distances and ages of classical, old, globular clusters. Such age determinations are done by comparing colour-magnitude diagrams of the clusters with theoretical isochrones. To make this comparison the distances of the clusters must be known. The ages determined depend sensitively on the distances adopted. A 10 percent change in adopted distance results in a 20 percent change in derived age.

1.2. *Basic Ideas on Stellar Pulsation*

Pulsating variables have been, and continue to be, of great importance for our understanding of stellar evolution generally and for globular clusters in particular. One can see this in a general way by considering why some stars pulsate and others do not. Ritter (1879) showed that the natural frequency of a homogeneous sphere performing adiabatic radial pulsations is given by,

$$(1/P^2) = (g/R) \times constant, \tag{1.1}$$

where P is the period of pulsation, g the surface gravity of the star, and R the star's radius.

Since $g \propto M/R^2$ where M is the mass of the star † we have

$$(1/P^2) = (M/R^3) \times constant \tag{1.2}$$

† Throughout M is used for a mass, $M_\odot$ for the mass of the sun and M with any other subscript for an absolute magnitude (i.e. M_V is the visual absolute magnitude).

and

$$P\sqrt{\rho} = Q \tag{1.3}$$

where ρ is the density of the star and Q is a constant.

Later work showed that this relation is approximately true for models of real stars with ρ being now defined as the star's mean density. The pulsation constant, Q, can be calculated for various stellar models. Within a given class of variable, the classical Cepheids † say, Q varies only slightly with stellar mass.

It is useful to see in a general way how the pulsation equation can be put in a form containing only observables. Stefan's law gives,

$$L = 4\pi R^2 \sigma T^4 \tag{1.4}$$

where L is the luminosity, T the effective temperature, and σ is Stefan's constant. Also it is known (e.g. from stellar evolution tracks) that the mass, M, can be expressed as a function of L, T and the chemical abundance (simplified here to [Fe/H]). Thus we have in general that the period is a function of L, T and [Fe/H], i.e.;

$$P = F_1(L, T, [Fe/H]). \tag{1.5}$$

For a class of stars (e.g. the Cepheids in the solar neighbourhood) where [Fe/H] is approximately constant, evidently,

$$P = F_2(L, T). \tag{1.6}$$

Since we can generally replace T by a colour (C), this is a period-luminosity-colour (PLC) relation.

The $P\sqrt{\rho}$ (or PLC) relation applies in principle to any star (with a suitable value of Q). However stellar pulsation is only found amongst quite restricted groups of stars. This is of great practical importance since in many cases we can deduce the properties of a pulsating star merely by knowing from its light variations that it belongs to a particular pulsation class.

Eddington (1917, 1918a,b) developed the basic theory of pulsating stars showing why one would expect some to pulsate and others not. Any star experiencing some transient external or internal disturbance will start to pulsate. However, Eddington showed that for most stars such pulsation would be damped out very rapidly (i.e. on a time scale of a few thousand years, which is obviously very short in terms of stellar evolution).

The size of a star of a given structure is determined essentially by the flow of energy (heat) outwards through the star. To have sustained radial pulsations of the star, this heat flow must vary in a periodic fashion. One obvious way for this to happen is for the energy generation at the centre of the star to vary periodically. This method of producing pulsating stars is not of importance for any of the variables that we shall be considering. Eddington realized that one could also get self-sustaining pulsations if the opacity of stellar material at some suitable level in the star increased during the compression part of each cycle (that is when the star's temperature was rising) and decreased during the expansion phase releasing the energy absorbed during compression. In the case of classical Cepheids the layer in which helium can become doubly ionized, the He$^+$ ionization zone, is suitably placed to act as this 'valve'. The H-ionization zone is also important. Detailed studies based on these and analogous ideas have been able to define the areas in the HR diagram where variables stars are expected and such predictions are in reasonably good agreement with observations.

There are obviously two other questions regarding pulsations that immediately arise. (1)

---

† It is useful to take Cepheids as an example since these are the first class of variables that will be considered.

What determines the pulsation amplitude in any particular case? (2) What determines the mode of pulsation? (i.e. will the star pulsate in the fundamental mode or in some overtone?). Theoretical analyses of both these problems have led to increasingly complex computations involving the interaction of the pulsations with the detailed internal structure of the star. For instance as regards the pulsation mode, a major factor is the position of pulsation nodes with repect to the internal structure of the star.

In the case of Cepheids it is known (e.g. from work in the Magellanic Clouds (see for instance Beaulieu et al. 1995)) that whilst most such stars pulsate in the fundamental mode, some pulsate in the first overtone and a few are double mode pulsators. In the latter case, their light and velocity curves can be resolved into fundamental and first overtone components (examples of light curves are given in, for instance, Sterken and Jaschek 1996).

The requirement of having an appropriate ionization zone at a suitable depth within the star effectively confines the Cepheids to a relatively narrow band in the HR diagram (see e.g. Choisi et al. 1992). Within this band (the instability strip) a PLC relation applies. But because the strip is narrow, the PLC relation can with good approximation be replaced by a PL relation. This is one reason why the Cepheids are so important as distance indicators both within our Galaxy and for extragalactic work. The Cepheid PL relation was discovered from observations of the Cepheids in the Magellanic Clouds where they are all at approximately the same distance from us in each Cloud. The absolute calibration of the relation can be carried out in various ways. Fernie (1969) gives a history of the discovery and early calibration of the Cepheid PL relation. A discussion of the most recent calibration is given by Feast (1999a).

2. Cepheids in Young Globular Clusters

The term globular cluster is frequently restricted to a specific group of very old objects with rather similar and well defined characteristics. However, such objects must once have been young and we know of objects which will almost certainly evolve into something like the classical globular clusters. Young globular clusters are found in the Magellanic Clouds (satellite galaxies to our own). The Large Cloud (LMC) is at a distance of 55 kpc and the Small Cloud (SMC) is at about 66 kpc. In what at the time seemed quite a remarkable observation, Thackeray found that in the LMC cluster NGC1866, which is globular in appearance, the brightest stars were blue, as they are in a young cluster such as the Hyades, rather than red as in a classical globular cluster (Thackeray as reported by Baade (1951)). He also found that the cluster contained classical Cepheids. Cepheids are massive young objects and this confirmed the youth of NGC1866. That NGC1866 and some other objects should be considered as young globular clusters is shown by table 1 (from Mateo 1993). This lists the clusters, their present ages estimated from their colour magnitude diagrams, their present integrated visual absolute magnitudes and their estimated masses. The table also shows estimates, based on a current luminosity function and stellar evolutionary theory, of how bright the cluster will be in about 15Gyr (i.e. when it is of the same order of age as the present classical globular clusters). Since a typical classical (old) globular cluster has an absolute visual magnitude (M_V) of ~ -7.1 and a mass of $\sim 1.2 \times 10^5 M_\odot$ it seems rather certain that these clusters will eventually evolve into objects like present day "classical" globulars.

Two interesting general questions arise with regard to these young globulars. (1) The luminosity functions of clusters of all ages in the Magellanic Clouds and our Galaxy show a power-law increase to fainter cluster luminosities (Elson and Fall 1985). Did the cluster population in the halo of our Galaxy start off like this 10 to 15 Gyr ago? If so

TABLE 1. Young Globular Clusters in the Magellanic Clouds

Cluster	Age Myr	M_V	$M_V(15 Gyr)$	Mass $M_\odot$
NGC 2070	3	−10.9	−7.2	
NGC 330	10	−9.9	−6.2	
NGC 1850	50	−10.5	−7.5	6.10^5
NGC 1866	100	−9.5	−7.0	1.10^5
NGC 1783	1600	−8.4	−6.9	3.10^5
NGC 1978	2000	−8.5	−7.2	3.10^5
NGC 1835	12000	−8.7	−8.6	1.10^6

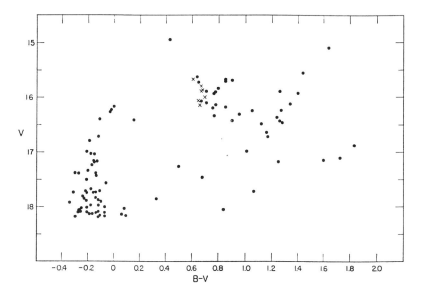

FIGURE 1. Colour-Magnitude diagram of NGC1866 from Arp and Thackeray (1967). The crosses are cluster Cepheids, the brightest of which is now known to be a non-member of the cluster (Welch et al. 1991).

presumably all the smaller clusters have disintegrated, their stars now forming part of the halo field population, with only the massive clusters being left behind. (2) Why have we so far found no young globular clusters in our own Galaxy? All the young clusters in our Galaxy are of quite low mass compared with globulars. One possible reason for this is that in our Galaxy the shearing effect of differential galactic rotation might be sufficient to prevent a large cluster from forming in the galactic disc.

There are of course small (open) clusters in our Galaxy of comparable age to clusters such as NGC1866. The advantage of a large cluster from the point of view of testing stellar evolution theories is its large stellar population. In particular, significant numbers of stars are present in the supergiant region which is only sparsely populated in open clusters. There has been a considerable amount of work on NGC1866 and its variables and on similar clusters, both observational and theoretical, in recent years (e.g. Chiosi et al. 1989, Brocato et al. 1989, Walker 1987, Welch et al. 1991), but some of the most interesting features were present in the early work of Arp and Thackeray (1967) and Arp (1967). The colour-magnitude (c-m) diagram in V and $(B-V)$ (figure 1) shows an evolved main-sequence and then a large gap as one moves to redder colours until the Cepheids are reached. These form the blue end of a supergiant sequence. Up to the 1960s this would have seemed a strange result. The general idea of the evolution of massive stars at that time was that stars would evolve off the top of the main sequence and move rather rapidly across the Hertzsprung gap to the red supergiant region. From there they would, in a somewhat unspecified way, become white dwarfs or perhaps supernovae. With a cluster of sufficient size one might expect to find some stars in the Hertzsprung gap, including Cepheids. One's natural expectation was to see stars strung out across the gap, some to the blue and some to the red of the Cepheids. However, evolutionary calculations carried out from the early 1960s (e.g. Iben 1964, Hofmeister et al. 1964) greatly modified this expectation. As before there was evolution from the top of the main-sequence to the red during a shell hydrogen-burning phase. This was however very rapid. Then after the star reached the red supergiant region there would be extended loops back to the blue during a core helium-burning phase. Thus the evolutionary tracks could pass through the Cepheid instability strip more than once. Details of the tracks vary according to the models used but the general features remain basically the same as in the early discussions. For stars of about the mass we are considering (about 5 solar masses) there will be three crossings of the instability strip. The first and third (both from blue to red) are expected to be quite rapid and we expect most or all of the observed stars (including Cepheids) to be found on the second crossing. NGC1866 provides a rather remarkable verification of some of these predictions. It is most unlikely that the distribution of the supergiants in the c-m diagram could be due to stars in the first-crossing phase. Also the position of the Cepheids with no supergiants to the immediate blue of them is really only compatible with them being on the blue end of the loop (i.e. at or near the end of the second crossing). This suggests that the loop in this case extends into the instability strip but not beyond (i.e. to the blue) of it.

Apart from the significance of this result for evolutionary theory it raises another important point. There is no particular reason why we should assume that the loop extends right up to the blue edge of the instability strip. Consequently there is a strong likelihood that the blue colour-limit of the Cepheids in this cluster is set, not by the limits of stellar stability against pulsations, but by the evolutionary tracks. The tracks may not allow stars in this cluster to fill the theoretically available instability strip. Since theory (e.g. Becker, Iben and Tuggle 1977) suggests that the extent to the blue of these supergiant loops is rather sensitive to metallicity, it would seem that whether the instability strip is fully populated or not will also depend on metallicity. In this case a PLC relation (with the colour properly corrected for metallicity effects) should still apply to the Cepheids. However, if the instability strip were only partially filled this would affect the PL relation which is a mean of the variables in the strip and might introduce an otherwise unexpected metallicity dependence into the PL relation. All these considerations follow from the study of Cepheids in a young globular cluster and would have been much more difficult to discover in any other way.

3. RR Lyrae Variables in Globular Clusters

3.1. *RR Lyrae Stars and the Oosterhoff Effect*

For the rest of this section of the book the term "globular cluster" will be taken in its conventional sense to mean a large, very old, system of stars. About 100 years ago astronomers (of whom Bailey, 1913, was the most prominent) started to find large numbers of variable stars in globular clusters. Most of these are what were at one time called "cluster variables" and are now commonly referred to as RR Lyrae stars after the bright (field) variable of this type. These stars are generally somewhat hotter than the sun (equivalent spectral type A to F), have periods in the range 0.2 to 1.2 day and light amplitudes between 0.2 and 2 mag. They were originally classified into three groups, a, b and c, the light amplitudes decreasing from type a to type c. It is now known that type a and b do not differ fundamentally from one another and they are classed together (type ab). These stars have asymmetrical light curves whereas the type c variables, with smaller amplitudes, have nearly sinusoidal light curves (see for instance the light curves in Sterken and Jaschek 1996). RR Lyraes belong to the horizontal branch (HB) phase of stellar evolution. A low-mass star ascends the Red Giant branch (RGB) for the first time whilst undergoing H-burning in a shell. At the RGB tip helium core-burning starts and the star drops rapidy to the zero-age horizontal branch (ZAHB). Its position on this branch depends on its mass. The spread of stars along the ZAHB is due to a range in the mass loss suffered by stars on the RGB. The stars then evolve off the ZAHB. The post-ZAHB tracks are generally to brighter luminosities and redder colours, but depend on the models used. As soon as shell-helium burning starts the star leave the HB region altogether and moves up the asymptotic giant branch (AGB).

Like the classical Cepheids, the He^+ ionization zone is the most important region in an RR Lyrae variable for maintaining the stellar pulsations. Whilst classical Cepheids in a given system such as the Magellanic Clouds almost certainly have only a rather small range in [Fe/H], the situation regarding RR Lyrae variables is quite different. Both in the field and in globular clusters they are found with a wide range of metallicities, from [Fe/H] ~ 0 to ~ -2.5 . †

Globular cluster c-m diagrams show that the edges of the RR Lyrae instability strip are sharply defined. Stars outside the strip do not pulsate and there is a separation in colour (or temperature) between ab and c type variables (see figure 2). The separation between the RRab and RRc stars is also seen in an amplitude - log period diagram (figure 3). These results lead to the interpretation of the ab stars as fundamental pulsators and the c types as first overtones. Early stellar models did in fact suggest that there should be such a division with the hotter stars pulating in the first overtone (Christy 1966).

One of the initially most puzzling, but also the most important, properties of globular cluster RR Lyrae stars is that they show the Oosterhoff dichotomy. This is a division of the clusters into two distinct groups (Oost I and II) with different mean periods and also different ratios in the numbers of ab to c types. This is shown in table 2 (data from Smith 1995) which gives the mean periods and the transition period between the ab and c types (P_{tr}) for the two groups as well as for a typical cluster from each group. Arp (1955) and Kinman (1959) showed that the Oost I clusters were more metal rich than the Oost II group and this has proved the key to understanding the effect. It also gives a hint that the absolute magnitudes of the RR Lyrae variables (and hence of the HBs) are a function of metallicity. For instance the Oosterhoff effect can be expressed as a dependence on metallicity of the transition period between ab and c types. However

---

† In the case of RR Lyraes in clusters their metallicities are taken to be that of their parent cluster except in the case of ω Cen where there is a range of metallicities (see section 3.5).

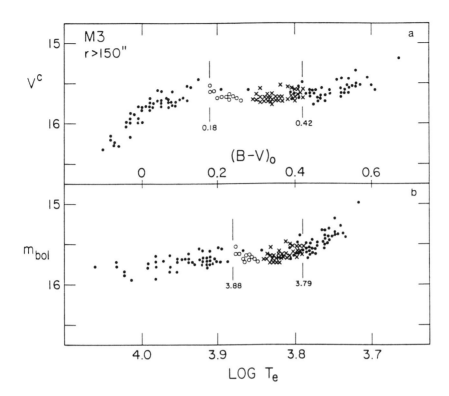

FIGURE 2. Colour-Magnitude and HR diagrams for the horizontal branch in the globular cluster M3 (Sandage 1990 figure 6). The RRab stars are marked as crosses and the RRc stars as open circles. The edges of the instability strip are marked.

TABLE 2. Properties of the Oosterhoff groups of globular clusters

Group	$\overline{P_{ab}}$	$\overline{P_c}$	P_{tr}	N(c)/N(ab,c)
M3 (Oost I)	0.56	0.32	0.45	0.16
M15 (Oost II)	0.64	0.38	0.57	0.48
Mean Oost I	0.55	0.32	0.43	0.17
Mean Oost II	0.64	0.37	0.55	0.44

Christy's (1966) theoretical work suggested that this transition period (P_{tr}) should be related to luminosity, viz.;

$$P_{tr} = 0.057(L/L_{\odot})^{0.6} \qquad (3.7)$$

where L is the luminosity of the RR Lyraes and $L_{\odot}$ is the solar luminosity. The way in which the luminosities of RR Lyraes vary with metallicity is currently of great interest. It effectively defines the relative ages of globular clusters of different metallicities (see below) and hence has important repercussions for our understanding of the evolution of the galactic halo.

It was at first rather puzzling that the Oosterhoff effect manifests itself as a dichotomy

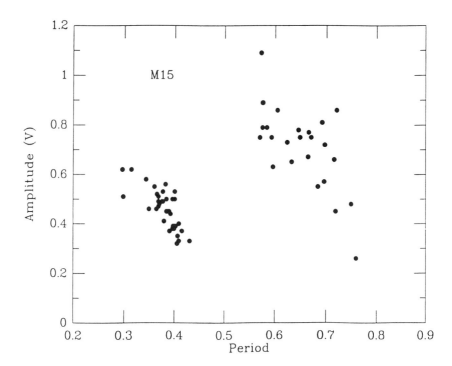

FIGURE 3. A Period-Amplitude diagram for RR Lyraes in the globular cluster M15 showing the division between the short period RRc stars and the longer period RRab stars (from Smith 1995, data from Bingham et al. 1984).

rather than a smooth progression of mean period with [Fe/H]. However, one sees the situation somewhat more clearly from figure 4 (Sandage 1993c). This plots the periods and metallicities of individual RR Lyraes in clusters. One sees that the dichotomy is due to a number of clusters with [Fe/H] $\sim$ -1.8. These clusters have rather few RR Lyraes and such variables as they do contain tend to have long periods. Sandage (1993a) attributes this to the fact that these clusters have rather blue HBs and the region of the RR Lyraes may contain a significant number of stars which have evolved back into the instability strip from a bluer part of the ZAHB. Such evolved RR Lyraes will be above the ZAHB and because they are brighter but of about the same temperature as the ZAHB variables they must also have larger radii and hence longer pulsation periods (from equation 1.2). Sandage, Katem and Sandage (1981) emphasised the fact that the Oosterhoff effect is not simply something that affects the average period of RR Lyraes in a cluster. Differences between individual RR Lyraes are seen in the two Oosterhoff groups. Sandage, Katem and Sandage investigated this in detail for the clusters M15 (Oost II) and M3 (Oost I). They found that the relations between amplitude, rise time of the light curve (i.e. the degree of asymmetry), and colour temperature were all distinctly different in the two clusters (see figures 5 and 6).

As already mentioned, the Oosterhoff effect leads one to suspect a relation between the absolute magnitude and the metallicity for RR Lyrae stars (and therefore for HB

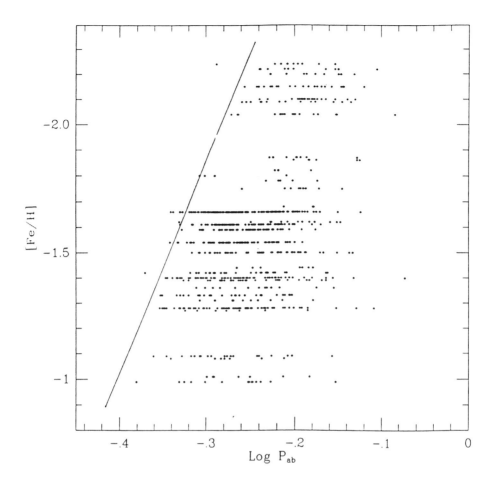

FIGURE 4. The distribution of the periods of RRab stars with metallicity for RR Lyraes in globular clusters. The line is the fundamental blue edge adopted by Sandage (from Sandage 1993c).

stars generally). It is usual to express this in the simplest possible form viz.;

$$M_V = \alpha[Fe/H] + \beta. \tag{3.8}$$

Considerable effort has gone into attempts to determine the constants α and β and no general agreement has yet been reached. A main driving force in these investigations is the desire to determine the relative and absolute ages of the globular clusters. For instance figure 11 of Sandage and Cacciari (1990) shows that if $\alpha \sim 0.19$ then there is an age difference of ~ 3 Gyr between clusters with $[Fe/H] \sim -0.7$ and those with $[Fe/H] \sim -2.2$, the more metal-poor ones being the older. Whereas if $\alpha \sim 0.39$ the trend is slightly in the opposite direction (see also Chaboyer, Demarque and Sarajedini 1996 and Reid 1998).

What follows is an attempt to sketch out the various ways in which estimates of α and β have been made and to indicate what might be the reasons for the discrepancies between

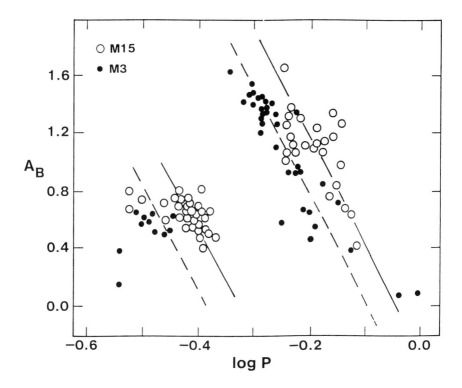

FIGURE 5. The relationships between the blue light amplitude and the period for RR Lyraes in the Oost I cluster M3 and the Oost II cluster M15 (from Smith 1995, data from Sandage, Katem and Sandage 1981).

the various results. It should be kept in mind that there is no fundamental reason why the relation between M_V and [Fe/H] should be linear.

3.2. *Sandage's Derivation of the M_V - [Fe/H] Relation*

Sandage (1993b) has derived an M_V − [Fe/H] relation from a consideration of the Oosterhoff effect. This derivation is semi-empirical, combining both observation and theory. The classical P$\sqrt{\rho}$ relation together with equation 1.4 above lead to,

$$\log P = 0.75 \log L - 0.5 \log M - 3 \log T + constant. \tag{3.9}$$

Pulsation studies of realistic models of RR Lyrae stars lead to a slightly modified relation together with a value for the constant (van Albada and Baker 1973). This is,

$$\log P = 0.84 \log L - 0.68 \log M - 3.48 \log T + 11.502. \tag{3.10}$$

The Oosterhoff effect shows there is a relation between P and [Fe/H]. Sandage's approach is to estimate this and also to seek for relations between [Fe/H] and mass (M) as well as between [Fe/H] and T. There is a range of periods at any metallicity (i.e. within any given cluster). Sandage's method is to confine discussion to the fundamental blue edge (FBE). This is essentially the locus of the shortest period RRab stars as a function of metallicity. There are certain advantages to doing this. It is possible to to establish a

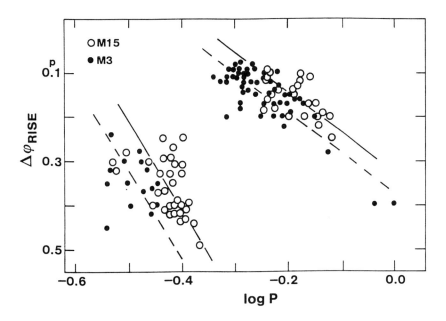

FIGURE 6. The relationships between the rise-time (the fraction of the light cycle from minimum to maximum) and period for RR Lyraes in the Oost I cluster M3 and the Oost II cluster M15 (from Smith 1995, data from Sandage, Katem and Sandage 1981).

P - [Fe/H] relation at the FBE (see fig 4). In addition the theoretical work of Christy (1966) (equation 3.7 above) led to the expectation of a period -luminosity (and hence a period - [Fe/H] relation) at the FBE. Furthermore, double mode RR Lyraes †, whose light curves show that both the fundamental and the first overtone are excited, occur only near the FBE. This was shown clearly by Sandage, Katem and Sandage (1981) in the case of M15, and of course might be reasonably anticipated from the distribution of ab and c types (in, e.g. figure 2). The RRd variables are important because it is possible from the ratio of their two periods to estimate their masses (see below) and hence this is one way to a mass - [Fe/H] relation at the FBE.

¿From a discussion of the period - [Fe/H] distribution in field and cluster RR Lyraes (e.g. figure 4) Sandage adopts

$$\log P = -0.122[Fe/H] - 0.500. \qquad (3.11)$$

In a similar way, after a discussion of the relation between colours and temperatures for the RR Lyraes he derives at the FBE the relation,

$$\log T = 0.012[Fe/H] + 3.865. \qquad (3.12)$$

In this case some uncertainty is introduced because of uncertainties in the corrections of the colours for interstellar reddening as well as the problem of calibrating the colour -temperature relation, taking into account any effects of metallicity on this relation.

† double mode RR Lyraes are called RRd variables.

Finally Sandage uses models of HB stars (Dorman 1992) to obtain,

$$\log M = -0.059[Fe/H] - 0.288. \qquad (3.13)$$

In addition to this one needs to adopt bolometric corrections (BC) for stars in this temperature range and Sandage uses,

$$M_V = M_{bol} - 0.06[Fe/H] - 0.06. \qquad (3.14)$$

This relation is derived by Sandage and Cacciari (1990) and is based on a theoretical relation between BC and metallicity (Kurucz models) at the mean temperature of RR Lyrae variables. Putting these various equations together leads to,

$$M_{bol}(RRab)_{FBE} = 0.36(\pm 0.12)[Fe/H] + 1.00(\pm 0.10) \qquad (3.15)$$

and,

$$M_V = 0.30[Fe/H] + 0.94. \qquad (3.16)$$

How reliable is this important result? Sandage's derivation of the log P and log T - [Fe/H] relations is essentially empirical. One might like to have more complete data but it seems unlikely that one would wish to significantly change these relations on the basis of the existing data. The M - [Fe/H] relation is, however, an entirely theoretical one, coming from evolutionary models. Some check is possible on this relation (as mentioned above) by combining observations of the period ratios in RRd variables with model predictions. If one combines observations of RRd variables in M3 and M15 with models (Kovács et al. 1992) as suggested by Sandage one would obtain,

$$M_V = 0.39[Fe/H] + 1.07 \qquad (3.17)$$

i.e. an even steeper slope.

3.3. *RR Lyrae Luminosities from Parallaxes*

A fundamental way to calibrate a M_V - [Fe/H] relation for RR Lyrae variables is to use trigonometrical parallaxes. Globular clusters are too far away for their parallaxes to have been directly measured. However, there are a considerable number of field RR Lyrae variables with measured parallaxes in the Hipparcos (ESA 1997) catalogue. These parallaxes are small and it is necessary to consider carefully how to analyse them in order to avoid the introduction of statistical bias into the result. A number of recent papers on the use of Hipparcos parallaxes in various connections suggest there is considerable misunderstanding regarding the origin of such bias and the way it can be avoided. Since this has relevance to many problems it seems worth considering it in some detail.

An absolute magnitude (M_V, say) is derived from the apparent magnitude corrected for interstellar extinction (V_o) and the measured parallax (π) by the usual relation,

$$M_V = V_o + 5log\pi + 5. \qquad (3.18)$$

Suppose we have a set of measured parallaxes, all with the same uncertainty (σ_π), of stars all of the same absolute magnitude. If one selects all those with some given value of the measured parallax then the sample will contain stars whose parallax has been over-estimated and stars for which it has been underestimated. The stars with overestimated parallaxes are actually further away than the measured parallax indicated. If the stars observed are uniformly distributed in space a little consideration shows that there will be more of them in the sample than there are stars with underestimated parallax. Thus the mean parallax will be an overestimate of the true mean parallax and the mean value of

M_V derived from it will be fainter than the true value. This effect is called Lutz-Kelker Bias (Lutz and Kelker 1973). In practice the effect may be very significant. For instance if $\sigma_\pi/\pi = 0.1$ then the absolute magnitude derived needs correcting by $\Delta M_V = -0.11$. If $\sigma_\pi/\pi = 0.175$ then ΔM_V is –0.43. Formulae have been given with which to handle different space distributions of the objects concerned (e.g. Hanson 1979) and the confidence limits to be placed on the corrections has been discussed (Koen 1992). Bias can arise in a variety of astronomical contexts. Its importance was first noted by Eddington (1913, 1940) and some aspects of it were discussed in detail by Malmquist (1920).

The bias problem is compounded if individual values of M_V derived from equation 3.18 are combined by weighting according to $1/\sigma_M^2$, where σ_M is the standard error of M_V. Since $\sigma_M = 2.17\sigma_\pi/\pi$ this means that stars with overestimated parallaxes are given higher weight than those with underestimated ones, thus increasing the bias further †. The bias problem is not of course something that is inherent in the measuring process or in the parallax catalogue, unless the stars are selected for the catalogue on the basis of their measured parallax. The bias simply results from the way the data are handled. If one has a group of stars for which one wishes to derive absolute magnitudes from observed parallaxes without knowing anything about the relationship, if any, between the individual absolute magnitudes, then there is probably no alternative but to proceed in the standard way and apply Lutz-Kelker type corrections to the individual values. However, if it is known that all the stars are of the same absolute brightness and we simply wish to derive the best estimate of this; or if the relative absolute magnitudes are known, because these are related to some other measured property of the star (such as the period of a variable or its metallicity), then we can avoid the bias problem.

The Lutz-Kelker problem arises because the mean parallax of a group of stars each with the same measured parallax is biased. However, suppose we knew the true parallaxes of the stars. Then the measured parallaxes of a set of stars all with the same true parallax would be distributed symmetrically (presumably in a Gaussian fashion) about the true parallax and the mean of the measured parallaxes would be an unbiased estimate of the true parallax. Consider the simple case when all the stars have the same (but initially unknown) absolute magnitude. Then all the stars of the same V_o must be at the same distance, i.e. have the same (true) parallax. Thus the mean parallax, $\bar{\pi}$, (weighted if necessary) of these stars is an unbiased estimate of the true parallax. Thus we can obtain an unbiased estimate of $10^{0.2M_V}$ from,

$$10^{0.2M_V} = \bar{\pi}.10^{0.2(V_o+5)}. \tag{3.19}$$

More generally we can use an equation like 3.19 with individual observed parallaxes and each estimate of $10^{0.2M_V}$ weighted proportional to the square of the reciprocal of $\sigma_\pi.10^{0.2(V_o+5)}$ (see, e.g. Feast and Catchpole 1997, Koen and Laney 1998) to obtain the best unbiased estimate of $10^{0.2M_V}$.

Clearly if the coefficient α in equation 3.8 is known or assumed, a value of β (or rather $10^{0.2\beta}$) can be obtained from the parallaxes of RR Lyrae stars using the equation,

$$10^{0.2\beta} = \pi.10^{0.2(V_o-\alpha[Fe/H]+5)} \tag{3.20}$$

in the same way. It is possible in addition to take into account any intrinsic scatter in equation 3.8 (see Koen and Laney 1998). Taking α as 0.3 one then obtains from the Hipparcos parallaxes of RR Lyrae stars (Koen and Laney 1998),

$$M_V = 0.3[Fe/H] + 0.84 \tag{3.21}$$

† Note that in principle this does not apply if the bias-corrected parallaxes are used in equation 3.18.

with an uncertainty, σ_M, of ± 0.22 mag. Gratton (1998) has pointed out that one can also use this method with the parallaxes of HB stars other than RR Lyraes in the Hipparcos catalogue. A rediscussion of his data, again with the value of α assumed, leads (Koen and Laney 1998) to,

$$M_V = 0.30[Fe/H] + 1.07 \tag{3.22}$$

and $\sigma_M = \pm 0.10$ mag. Note, however, that some of these HB stars, especially the bluer ones, have had large corrections applied to them for the fact that the HB is far from horizontal and what is required is the absolute magnitude at the colour of the RR Lyrae variables. The red HB star with the best determined parallax (HD17072) may actually be a first giant branch star rather than on the HB and was omitted in deriving equation 3.22.

3.4. *Other Determinations of RR Lyrae Luminosities*

One can in principle determine the distance and hence the absolute magnitude of a radially pulsating star by the method of pulsation parallaxes pioneered by Baade and Wesselink and usually known by their names. This method can be applied in various forms. A common way of applying it is in essence as follows. Assume we know, for the type of variable under consideration, the relation between colour and surface brightness at the wavelengths in which we are interested. This, together with the apparent magnitude, gives us directly the angular diameter of the star and we can measure the change of this during the pulsation cycle. On the other hand, integration of the radial velocity curve gives the variation in the linear diameter through the cycle. These two result together then give the star's distance.

In principle this is an attractive method and can give results with a high degree of internal consistency. However, one needs to be aware that in application there are a number of uncertainties which could give rise to systematic errors. (1) In the case of the RR Lyraes the surface brightness-colour relation relies on stellar models (e.g. Jones et al. 1992). It is thus not entirely empirical (see also below). (2) One has to assume some value for the limb darkening of the star and this is not known directly. (3) It is assumed that the absorption lines measured for the radial motion of the star's atmosphere are formed in the same photospheric level which gives rise to the observed brightness and colour variations of the star. In the case of RR Lyrae stars we see doubling of the absorption lines at certain phases in the pulsation cycle. Evidently at these times we are seeing lines formed at different depths in the atmosphere and the radial velocity changes cannot easily be used with integrated photometry to derive a distance. It is usual to omit such phases in the Baade-Wesselink analysis but clearly this adds uncertainty.

Whilst one might have reservations about the absolute calibration of Baade-Wesselink type analyses, it has been generally hoped that relative luminosities could be reliably obtained. That is, for RR Lyraes one might hope to derive a satisfactory value for the slope, α, of equation 3.8. This of course is not absolutely certain. The latest available data for RR Lyrae variables in the galactic field is shown in figure 7 (Fernley et al. 1998a). These authors derive,

$$M_V = 0.20[Fe/H] + 0.98 \tag{3.23}$$

with an rms scatter of ± 0.12 mag †. This is steeper than earlier results from this group but not as steep as required by Sandage (0.30). There has been some discussion of how to correctly determine the slope from the observations (Feast 1997). This depends on

---

† In deriving this relation Fernley et al. (1998) omit the star SS Leo (which is marked in figure 7) on the grounds that they believe it to have evolved significantly from the ZAHB.

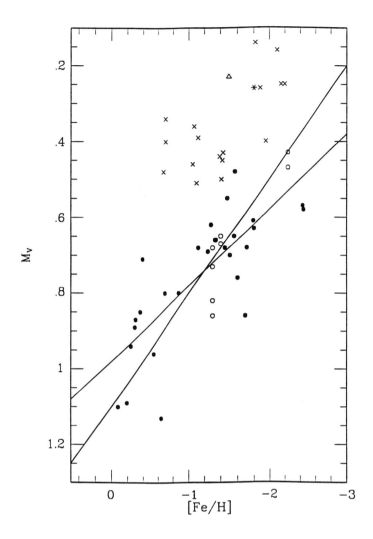

FIGURE 7. The relation between M_V and [Fe/H] for RR Lyraes. Filled circles, Baade-Wesselink results for field variables; Open circles, Baade-Wesselink results for variables in clusters; Crosses, results from clusters with main-sequence distances; The asterisk, result from LMC clusters with the Cepheid distance to the LMC. The open triangle is the Baade-Wesselink result for the field variable SS Leo. (See text for details).

selection effects and on how much of the scatter in the relation is intrinsic. The lines shown in figure 7 correspond to equation 3.23 and the best fit to the Baade-Wesselink results for field RR Lyraes taking α as 0.30.

Although some of these issues remain unresolved, an apparently clear indication of a low slope for the relation seemed to have come from HST work on the cm diagrams of globular clusters in M31 (Fusi Pecci et al. 1996). Here, with all the clusters at approximately the same distance and with a range of metallicities, one might expect to solve this problem. Fusi Pecci et al. found that in M31 the apparent magnitude of the HB at the colour of

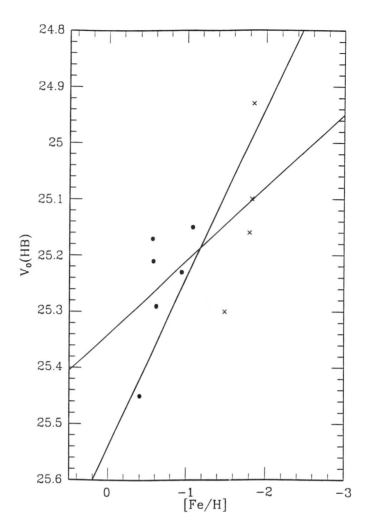

FIGURE 8. The apparent magnitude of the HB at RR Lyrae colours in M31 clusters. The crosses are the clusters with HBs which extend at least as far to the blue as the instability strip. The lines are the best fits to the data adopting slopes of 0.13 and 0.3 (see text for details).

RR Lyrae variables showed only a rather slight variation with [Fe/H]. Their results lead to,

$$M_V = 0.13[Fe/H] + constant. \tag{3.24}$$

The uncertainty in their slope is ±0.07. However, an examination of the cm diagrams of the clusters used shows that only three of them (the clusters, G105, G219 and G351) have HBs which extend to the colours of RR Lyraes or bluer. The others show only stubby red HBs. To these data one can now add two more M31 clusters (Holland et al. 1997), G302 which has a good HB and G312 which has only a stubby red branch. Figure 8 shows the available data on M31 clusters.

Whilst the sum total of M31 clusters as it stands points to a shallow overall slope for the

$M_V - [Fe/H]$ relation, some caution is need in interpreting these results. The clusters with HBs extending to, or beyond, the RR Lyrae gap are marked in figure 8. These are clusters in which RR Lyraes would actually be expected. Clusters with stubby red branches have few if any RR Lyraes and to that extent the positions for the HBs of these clusters in figure 8 are extrapolations of the red branch into regions where no stars in the clusters are actually found. If we limit ourselves to the clusters with good HBs it is clear that the data are not sufficient to give a satisfactory value for the $M_V - [Fe/H]$ slope and a steep slope (~ 0.3) for these metal-poor clusters is by no means ruled out. If (see figure 7) we restrict ourselves to $[Fe/H] < -1.0$ then the field RR Lyrae variables with Baade-Wesselink distances are not sufficiently well distributed in [Fe/H] to give a definitive value of α either. RR Lyraes more metal rich than this limit are found in the field but not generally in globular clusters. It is therefore not entirely clear what the significance of these metal rich field stars are for the globular cluster problem.

The group which has carried out much of the work on Baade-Wesselink analyses of field RR Lyrae variables has also applied this technique to two RR Lyraes in the cluster M92, two in M5 and four in M4 (Liu and Janes 1990, Storm et al. 1994). The results for these cluster RR Lyraes are shown in figure 7. Again the results are not sufficient to determine α with any accuracy. Nevertheless they would be consistent with a slope of ~ 0.3 either considered on their own or taken together with the field stars with [Fe/H] less than -1.0. Cohen (1992) has also carried out Baade-Wesselink analyses for RR Lyraes in M5 and M4. Her results are 0.18 mag fainter than those just discussed but the difference between the two clusters is closely the same for both groups. The difference in the absolute values is no doubt partly a reflection of the somewhat different methods of analysis used and perhaps shows the sensitivity of the Baade-Wesselink method to the precise way the data are handled.

Some additional support for a value of α near 0.3 comes from considering the M_K-logP relation for RR Lyrae variables (where M_K is the absolute magnitude at K (2.2 microns)). Longmore et al. (1986) and Longmore et al. (1990) find that within a given globular cluster this relation has a small scatter and that the slope $\Delta M_K / \Delta \log P$ is closely the same for clusters of a wide range of metallicities (the mean value of the slope is -2.31 ± 0.06). The scatter in the relation remains small even for the cluster ω Cen where there is a wide range of metallicities amongst the RR Lyraes. In addition the Baade-Wesselink results for RR Lyraes in the general field, which also cover a wide range of metallicities, show a M_K - log P relation with a closely similar slope (-2.33 ± 0.20, Jones et al. 1992). These results suggest that the zero-point of the M_K - log P relation is rather insensitive to metallicity. If this is so the relation can be used to derive the relative distances of globular clusters of different metallicities and to use these to derive the slope of the M_V - [Fe/H] relation. In this way Longmore et al. (1990) find values of α in the range 0.32 to 0.40 depending on the way the data are treated. Thus these data suggest a steep M_V - [Fe/H] slope at least for the metallicity range over which RR Lyraes are found in globular clusters ($[Fe/H] \le -0.9$). There is unfortunately an uncertainty in this result since a smaller value of α could be obtained if the reddening law were unusual for the two least metal-poor clusters in the Longmore et al. sample.

The above discussion has been focussed on the determination of α. We now consider further the absolute calibration of equation 3.8.

An entirely independent way of deriving the absolute magnitudes of RR Lyrae stars in globular clusters is by fitting subdwarfs, having known metallicity and with Hipparcos parallaxes, to the main sequences of globular clusters. This has now been done for several galactic globular clusters. Gratton et al. (1997) derive $M_V(HB)$ for eight globular clusters, Reid (1997) for the mean of three clusters, Reid (1998) for 7 clusters, Pont et

TABLE 3. Inferred $M_V(RR)$ at $[Fe/H] = -1.8$

Method	$M_V(RR)$	Remarks
LMC Clusters	0.26 ± 0.1	LMC Cepheid distance
Galactic Globulars	0.30 ± 0.04 (int.)	main sequence fits
Field RR Lyraes	0.30 ± 0.22	Parallaxes
RRd variables	~ 0.3	Kovács and Walker (1998)
Sandage calibration	0.4	
HB parallaxes	0.47 ± 0.12	excludes HD17072
Baade-Wesselink	0.62	$\alpha = 0.20$
Baade-Wesselink	0.45	McNamara (1997) $\alpha = 0.287$
Statistical parallaxes	0.62 ± 0.10	Tsujimoto et al. 1998
Statistical parallaxes	0.73 ± 0.17	Fernley et al. 1998a

al. (1998) for M92. The results are plotted in figure 7. Also shown is the mean absolute magnitude of RR Lyraes in LMC clusters if the LMC modulus derived by Feast and Catchpole (1997) from Cepheids is adopted together with the data on RR Lyraes in LMC globular clusters (Walker 1992). The result for the LMC is clearly consistent with the results from main sequence fitting but the Baade-Wesselink results are about a third of a magnitude fainter. This implies a difference in distance scale of ~ 16 percent and a difference in age estimates of globular clusters of ~ 33 percent. The scatter amongst the globular cluster results in figure 7 precludes an accurate estimate of α but the results are again not inconsistent with $\alpha \sim 0.3$ (see also Reid 1997, 1998).

The position with respect to the value of α thus remains somewhat unsatisfactory. The M31 slope depends on extrapolating stubby red HB of clusters to regions in which stars in these clusters are not found. If one omits these relatively metal-rich clusters and also Baade-Wesselink determinations for metal-rich field RR Lyraes on the ground that such stars are not found in globular clusters, then the data is not inconsistent with a steep slope and this is supported by the M_K - log P relation discussed above. Possibly further work in M31 will throw more light on this problem and indicate whether, for instance we need to consider a change of α at around [Fe/H] of –1.0.

As regards the absolute calibration, in addition to the methods described above one can estimate RR Lyrae absolute magnitudes using radial velocities and proper motions in the method of statistical parallaxes. This has recently been done for field RR Lyraes by two groups. Fernley et al. (1998b) obtained $M_V = 0.77 \pm 0.17$ at a mean $[Fe/H] = -1.66$ whilst Tsujimoto et al. (1998) obtained $M_V = 0.69 \pm 0.10$ at a mean $[Fe/H] = -1.58$.

Table 3 shows the absolute magnitudes for RR Lyraes derived in different ways. Where a value of α has not been used in deriving the published results these have been converted to the expected value at $[Fe/H] = -1.8$ by using a value of $\alpha = 0.30$. In addition to values already discussed, table 3 includes an estimate by Kovács and Walker (1998) from double mode RR Lyraes. This is essentially a somewhat different application of Sandage's method. They do not give a precise estimate and the value in the table is estimated from their work. The first three entries in this table, via LMC Cepheids, via main-sequence fitting, and via parallaxes, are all empirical determinations which seem soundly based. They agree very closely with one another at an absolute magnitude of ~ 0.28 mag . The RRd result of Kovács and Walker also agrees with these results. The other results range from 0.4 mag to 0.7 mag. Sandage's absolute calibration rests on the theoretical models he adopts for the masses. The result from HB stars has been discussed above. As regards the Baade-Wesselink result, McNamara (1997) has pointed out that adopting a new optical surface brightness calibration based on recent Kurucz models will lead to a brightening of the derived Baade-Wesselink absolute magnitude. His result leads to $M_V = 0.45$ at $[Fe/H] = -1.8$ (see table 3).

Regarding the statistical parallaxes, two reasons why they might differ from other results have been suggested:

(1) The field RR Lyraes may be fainter than those in globular clusters. This could either be due to the fact that on the average they are closer to the ZAHB than those in clusters (Chaboyer et al. 1998) or else that they are different in detailed chemical composition from cluster stars at a given overall metallicity (Reid 1998, Sweigart 1997). The latter explanation has some support from the fact that the detailed chemical composition of cluster red giants is different from those of similar metallicity in the field. This difference could imply more mixing in the cluster stars with dredge up of additional helium and an increase in HB luminosity. However, the Baade-Wesselink results for cluster and field stars do not seem to show evidence for a field/cluster difference although the sample is small (figure 7). Nor do the parallaxes of field RR Lyrae (table 3) give results differing from those derived by other methods for the cluster stars though the standard error of the parallax solution is quite large.

(2) The galactic model adopted in the statistical parallax analysis is possibly too simple (Feast 1998). In the analysis a classical halo model was adopted by both groups of workers. However, it is now thought that a significant part of the halo population is made up of separate streams probably from infalling satellite galaxies. If some streams are seen preferentially in one direction and others in another, this could affect the statistical parallax result. Martin and Morrison (1998) show how the derived mean model of the halo changes when the RR Lyrae scale is varied.

3.5. *The Width of the HB*

So far we have not considered whether the [Fe/H] -M_V relation (and the HB itself at a given metallicity) has any intrinsic scatter. However, theory suggests that stars evolve rapidly from the RGB tip to the ZAHB and then evolve away from it sufficiently slowly for a significant number of post-ZAHB stars to be found in the general HB region of the HR diagram. We might therefore expect to find a mixture of ZAHB stars and evolved HB stars in a cluster and this would lead to an intrinsic width to the observed HB. Sandage (1990) showed that there was indeed such a spread (e.g. figure 9). Ones first thought might be that this spread was due to observational scatter. However, the RR Lyraes show that this is not so. The scatter is real. Pulsating stars of the same mass and temperature must, from the pulsation equation 3.10, show a relation between luminosity and period. Thus in a cluster one would expect that if there is a range of magnitudes amongst RR Lyraes of the same temperature (or colour), there would be a relation between their brightness and their period. The brighter RR Lyraes should have longer periods since at a given temperature the brighter variables must be bigger and thus pulsate with a longer period (the $P\sqrt{\rho}$ relation). Sandage showed in detail that this was so in a number of clusters. He also showed (figure 10) that the width of the HB increased with increasing metallicity.

It will be evident from this that the application of the M_V -[Fe/H] relation, particularly to individual variables (for instance in the field), may not be straight forward. This is especially so for the metal-rich RR Lyraes which are found in the field but not in clusters. If the trend found by Sandage (figure 10) for the widths of the cluster HBs continues to higher metallicities then the metal-rich field RR Lyraes will have a considerable spread in luminosities at a given [Fe/H]. It seems possible that this might lead to selection effects in the RR Lyraes chosen for Baade-Wesselink studies and this could

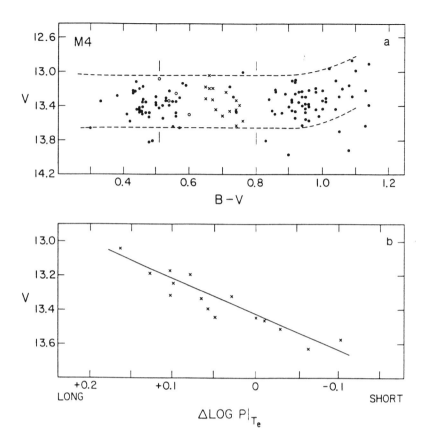

FIGURE 9. (a) The HB of the cluster M4 showing the spread in magnitude at a given colour (crosses, RRab stars; open circles, RRc stars). (b) The correlation of the apparent magnitudes of the RRab stars in the cluster with the deviations from a period-temperature relation, read at fixed temperature (from Sandage 1990).

possibly contribute to the low slope of the M_V - [Fe/H] relation which is found when these stars are included (see section 3.4). On the other hand (Skillen et al. 1993) argue that the field metal rich RR Lyraes come from a population without a blue HB and that therefore there can be no evolved RR Lyraes in this population. However, the question of whether evolved metal-rich RR Lyraes are expected depends on the detailed shapes of the evolutionary track at these metallicities. For instance the post-ZAHB tracks of Dorman (1992) clearly loop to the blue. His table 12 and fig 11 show for instance that a $0.59 M_\odot$ star of $[Fe/H] = -0.47$ starts as a red ZAHB star but loops above the ZAHB to the temperatures of the instability strip.

It is interesting to note that, if the lower envelope of the HB is taken as the ZAHB and this is assumed to follow the M_V -[Fe/H] relation of Sandage (i.e. with $\alpha = 0.3$ in equation 3.8) then, because of the increase in HB width with metallicity, the M_V of the upper envelope has a smaller effective value of α, ~ 0.19 for [Fe/H] ≤ -1.6, and < 0 for [Fe/H] ≥ -1.5.

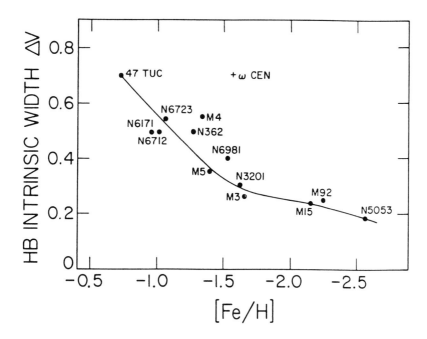

FIGURE 10. The width of the HB as a function of metallicity (from Sandage 1990). The anomalous cluster ω Cen is discussed in the text.

An example of the complications that can arise is shown by the globular cluster ω Centauri. This cluster is unique in having stars of a wide range in metallicities. Initially this seemed an ideal place to derive an M_V - [Fe/H] relation. Yet there is essentially no correlation between V and [Fe/H] for the RR Lyraes in this cluster (figure 11). Gratton et al. (1986) explain this as due to the fact that most of the metal-poor RR Lyraes have evolved from blue HB stars. The evolutionary tracks they use indicate that such stars will have about the same luminosity independent of metallicity. The most metal-rich RR Lyraes in the cluster are fainter than the others and are supposed to be near the ZAHB. These considerations show what care is necessary in constructing an M_V - [Fe/H] relation from field RR Lyraes or using it for field stars or in clusters with only a few variables. In the latter case, however, one can obtain some idea of the ZAHB from the constant stars on either side of the instability strip.

3.6. *The Masses of the RR Lyrae Stars*

The calibration of the luminosities of the RR Lyrae stars can be seen as, in a sense, a problem of mass determination. If one takes a pulsation equation such as equation 3.10 above, then in Sandage's method one adopts empirical values for P and T and uses theoretical masses to derive L. On the other hand if one can derive L in some empirical way then the equation can be used to derive a mass. Table 4 lists some estimated masses at typical metallicities. These were obtained in the following way (numbering as table 4).

(1) Evolutionary masses adopted by Sandage (1993b) from Dorman (1992).

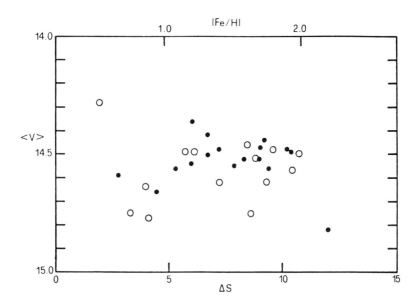

FIGURE 11. The relation between metallicity and apparent magnitude for RR Lyraes in ω Cen (from Gratton et al. 1986). ΔS is a measure of metallicity. The different symbols are not of significance for the present discussion.

TABLE 4. RR Lyrae Mass Estimates (solar masses).

[Fe/H]	Evolution (1)	RRd(1) (2)	RRd(2) (3)	LMC (4)	Stat. Par. (5)	B-W (6)
−1.0	0.59	0.58		0.69		0.50
−1.66					0.44	
−2.0	0.68	0.72		0.80		0.52
−2.15			∼ 0.8			

(2) Masses from Sandage (1993b) using the double mode results of Kovács et al. (1992).
(3) The result for double mode RR Lyraes is sensitive to changes in relative abundance of the elements (Kovács and Walker 1998). Their recent calculations show that the RR Lyrae mass should be near $0.8 M_\odot$ at the metallicity of M15 ([Fe/H] = −2.15).
(4) The result for the Sandage relation with the luminosities increased to agree with the RR Lyrae luminosities in LMC globular clusters and taking the LMC modulus to be 18.70 (Feast and Catchpole 1997).
(5) The result using the luminosity derived by Fernley et al. (1998b) from statistical parallaxes.
(6) Result from Baade-Wesselink luminosities adopting the Fernley et al. (1998a) solution.

The Baade-Wesselink and the statistical parallax results give particularly low masses ($\sim 0.5 M_\odot$). The RR Lyraes will eventually evolve, after further mass loss, into white dwarfs. Sandage (1993b) summarizes the evidence for white dwarf masses. For the general field this is believed to be close to $0.6M_\odot$. Renzini et al. (1996) suggest that the masses of white dwarfs in globular clusters can be estimated rather robustly at $0.51 \leq M_{wd} \leq 0.55M_\odot$ and they adopt $M_{wd} = 0.53 \pm 0.02M_\odot$ (see also Reid 1996). Since there is likely to be significant mass loss on the AGB (see sections 4.1 and 4.3), the low luminosities and masses implied by the statistical parallaxes and the Baade-Wesselink analyses are difficult to accept unless there is a systematic error in the adopted temperature scale.

4. Mira and Semiregular (SR) Variables in Globular Clusters

4.1. *Miras in Globular Clusters*

Mira variables are cool, giant, pulsating stars with periods in the range $\sim$100 to $\sim$500 day. The period range extends to 1000 or 2000 day if we include OH/IR variables which are generally too faint (due to their low temperatures and to dust shell obscuration) to be readily detected at optical wavelengths. Miras have large amplitudes at optical wavelengths. The General Catalogue of Variable Stars defines long period variables as Miras if their visual light amplitudes exceed 2.5 mag. However, they can have visual amplitudes of 6 mag or more. This large light range in the visible is partly due to the low temperature of the stars, so that the optical region is in the short-wavelength tail of the energy distribution. Thus relatively small changes in temperature during the light cycle result in large changes of flux in the optical region. In addition the spectra of Miras are dominated by strong bands of diatomic molecules (especially TiO in oxygen-rich Miras). These strengthen toward temperature minimum and their blocking effect further increases the light range at optical wavelengths. In the case of normal Miras most of the energy is radiated in the 1.0 to 3.5 micron region of the spectrum. A good deal of our current understanding of Miras is due to the fact that there has been, in relatively recent times, extensive monitoring of many of them in the infrared. The infrared amplitudes are modest, typically about one magnitude or less in the K band (2.2 micron) and this, together with the fact that the energy distribution generally peaks at about 1.5 micron, makes it possible to use infrared observation to derive mean bolometric luminosities of good accuracy. The large light amplitudes in the visible can be thought of as due to large variations of the bolometric correction round the light cycle.

In the galactic field there are three broad classes of Miras, oxygen-rich Miras whose optical spectra are, as just mentioned, dominated by absorption bands of the TiO molecule (M-type Miras), oxygen-rich Miras with strong enhancements of the s-process elements which show strong bands of ZrO (S-type Miras) and Miras in whose atmospheres the ratio of C/O is greater than unity and whose spectra are dominated by molecules containing carbon (C_2 etc.) (C-type Miras). Of these three groups only the first is found in globular clusters and we shall not concern ourselves here with the other two groups. Miras are losing mass. Far infrared observations (at 10 microns and longer wavelengths) show that many Miras are surrounded by cool dust shells. Mass-loss rates, derived from far infrared fluxes or measurements of CO lines in the mm region are in the range $\sim 10^{-7}$ to $\sim 3 \times 10^{-5} M_\odot yr^{-1}$ and up to $\sim 10^{-4} M_\odot yr^{-1}$ for extreme OH/IR Miras (see, e.g. Whitelock 1990, Whitelock et al. 1994).

A characteristic of Miras is that at certain phases in their light cycles they show Balmer lines of hydrogen, and sometimes other atomic lines, in emission. We know that these emission lines are being excited deep in the atmosphere of the star because high resolution spectra show that they are overlaid by absorption lines from the TiO bands. The

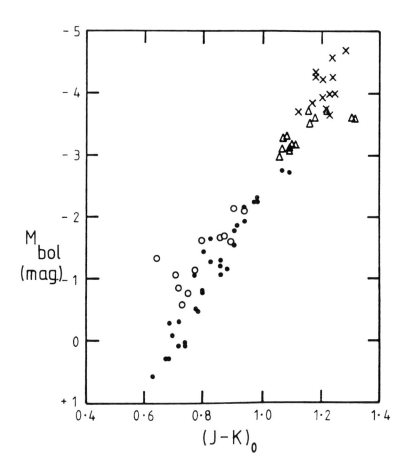

FIGURE 12. The giant branch of 47 Tuc together with Miras from various clusters; Crosses, Miras; Triangles, 47 Tuc non-Mira variables; Open circles, 47 Tuc AGB stars; Filled circles, 47 Tuc RGB stars (from Feast and Whitelock 1987).

emission lines are generally believed to be excited by shock waves generated by the star's pulsation. The radial velocity curve derived from infrared CO lines also show the characteristics, including doubling of the lines at certain phases, that are expected from the period passage of shock waves through the atmosphere (e.g. Hinkle et al. 1982).

Miras are found in relatively metal-rich globular clusters. Their importance can be judged from figure 12. This shows an HR diagram (Feast and Whitelock 1987) with the known cluster Miras together with the RGB/AGB of 47 Tuc, a cluster which contains three Miras. The exact relative positions of the Miras from different clusters in this figure depends on the relative distances of the clusters concerned which is still somewhat uncertain. This will not, however, affect the qualitative conclusion that Miras lie at the tip of the giant branch of the clusters in which they occur. They are thus the bolometrically brightest and the coolest stars in globular clusters. In this figure the infrared colour index, (J–K), is taken as a monotonically varying temperature index. This is probably satisfactory over the temperature range of the stars concerned in this plot (but see below). The (B–V)

TABLE 5. Asymmetric drift (V) and total velocity dispersion (σ_T) for galactic Miras

Period Range	Period Mean	V	σ_T	No. of Stars
< 145	131	-20 ± 13	81	22
145 - 200	179	-98 ± 22	180	46
200 - 250	225	-48 ± 9	101	71
250 - 300	270	-20 ± 10	88	77
300 - 350	324	-19 ± 6	69	83
350 - 410	382	-10 ± 8	58	54
> 410	454	-2 ± 8	50	35
day	day	km.s^{-1}	km.s^{-1}	

colour is a poor temperature indicator for these stars since the flux at these wavelengths is severely distorted by the effects of TiO absorption. In fact in a cluster like 47 Tuc the reddest giants in (B–V) are not the coolest. The giant branch of this cluster in a V,(B–V) plot turns back to bluer (B–V) colours near its top (Arp, Brueckel and Lourens 1963). Miras lie at the tip of the AGB and are therefore in the last stage before the rapid evolution of the star to the left in the HR diagram, the (probable) ejection of a planetary nebula shell and the descent to the white dwarf stage. Thus Miras are crucial for an understanding of the way in which AGB evolution is terminated. The fact that they are losing mass presumably has something to do with this. But the mass-loss process is not yet entirely understood and it has so far proved impossible to predict mass-loss and mass-loss rates from the basic theory of the evolution of AGB stars. Understanding these stars observationally is therefore vital to our understanding of stellar evolution.

Clearly we need to know the brightness and effective temperatures (or equivalently the radii) of Miras since these define the AGB tip and are the parameters any complete theory must reproduce. These questions will therefore be dealt with in some detail.

Near the top of the AGB, low-mass stars undergo thermal pulsing (see, e.g. Iben and Renzini 1983). The Miras belong to this thermal pulsing stage and it seems likely that stars only manifest Mira characteristics during the brightest phases of each thermal cycle (Feast 1989b). The pulsation of Miras is probably driven by energy stored and released each pulsation cycle by a hydrogen ionization zone in the star.

Miras in the galactic field provide important information on the properties of these stars which we can then apply to those in globular clusters. Field Miras show a correlation between their kinematic properties (asymmetric drift and velocity dispersion) and their periods (see Feast 1963, Feast, Woolley and Yilmaz 1972, and table 5 from Feast 1989a). This table shows rather clearly that the Mira period-sequence cannot be simply interpreted as an evolutionary sequence (i.e. a Mira gradually evolving from the shortest period to the longest period) though this has at times been suggested. Instead, since age and chemical composition (or for evolved stars, initial mass and chemical composition) are correlated with galactic kinematics, the Mira sequence is best considered as a sequence of stars of different initial masses and/or chemical compositions. The evidence from the globular clusters suggests that a Mira is at the AGB tip of the population to which it belongs. Of course this does not preclude some change of period during the time the star remains in the Mira instability strip. However, the galactic kinematics suggest that the change in log P (Δ log P) is relatively small. Data for Miras in clusters reinforce this conclusion. In any cluster containing more than one Mira the range of periods is small indicating little evolution with period (e.g. In 47 Tuc the Miras have periods of

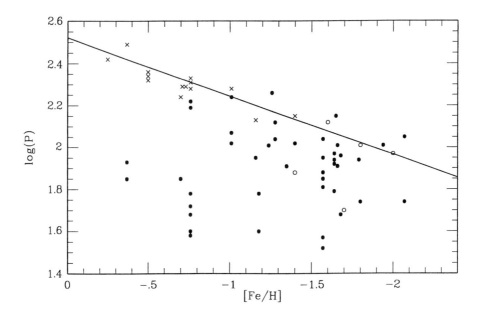

FIGURE 13. The period-[Fe/H] diagram for Miras and SR variables in clusters. Crosses, Miras; Filled circles, SR variables; Open circles, metal-poor SR variables in the galactic field.

212, 203, and 192 day; in NGC6637, 196 and 195 day; in NGC6356, 231, 220, 207, and 220 day).

It has long been known (Feast 1981) that there is at least a rough correlation between the periods of Miras in globular clusters and the cluster metallicity. The exact slope of this relation is however somewhat difficult to derive with any precision. By including the SRd variables in globular clusters (see section 4.2) to extend the correlation to shorter periods and lower metallicities, Feast (1992) found that the data could be reasonably represented by the equation,

$$[Fe/H] = 3.6 \log P + constant \tag{4.25}$$

Figure 13 shows an [Fe/H] -log P plot which in addition to Miras contains SR variables in clusters as well as some SR variables from the general field of the galactic halo. The bulk of the data for this figure is that used by Frogel and Whitelock (1998). A few additional variables have been added including the Mira in the SMC cluster NGC121. As we shall see in section 4.2 there is evidence that a red giant SR variable evolves with increasing period, i.e. vertically upwards in figure 13. Thus the termination of the AGB as a function of metallicity is plausibly represented by the upper envelope to the points in this figure. The line shown has a line of slope 3.6 as previously derived. Individual field Miras cannot be entered in a diagram such as figure 13 since the complexity of their atmospheres is such that little success has so far been achieved in obtaining abundances for them from direct spectroscopic analyses.

As mentioned above Miras show a good correlation of mass-loss rates with pulsation periods and amplitudes. Whitelock et al. (1987) found for simultaneous infrared observations from IRAS in the far infrared and from the ground in the near infrared that the mass in a (field) Mira dust shell could be expressed by the equation;

$$logM = 2.17 \log P + 1.88\Delta L - 13.38 \qquad (4.26)$$

where ΔL is the pulsation amplitude at 3.4 microns. These results strongly suggest that pulsation drives the mass loss. Probably there is a two stage process. Pulsation raises matter to a sufficient height above the star that dust formation can take place. The dust particles are then ejected from the star by radiation pressure (see for instance, Wood 1979). The results are consistent with the idea that Miras evolve with increasing amplitude and a small change in log P.

The kinematics of field Miras (Table 5) show that those with $P \sim 179$ day have an asymmetric drift of 98 ± 22kms^{-1}. This is distinctly less than that of classical halo objects †. However, it is greater than that of the thick disc (~ 30kms^{-1} (e.g. Sandage and Fouts 1987)). This strongly suggests that Miras of this period belong to a population intermediate between the halo proper and the thick disc. We can perhaps call this an intermediate halo population.

It has been customary since the work of Zinn (1985) to divide the globular clusters into 'halo' and 'disc' subgroups. The Miras suggest a somewhat more complex situation. For instance amongst the disc clusters are NGC 6712 (with a Mira of period 191 day), NGC6553 (with a Mira of period 265 day) and NGC6927 (with (probably) a Mira of period 310 day). Over this period range the kinematics of the field Miras change rapidly. This suggests that the clusters too cannot be classified into one homogeneous ('disc') population. If we take the kinematics of the field Miras as a guide the group of clusters containing Miras of period near 200 day (47 Tuc, NGC 6712, NGC6637, NGC6356) belong to a population with a larger asymmetric drift (~ 70kms^{-1}) than the standard thick disc (Feast 1989a)‡.

Rudimentary period-luminosity relations for Miras at visual and infrared wavelengths have long been known through statistical parallax analyses (e.g. Feast and Clayton 1969, Robertson and Feast 1981). But it was not until extensive infrared observations were made of Miras in the LMC where they are all at about the same distance from us that it became clear that there was a narrow PL relation in the infrared (at K, 2.2microns) and in M_{bol} (figure 14).

The relations found by Feast et al. (1989) from the LMC Miras together with a distance modulus of 18.70 for the LMC leads to,

$$M_K = -3.47 \log P + 0.78 \qquad (4.27)$$

with a scatter of $\sigma_M = 0.13$ mag,
and,

$$M_{bol} = -3.00 \log P + 2.65 \qquad (4.28)$$

with a scatter of $\sigma_M = 0.16$ mag.

If the Miras are evolving through an instability strip of finite width (or possibly if there is a range of metallicities at a given initial mass) we might expect that the PL relation is a first approximation to a PLC relation (see section 1.2). In fact Feast et al. find evidence for a PLC relation and derive (with a LMC modulus of 18.70),

$$M_{bol} = -4.32 \log P + 2.37(J - K)_0 + 2.93 \qquad (4.29)$$

† The exact value for halo objects is still somewhat uncertain (see, e.g. Martin and Morrison 1998) but it is generally believed to be close to the circular velocity of galactic rotation at the sun. A recent value for this (231 ± 15kms^{-1}) may be deduced from the kinematics of Cepheids (Feast and Whitelock 1997).

‡ Sandage and Fouts 1987 find evidence that the velocity dispersion perpendicular to the galactic plane is correlated with [Fe/H] within the thick disc.

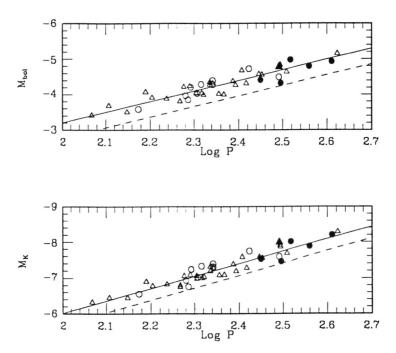

FIGURE 14. The period-luminosity relations for Mira variables. Open triangles, LMC Miras; Open circles, galactic globular cluster Miras; Closed circles, Miras with companions of known spectroscopic parallax; Close triangle, R Leo using its trigonometrical parallax. The solid line is the LMC relation, the dotted lines show other relations that have been suggested but which do not fit the data. (See text for further discussion) (from Whitelock et al. 1994).

although the scatter about this equation ($\sigma_M = 0.13$ mag) is only slightly less than that about the PL relation in M_{bol} ¶.

The calibration of the Mira PL relations, including the use of Hipparcos parallaxes of Miras, has been reviewed recently by Feast and Whitelock (1998). Since the emphasis here is on Miras in globular clusters the direct calibration of the Miras in 47 Tuc is of particular importance. This can be made (see Feast and Whitelock 1998) using the distance to 47 Tuc derived by Reid (1998) from main sequence fitting to subdwarfs with Hipparcos parallaxes. One then obtains absolute magnitudes which are 0.09 ± 0.15mag brighter than those given by equations 4.28 and 4.29. Clearly the agreement is satisfactory.

In using the PL relations one has to consider whether they might not be subject to metallicity effects (i.e. if the metallicities of the calibrating and programme stars are different at the same period). That there are no very large differences in luminosity between Miras in the LMC field and those in globular clusters in our Galaxy was demonstrated by Whitelock et al. (1994). Figure 14, in which globular clusters are shown as open circles, is from their paper. The PL relation shown by Miras in galactic globular clusters is evidently very similar (and probably the same) as that shown by the LMC Miras.

¶ Note that the comparison of this relation with the theoretical pulsation relation in Feast et al. depended on a $logT - (J - K)_0$ relation which is no longer thought appropriate (see below).

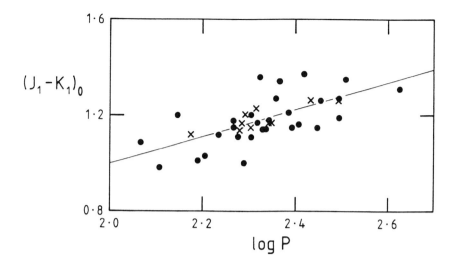

FIGURE 15. The $(J - K_0) - logP$ relation for Miras. filled circles, LMC Miras; crosses, Miras in galactic globular clusters (from Feast et al. 1989).

The zero-points of the LMC and the cluster relations have been revised since this figure was published. But as indicated above when comparing the zero-point derived from the Cepheid distance to the LMC and the subdwarf distance to 47 Tuc, these zero-points remain the same within the uncertainties. The slope of the cluster PL shown in figure 14, depends on the relative distances of clusters with Miras of different periods, that is of different metallicities. The cluster distances used by Whitelock et al. (1994) in producing figure 14 were obtained from HB magnitudes assuming an $M_V - [Fe/H]$ relation of the form,

$$M_V = 0.15[Fe/H] + constant \tag{4.30}$$

This is shallower than the 'Sandage' slope discussed earlier. However, as indicated there, the slope in the higher metallicity range is not well determined and might be less than 0.30. In any case adopting any slope in the range ~ 0.3 to ~ 0 will affect the present comparison very little.

There are slight infrared colour differences between LMC Miras and Miras of the same period in the general field of our Galaxy (see Whitelock et al. 1994, Glass et al. 1995). If we adopt the theoretical treatment of Wood (1990) this is in the sense that at a given period the LMC Miras may be slightly more metal poor and slightly brighter than those in the field of our Galaxy (see also Feast 1996). However Miras in galactic globular clusters have closely the same $(J - K)_0$ colours at a given period as those in the LMC. This is shown in figure 15 (from Feast et al. 1989). The line in this figure is the mean LMC relation and is given by,

$$(J - K)_0 = 0.56 \log P - 0.12 \tag{4.31}$$

As can be seen in figure 15 the spread in $(J - K)_o$ colours at a given period is much greater for the LMC Miras than for those in galactic globular clusters. This result has not been adequately explained. It seems rather unlikely that it could be due to either uncertainties in the reddenings or in the measured colours of the LMC variables. One

TABLE 6. A Comparison of RGB-tip and Mira Luminosities

[Fe/H]	M_{bol} RGB-tip	M_{bol} Mira	$\overline{P}$ day
0	−4.04	−4.90	312
−0.5	−3.95	−4.48	227
−0.7	−3.91	−4.32	200
−1.0	−3.85	−4.07	165
−1.5	−3.76	−3.65	120
−2.0	−3.65	−3.23	87

possibility is a spread of metallicities at a given period in the LMC. But any explanation has also to account for the fact that the cluster Miras fit the mean LMC relation very well.

Having discussed the calibration of Mira luminosities we can now re-examine their position in the HR diagram. Table 6 gives the results of Salaris and Cassisi (1998) for the luminosity of the theoretical RGB tip as a function of metallicity. According to their results the age of the population has only a small effect on these luminosities for ages larger than a few Gyr and they should thus be applicable to globular clusters. Their theoretical luminosities have been increased by 0.09 mag to bring them onto the empirical Cepheid scale (Feast 1999c). Also shown in the table are the corresponding luminosities for Miras. These were found by combining the PL relation (equation 4.29) with the P -[Fe/H] relation (equation 4.25). The zero-point for the latter equation was determined by adopting [Fe/H] = −0.76 at P = 202 day from the mean of the 47 Tuc Miras. The table shows that in the period range 312 to 165 day, the range over which most clearly established cluster Miras are found, their luminosities are distinctly above the RGB tip. Thus there can be little doubt that these stars are on the AGB (and at its tip). At the short period end of the distribution which may be represented by some SR variables in metal poor clusters (see section 4.2), the predicted luminosities are fainter than the RGB tip. One has the choice of concluding that these metal-poor variables fall on the RGB, but not at its tip, or that the AGB in these systems terminates below the RGB tip. The first of these two alternatives seems rather arbitrary. Further work on variables in metal poor systems should help clarify this matter.

To define the AGB tip one evidently needs not only the luminosities of Miras but also their effective temperatures, or equivalently, and in a sense more fundamentally, their radii. The effective temperature, or surface brightness, can be obtained rather directly from a measure of the angular diameter of a star together with an apparent magnitude. The problem of the determination of the angular diameters of Mira variables is closely related to the problem of their pulsation modes. Are Miras pulsating in the fundamental or the first overtone? This problem has been much debated for many years.

Globular cluster are too distant for direct measurement of the angular diameters of their red giants or Miras. However, many red giants and Miras in the galactic field have had their angular diameters measured. The first measurements were made from lunar occultations but more recently a number of groups have been making observations using ground-based interferometery (see, e.g Haniff 1995). It is also possible to use the fine guidance sensors on the HST (which are interferometers) to measure Mira angular diameters (Lattanzi et al. 1997).

Nevertheless there are considerable problems in the angular diameter and effective temperature estimates for Miras. A main reason for this is that the angular diameter of a

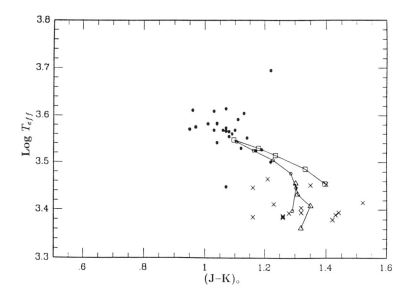

FIGURE 16. The $logT_{eff} - (J - K)_0$ diagram for Miras (crosses) and non-Miras (filled circles). The symbols joined by lines are model results from Bessell et al. (1989). The diagram is from Feast (1996). Note that T_{eff} is the same as T used in the text.

Mira is a strong function of the wavelength at which the measurements are made. This is the result of the extreme depth of the star's atmosphere. The diameter of a Mira measured in the visible at the wavelength of one of the strong TiO absorption bands is about twice that measured outside the band in a nearby region (e.g. Bonneau et al. 1982). Thus we need to ask which diameter interests us. In particular which diameter would correspond to predictions of AGB evolution and which would be appropriate to studying the star's pulsational properties? This is a complex problem. However, as at least a first approximation it would seem that the appropriate diameter is that measured in a spectral region where most of the radiation of the star is being emitted, that is at about 1.6 micron. Models (e.g. Bessell et al. 1996) indicate that at about this wavelength we would be measuring the diameter of the stellar photosphere. Comparison of angular diameters measured at about 1.6 microns with those measured at visual wavelengths and correct to 'photospheric' using model atmospheres shows good agreement (Feast 1996). This is an indication that the atmospheric models are satisfactory at least to a first approximation.

Figure 16 shows a plot of effective temperatures derived from angular diameter measurements plotted against $(J - K)_0$ (Feast 1996). For the constant or low amplitude (SR) variables the results are probably rather reliable. For the Miras there are at least three problems besides the question of correction to photospheric diameter just discussed. The first is that we can expect a significant scatter since the effective temperature varies significantly round the light cycle (variations of $\Delta logT \sim 0.1$ are to be expected). Secondly a variable will have usually been measured near maximum light since the large light range of the variables makes them difficult to measure near minimum. Thus it seems likely that the mean temperatures may be biased too high (but note that the values of

$(J - K)_0$ used in the plot correspond to the phase at which the diameter observations were made). Thirdly, the hope that there might be a useful log T - $(J - K)_0$ relation for Miras may be illusory. Predictions of models are plotted in figure 16 and these show that at low temperatures the temperature is nearly independent of the colour. The available data also show a considerable scatter for Miras in the log T - log P plane.

Figure 16 does however suggest a rather steep decline of log T with increasing $(J - K)$ in the region of relevance for the Miras. Feast (1996) obtained,

$$\log T = -0.590(J - K)_0 + 4.194 \qquad (4.32)$$

That a rather steep relation is required in the case of globular cluster Miras, at least, is shown by the following considerations. Wood (1990) has parameterized the loci of AGBs by the relation,

$$M_{bol} = 15.7 \log T + 1.884 \log z - 2.65 \log M + constant \qquad (4.33)$$

where z is the metal abundance. One can combine this with a pulsation equation,

$$\log P = 1.5 \log R - 0.5 \log M + constant \qquad (4.34)$$

to obtain,

$$\log T = -0.130 \log P - 0.073 \log z + 0.038 \log M + constant. \qquad (4.35)$$

This is quite a useful equation since the dependence on mass (M) is negligibly small. Putting this equation together with the period-colour relation (equation 4.31) and equations 4.25 leads to,

$$\Delta \log T / \Delta(J - K) = -0.70, \qquad (4.36)$$

which, given all the uncertainties, is satisfactorily close to the slope given by equation 4.32.

Linear diameters can be obtained by combining angular diameter measurements (correct to photospheric as discussed above) with a PL relation (Feast 1996) or by combining the angular diameters with individual parallaxes from the Hipparcos catalogue (van Leeuwen et al. 1997, Feast 1999b, Whitelock et al. 1999). In both case one find that Miras in the field with periods close to 300 day have linear radii of about $400R_\odot$. There are no directly estimated radii for Miras with periods as short as those typical of the globular cluster variables. A rough extrapolation, taking into account that a 200 day globular cluster Mira is likely to be somewhat less massive than a 300 day Mira in the general field leads to an estimated radius for the 200 day globular cluster variables of $\sim 250R_\odot$. Thus one concludes from the discussion of this section that at the top of the AGB in a cluster like 47 Tuc the stars have luminosities of $\sim 4 \times 10^3 L_\odot$ and radii of $\sim 250R_\odot$, implying effective temperatures of $\sim 3000K$.

4.2. Semiregular (SR) Variables

Low amplitude SR variability occurs toward the top of globular cluster giant branches (e.g. Lloyd Evans and Menzies 1973) and in metal-rich clusters the SRs lie just below the Miras (figure 12). They seem likely to be AGB stars since, as we shall see, there is evidence that in the metal rich clusters they evolve into the cluster Miras. However, they are not all above the RGB tip so that it is not possible to be quite certain of this. SR variables either in clusters or in the general field have been less completely studied than the Miras. Their low light amplitudes and lack of regularity make their study less attractive especially using photographic techniques. However, a number of infrared studies of field stars of this type have been made in recent times (e.g. Kerschbaum and Hron 1994). Given the lack of regularity in many of the variables, it has not always

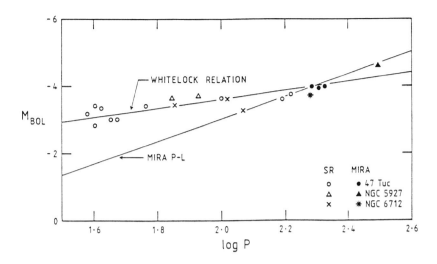

FIGURE 17. The SR and Mira variables in meta-rich globular clusters (from Whitelock 1986). The Mira PL is shown. The SRs define a line of shallower slope (the Whitelock relation) (from Feast 1989a).

been clear as to how much of the observed variability should be attributed to pulsation. Schwarzschild (1975) suggested that large convection cells are present in the atmospheres of red giant and supergiant stars and that the changes in these could produce low level light variations. Since the turn-over time scale for a typical cell is ~ 200 day one expects light varations on time scales of this length. This is indeed the order of periodicity or quasi-periodicity in many SR variables. Recent interferometric observations have provided support for Schwarzschild's proposal at least in the case of α Ori (a massive supergiant irregular variable). In the case of Mira variables, interferometry often shows the stars to be non-circularly symmetrical. This is probably due to unresolved surface structure (see Haniff 1995 for a review).

Nevertheless, if we confine attention to SR variables for which periods can be obtained with some degree of confidence, then there is good evidence that the light variations are due primarily to pulsation. We should of course anticipate that less regular variations might be superimposed on pulsational light variations due to the Schwarzschild effect. In the case of field stars the evidence for pulsation comes from a discussion of the available angular diameter measurements. These can be combined with Hipparcos parallaxes to obtain linear diameters and the results compared with the predictions for the pulsational radii of low mass stars as already mentioned in section 4.1. Feast (1999c) and Whitelock et al. (1999) conclude that;

(1) The SR variables with reasonably well determined periods in the field do in general have radii and periods which are consistent with the pulsation hypothesis.

(2) The SRs and the Miras are pulsating in the same mode which, on conventional theory is the first overtone.

The situation regarding SR variables in globular clusters was considerably clarified by Whitelock (1986) who derived bolometric magnitudes of SR and Mira variables in clusters from infrared observations. Her results for metal-rich clusters are shown in the figure 17. It is clear that in any cluster the luminosities of the SRs increase with the period and that these stars define a sequence terminating at the position of the Miras in the cluster. Thus the sequence defined by the 'Whitelock' line can be taken as an evolutionary sequence. Stars evolve along this line with increasing period, amplitude and luminosity until they reach the Mira stage. As discussed in section 4.1 the Mira PL relation is a sequence of AGB tips. The Whitelock line makes clear how stars evolve onto the Mira PL relation along lines of lower slope in a diagram such as figure 17.

The Whitelock relation is important for a number of reasons. For one, it clarifies our understanding of SR variables in the general field. Wood and Sebo (1996) find in the LMC that in a luminosity-log P plot the SRs lie above the Mira PL relation (i.e. at a given period the SRs are brighter). A similar result was recently obtained by Bedding and Zijlstra (1998) for SRs in the general galactic field using Hipparcos parallaxes to derive absolute magnitudes. In the Galaxy the kinematics of the field SRs show them to be predominently old disc stars (Feast, Woolley and Yilmaz 1972) and therefore probably more metal rich and/or younger that those in 47 Tuc. Nevertheless, they can be seen as analogous to the 47 Tuc type SRs and on evolutionary tracks of shallow slope which will end on the Mira PL relation; in their case at longer periods than the Miras in globular clusters. A 'Whitelock' line such as that shown in figure 1 of Bedding and Zijlstra (1998) connects the SR variables with Miras whose periods are about a year. These also have old disc kinematics so that one can naturally regard the latter as evolutionary products of the former. Wood and Sebo on the other hand suggest that the SRs and the Miras define two nearly parallel sequences which they take to be fundamental and first overtone pulsation sequences. However, the Whitelock line shows that at least in the case of globular clusters this cannot be so. In addition the Wood/Sebo picture would have SRs of period near 100 day as overtone pulsators of Miras with periods near 200 day. In our Galaxy at any rate this cannot be true since 100 day SRs have old disc kinematics whereas 200 day Miras have intermediate halo kinematics (Feast, Woolley and Yilmaz 1972 and table 5 above). The Whitelock relation is also of interest since, in combination with theory, it should place some constraints on mass-loss rates on the AGB. Wood's parameterization of the AGB (equation 4.33) and the pulsation equation (4.34) can be combined to eliminate the temperature. One then obtains,

$$M_{bol} = -2.036 \log P + 0.73 \log z - 2.049 \log M + constant \qquad (4.37)$$

(Feast 1996). This evidently predicts that at constant mass and z an evolutionary track should be,

$$M_{bol} = -2.036 \log P + constant. \qquad (4.38)$$

 Figure 18 shows the data for one cluster (47 Tuc). In this way we need not be concerned with the absolute calibration of the luminosities or the value of z. The two regressions are shown, these have slopes ($\Delta M_{bol}/\Delta log P$) of –1.13 and –1.39. These are both distinctly different from that predicted by equation 4.38. However, this is expected if there is significant mass-loss as a star ascends the AGB. Putting the slope, –1.39, into equation 4.37 yields,

$$\Delta \log M = 0.22 \Delta M_{bol} \qquad (4.39)$$

for the change of mass with luminosity on the AGB. For the range of M_{bol} covered by the variables in 47 Tuc ($\sim$ 0.8mag) we have $\Delta log M \sim -0.17$ or a decrease in mass by a factor of 1.5 over this luminosity range. According to Wood (1990) a star ascends the AGB at a rate of one magnitude in $\sim 1.2 \times 10^6$ year. Thus this result implies an average mass loss rate of about $3 \times 10^{-7} M_\odot yr^{-1}$. This is in the range of mass-loss rates for late type stars with optically thin shells (Knapp and Morris 1985). Whilst this value for the mass loss does not seem too unrealistic (e.g from a mass of, say,$\sim 0.8 M_\odot$ on the lower

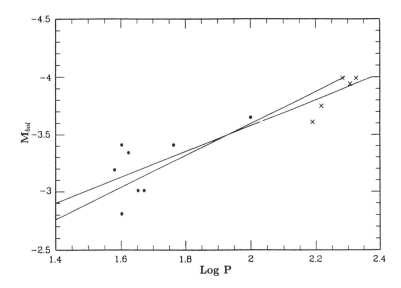

FIGURE 18. The Whitelock relation for 47 Tuc variables (crosses, Miras; filled circles, SR variables). The two regressions are shown (from Feast 1996).

AGB to $\sim 0.5 M_\odot$ at the tip) it must be remembered that this is a very rough estimate at the present time relying, amongst other things, on the correct parameterization of the AGB (equation 4.33).

Since the Whitelock relation is an evolutionary track it provides an opportunity to test the metallicity dependence of the AGB tracks predicted by equation 4.33. Unlike the Miras the SRs are on the whole of sufficiently high surface temperature that a standard $logT - (J - K)_0$ relation (Bessell et al. 1983, Whitelock 1986) may be used to derive values of log T for them. When this is done one finds that at a given period the SRs in metal-rich globular clusters are significantly cooler than those in metal-poor clusters (figure 19). However a plot of $logT + 0.073 \log z$ against log P (figure 20) shows no significant separation of stars of different metallicities at a given period verifying the metallicity term in equation 4.35. This test is possible because the mass term in equation 4.35 is negligible.

We can now return to a discussion of figure 13; the relation between [Fe/H] and log P. Besides globular cluster stars, this plot contains five SR variables from the general field whose chemical abundances indicate membership of the galactic halo. Values of [Fe/H] for these stars are from Preston and Wallerstein (1963), Leep and Wallerstein (1981), Luck and Bond (1985) and Giridhar et al. (1998). In cases where more than one determination has been published means have been taken. Stars of this type in the general field are classified as SRd variables although as used in the General Catalogue of Variable Stars this term is applied to a rather heterogeneous group of objects. Some field SRd stars and some SR stars in clusters have strong Balmer emission lines at certain phases and have been suggested as the continuation of the Mira sequence to shorter periods and lower metallicities (Feast 1965, 1973, Preston 1967). However, the available data both in clusters and the general field make it difficult to distinguish very satisfactorily between metal-poor variables with Balmer emission at some phase and those without emission at

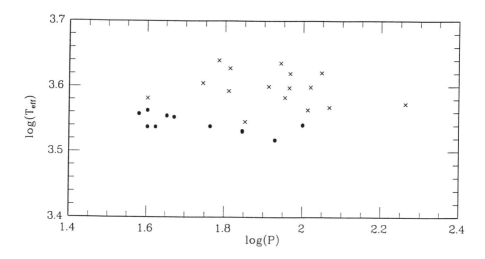

FIGURE 19. The relation between $logT_{eff}$ and log P for SR variables in globular clusters. The crosses are for metal-poor clusters ($[Fe/H] < -1.1$) and filled circles for metal-rich clusters ($[Fe/H] \geq -0.8$) (data from Whitelock 1986).

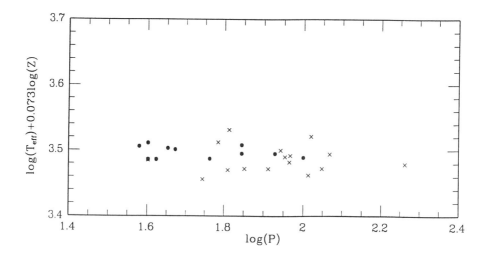

FIGURE 20. As figure 19 but with the factor 0.073[Fe/H] added to the ordinate.

any phase. Thus no attempt is made to distinguish between the two groups here.
Figure 13 shows that the maximum period to which the Mira and SR variables extend is a function of metallicity. The relation of Mira period to metallicity found previously (equation 4.25) for metal-rich clusters fits the upper envelope of figure 13 at all metallicities. Since we have just seen that SR variables in clusters evolve with increasing period (i.e. vertically upwards in figure 13) the upper envelope in this figure gives the period of

a star at the tip of the AGB. Provided there is no large difference in the (pulsational) masses at the tip of the AGB as a function of metallicity in globular clusters (and a large difference seems unlikely unless very different amounts of mass are lost in the post-AGB evolution of stars of different metallicity), equation 4.25 together with the pulsation equation 4.34 and the previous discussion of Mira radii shows (very roughly) that over the range of cluster metallicities from [Fe/H] ~ -0.5 to ~ -2.0 the stellar radius at the AGB-tip decreases by a factor of ~ 2 from $\sim 250R_\odot$ to $\sim 125R_\odot$.

Both in the case of the SRs, and of the Miras discussed in the previous section, the numerical results are subject in many cases to considerable uncertainty. Nevertheless, they suggest that we are at least beginning to understand the place of these stars in stellar evolution and in particular in globular cluster evolution, to obtain from them clues to stellar evolution which cannot be obtained in other ways, and to derive numerical results against which to test theory.

5. Acknowledgements

Drs Frogel and Whitelock kindly provided data used in their 1998 paper. Some of the new work reported in this section has been done in collaboration with Dr Whitelock and I am grateful to her for the use of unpublished results and for many helpful discussions.

REFERENCES

Arp, H.C. 1955, AJ, 60, 317

Arp, H. 1967, ApJ, 149, 91

Arp, H.C., Brueckel, F. and Lourens, J. v. B. 1963, ApJ, 137, 228

Arp, H. and Thackeray, A.D. 1967, ApJ, 149,

Baade, W.A. 1951, Pub. Obs. Univ. Michigan, X, 7

Bailey, S.I. 1913, Harv. Coll. Obs. Ann. 78, No.1

Beaulieu. J.P. et al. 1995, A&A, 303, 137

Becker, S.A., Iben, I. and Tuggle, R.S. 1977, ApJ, 218, 633

Bedding, T.R. and Zijlstra, A.A. 1998, ApJ, 506, L47

Bessell, M.S., Wood, P.R. and Lloyd Evans, T. 1983, MNRAS, 202, 59

Bessell, M.S., Brett, J.M., Scholz, M. and Wood. P.R. 1989, A&A, 213, 209

Bessell, M.S., Scholz, M. and Wood, P.R. 1996 A&A, 307, 481

Bingham, E.A., Cacciari, C., Dickens, R.F. and Fusi Pecci, F. 1984, MNRAS, 209, 765

Bonneau, D., Foy, R., Blazit, A. and Labeyrie, A. 1982, A&A, 106, 235

Brocato, E., Buonanno, R., Castellani, V. and Walker, A.R. 1989, ApJ. Sup. 71, 25

Chaboyer, B., Demarque,P., Kernan, P.J. and Lawrence M.K. 1998, ApJ, 494, 96

Chaboyer, B., Demarque, P. and Sarajedini, A. 1996, ApJ, 459, 558

Chiosi, C., Bertelli, G., Meylan, G. and Ortolani, S. 1989, A&A, 219, 167

Chiosi, C., Wood, P., Bertelli, G. and Bressan, A. 1992, ApJ, 387, 320

Christy, R. F. 1966, ApJ, 144, 108

Clayton, M.L. and Feast, M.W. 1970, MNRAS, 146, 411

Cohen, J.G. 1992, ApJ, 400, 528

Dorman, B. 1992, ApJ. Sup. 81, 221

Eddington, A.S. 1913, MNRAS, 73, 359

Eddington, A.S, 1917, The Observatory, 40, 290

Eddington, A.S. 1918a, MNRAS, 79, 2

Eddington, A.S. 1918b, MNRAS, 79, 177

Eddington, A.S. 1940, MNRAS,100, 354

Edmonds, P.D. and Gilliland, R.L. 1996, ApJ, 464, L157

Elson, R.A.W. and Fall, S.M. 1985, PASP, 97, 692

ESA, 1997, The Hipparcos Catalogue, ESA SP-1200

Feast, M.W. 1963, MNRAS, 125, 367

Feast, M.W. 1965, The Observatory, 85, 16

Feast, M.W. 1973, in Fernie, J.D. ed. Variable Stars in Globular Clusters and in Related Systems, Reidel, Dordrecht, p. 131

Feast, M.W. 1981, in Iben, I. and Renzini, A. eds. Physical Processes in Red Giants, Reidel, Dordrecht, p.193

Feast, M.W. 1989a, in Schmidt, E.G. ed. The Use of Pulsating Stars in Fundamental Problems of Astronomy, Cambridge University Press, p.205

Feast, M.W. 1989b, in Johnson, H.R. and Zuckerman, B. eds. 1989b, Evolution of Peculiar Red Giant Stars, Cambridge University Press, p.35

Feast, M.W. 1992, in Bergeron, L., ed. Highlights of Astronomy vol.9 Kluwer, Dordrecht, p.613

Feast, M.W. 1996, MNRAS, 278, 11

Feast, M.W. 1997, MNRAS, 284, 761

Feast, M.W. 1998, Mem. Soc. Ast. It., in press

Feast, M.W. 1999a, in New Views of the Magellanic Clouds, IAU Symp. 190, ASP. Conf. Ser. in press

Feast, M.W. 1999b, in Asymptotic Giant Branch Stars, IAU Symp. 191, in press

Feast, M.W. 1999c, in The Stellar Content of the Local Group, IAU Symp. 192, in press

Feast, M.W. and Catchpole, R.M. 1997, MNRAS, 286, L1

Feast, M.W., Glass, I.S., Whitelock, P.A. and Catchpole, R.M. 1989, MNRAS, 241, 375

Feast, M.W. and Whitelock, P.A. 1987, in Kwok, S. and Pottasch, S.R. eds. Late Stages of Stellar Evolution, Reidel, Dordrecht, p.33

Feast, M.W. and Whitelock, P.A. 1997, MNRAS, 291, 683

Feast, M.W. and Whitelock, P.A. 1998, in Caputo, F. and Heck, A. eds. Post-Hipparcos Cosmic Candles, Kluwer, Dordrecht in press

Feast, M.W., Woolley, R.v.d.R. and Yilmaz, N. 1972, MNRAS, 158, 23

Fernie, J.D. 1969, PASP, 81, 707

Fernley, J. et al. 1998a, MNRAS, 293, L61

Fernley, J. et al. 1998b, A&A, 330, 515

Frogel, J.A. and Whitelock, P.A. 1998, AJ, 116, 754

Fusi Pecci, F. et al. 1996, AJ, 112, 1461

Giridhar, S., Lambert, D.L. and Gonzalez, G. 1998, PASP, 110, 671

Glass, I.S., Whitelock, P.A., Catchpole, R.M. and Feast, M.W. 1995, MNRAS, 273, 383

Gratton, R.G. 1998, MNRAS, 296, 739

Gratton, R.G., Tornambe, A. and Ortolani, S. 1986, A&A, 169, 111

Gratton, R.G., Fusi Pecci, F., Carretta, E., Clementini, G., Corsi, C.E. and Lattanzi, M. 1997, ApJ, 491, 749

Haniff, C.A. in Stobie, R.S. and Whitelock, P.A., eds. Astrophysical Applications of Stellar Pulsations, ASP. Conf. Ser. 83, p.270

Haniff, C.A., Scholtz, M. and Tuthill, P.G. 1995, MNRAS, 276, 640

Hanson, R.B. 1979, MNRAS, 186, 875

Hinkle, K.H., Hall, D.N.B. and Ridgway, S.T. 1982, ApJ 252, 697

Hofmeister, E., Kippenhahn, R. and Weigert, A. 1964, Zs. f. Ap., 60, 57

Holland, S., Fahlman, G.G. and Richer, H.B. 1997, AJ, 114, 1488

Iben, I. 1964, ApJ, 140, 1631

Iben, I. and Renzini, A. 1983, Ann. Rev. Astron. Astrophys., 21, 271

Jones, R.V., Carney, B.W., Storm, J. and Latham, D.W. 1992, ApJ, 386, 646

Joy, A.H. 1949, ApJ, 110, 105

Kerschbaum, F. and Hron, J. 1994, A&A Sup., 106, 397

Kinman, T.D. 1959, MNRAS, 119, 538

Knapp, G.R., and Morris, M. 1985, ApJ, 292, 640

Koen, C. 1992, MNRAS, 256, 65

Koen, C. and Laney, C.D. 1998, MNRAS in press

Kovács, G. et al. 1992, A&A, 259, L46

Kovács, G. and Walker, A.R. 1998 preprint

Lattanzi, M.G., Munari, U., Whitelock, P.A. and Feast, M.W. 1997, ApJ, 485, 328

Leep, E.M. and Wallerstein, G. 1981, MNRAS, 196, 543

Liu, T. and Janes, K.A. 1990, ApJ, 360, 561

Lloyd Evans, T. and Menzies, J.W. in Fernie, J.D. ed. Variable Stars in Globular Clusters and in Related Systems, Reidel, Dordrecht, p.151

Longmore, A.J., Fernley, J.A. and Jameson, R.F., 1986 MNRAS, 220, 279

Longmore, A.J., Dixon, R., Skillen, I., Jameson, R.F. and Fernley, J.A., 1990, MNRAS, 247, 684

Luck, R.E. and Bond, H.E., 1985, ApJ, 292, 559

Lutz, T.E. and Kelker, D.H. 1973, PASP, 85, 573

Malmquist, K.G. 1920, Lund Medd.Ser.II,No. 22

Martin, J.C. and Morrison, H.L. 1998, AJ, 116, 1724

Mateo, M. 1993, in Smith, G.H. and Brodie, J.P. eds. The Globular Cluster - Galaxy Connection ASP Conf. Ser. 48, p.387

McNamara, D.H. 1997, PASP, 109, 857

Pont, F., Mayor, M., Turon, C. and Vandenberg, D.A. 1998, A&A, 329, 87

Preston, G.W. 1967, PASP, 79, 125

Preston, G.W. and Wallerstein, G. 1963, ApJ, 138, 820

Reid, I.N. 1996, AJ, 111, 2000

Reid, I.N. 1997, AJ, 114, 161

Reid, I.N. 1998, AJ, 115, 204

Renzini, A. et al. 1996, ApJ, 465, L23

Ritter, A. 1879, Wiedemanns Ann. 8, 172

Robertson, B.S.C. and Feast, M.W. 1981, MNRAS, 196, 111

Salaris, M. and Cassisi, S. 1998, MNRAS in press

Sandage, A.R. 1990, ApJ, 350, 603

Sandage, A.R. 1993a, AJ, 106, 687

Sandage, A.R. 1993b, AJ, 106, 703

Sandage, A.R. 1993c, in Nemec, J.M. and Matthews, J.M. eds. New Perspectives on Stellar Pulsation and Pulsating Variable Stars, Cambridge University Press, p. 3

Sandage, A and Cacciari 1990, ApJ, 350, 645

Sandage, A.R., Katem, B and Sandage, M. 1981, ApJ. Sup. 46, 41

Sandage, A.R. and Fouts, G. 1987, AJ, 92, 74

Schwarzschild, M. 1975, ApJ, 195, 137

Skillen, I., Fernley, J.A., Stobie, R.S. and Jameson, R.F. 1993, MNRAS, 265, 310

Smith, H.A. 1995, RR Lyrae Stars, Cambridge University Press

Sterken, C. and Jaschek, C. eds. 1996, Light Curves of Variable Stars, Cambridge University Press

Storm, J., Carney, B.W. and Latham, D.W. 1994, A&A, 290, 443

Sweigart, A.V. 1997 in Davis Philip, A.G. ed. Proc. Third Conf. on Faint Blue Stars, L.Davis, Schenectady, NY in press

Tsujimoto, T., Miyamoto, M., and Yoshii, Y. 1998, ApJ, 492, L79

van Albada, T.S. and Baker, N. 1973, ApJ, 185, 477

van den Bergh, S. 1993, in Djorgovski, S.G and Meylan, G. The Structure and Dynamics of Globular Clusters, ASP. Conf. Ser. 50, p.1

van Leeuwen, F., Feast, M.W., Whitelock, P.A. and Yudin, B. 1997, MNRAS, 287, 955

Walker, A.R. 1987, MNRAS, 225, 627

Walker, A.R. 1992, ApJ, 390, L81

Welch. D.L., Mateo, M., Coté, Fischer, P and Madore, B.F. 1991, AJ, 101, 490

Welch, D.L., Mateo, M. and Olszewski, E.W. 1993, in Nemec, J.M. and Matthews, J.M. eds. New Perspectives on Stellar Pulsation and Pulsating Variable Stars, Cambridge University Press, p.359

Whitelock, P.A. 1986, MNRAS, 219, 525

Whitelock, P.A., 1990, in Cacciari, C. and Clementini, G. eds. Confrontation between Stellar Pulsation and Evolution, ASP Conf. Ser.11

Whitelock, P.A., Feast, M.W. and Catchpole, R.M. 1991, MNRAS, 248, 276

Whitelock, P.A., Pottasch, S.R. and Feast, M.W. 1987, in Kwok, S. and Pottasch, S.R. eds. Late Stages of Stellar Evolution, Reidel, Dordrecht, p.269

Whitelock, P.A., et al. 1994, MNRAS, 267, 711

Whitelock, P.A. et al. 1999, in preparation

Wood, P.R., 1979, ApJ, 227, 220

Wood, P.R. 1990, in Mennessier, M.O. and Omont, A. eds. ¿From Miras to Planetary Nebulae, Editions Frontières, Gif-sur-Yvette, p.67

Wood, P.R. and Sebo, K.M. 1996, MNRAS, 282, 958

Zinn, R. 1985, ApJ, 293, 424

Born on 20 April 1943, **RAMÓN CANAL MASGORET** graduated in Physical Sciences at the University of Barcelona in 1965, where he obtained his doctorate in 1973 with a thesis directed by Evry Schatzman and Juan J. Orus.

He was Profesor Agregado (assistant lecturer) at the University of Barcelona (1979-82) and Professor at the University of Granada (1992-97) before occupying the chair in Astrophysics in the Physics Faculty of the University of Barcelona. He has also been associated with the Paris Institute of Astrophysics, where he worked for five years, the University of Chicago (1982-84), where he taught stellar evolution and supernovae, and the Max-Planck Intitute for Astrophysics (1985-91), where his work was dedicated to collapsed stars, supernovae and nucleosynthesis.

A pioneer of astrophysics in Spain, Ramón Canal participated in the I National Assembly of Astronomy and Astrophysics, celebrated in 1975 at Puerto de la Cruz (Tenerife), the first such meeting ever held in Spain.

He has directed seven doctoral theses and 14 master's dissertations. Among his former students are a number of university professors, CSIC researchers, and university lecturers.

A referee of mainstream scientific journals such as Astrophysical Journal and Astronomy and Astrophysics, he has over a thousand citations in the scientific literature.

He has been a member of the time allocation committee for the IUE (International Ultraviolet Explorer) of the European Space Agency (1982-85), a member of the IAC's CAT (Telescope Time Allocation Committe) for the Canarian Observatories (1984-86), Director of the Department of Astronomy and Metrology of the University of Barcelona (1987-94), serving member and President of the Advisory Committee (Mathematics and Physics) of the Spanish Advisory Commission for Research Activities, and a member (since 1992) of the World Institute of Science.

Ramón Canal was a founder member and the first President of the Spanish Astronomical Society (1992-96).

X–ray Sources in Globular Clusters

By RAMÓN CANAL

Department of Astronomy, University of Barcelona, 08028 Barcelona, SPAIN

The total mass of the globular cluster system of our Galaxy makes only $\sim 10^{-3}$ the mass of the Galactic disk. It contains, however, $\sim 20\%$ of all known low–mass binary X–ray sources, about one half of all binary pulsars, and more than a half of the millisecond pulsars in the Galaxy. Close binary systems containing neutron stars should thus form much more easily in the dense stellar environment of globular clusters than elsewhere in the Galaxy. In these lectures we first review the formation mechanism of neutron stars. Then, we present the evolutionary scenarios leading to the formation of binary X–ray sources and binary and millisecond pulsars in the Galactic disk and the Galactic bulge. We later discuss the specific mechanisms to form neutron star binaries in globular clusters. We end by discussing the open issues concerning the origin and evolution of X–ray sources and millisecond pulsars in globular clusters, and their relationship with the structure, dynamics and evolution of the clusters themselves.

1. Low–mass binary X–ray sources and millisecond pulsars

An early, unexpected result of X–ray astronomy was the discovery of several bright X–ray sources in globular clusters. Later on, searches for radio pulsars have produced many detections, especially of short period pulsars. We begin these lectures with a very schematic presentation of those two kinds of objects.

1.1. X–ray binaries

Luminous Galactic binary X–ray sources provided the first evidence of *neutron star binaries*, that is binary star systems containing neutron stars (Giacconi et al. 1971; Lewin et al. 1971; Schreier et al. 1972; Tananbaum et al. 1972). Those sources had luminosities L_x in the 2–20 keV range:

$$L_x \sim 10^{36} - 10^{38} \ erg \ s^{-1}$$

The low–luminosity end of this range corresponds to the detection limit of the early X–ray satellites for distances ~ 1 kpc. The high–luminosity end roughly coincides with the Eddington limit for a $\sim 1 \ M_\odot$ neutron star. The X–ray emission is powered by accretion of matter by the neutron star from its close stellar companion:

$$L_x = \frac{GM\dot{M}}{R} \tag{1.1}$$

and the above range of luminosities thus implies:

$$\dot{M} \sim 10^{-10} - 10^{-8} \ M_\odot \ yr^{-1}$$

for a typical M = 1.4 $M_\odot$, R = 10 km neutron star.

There are two main types of binary X–ray sources

(i) *High–mass binary X–ray sources* (HMBXS) or *class I sources* (Cen X–3 type sources). The companion of the neutron star has a mass M $\gtrsim$ 10 $M_\odot$, and it appears as a luminous, early–type star (O or early B). Also,

293

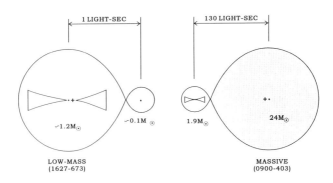

FIGURE 1. A comparison of the dimensions of LMBXS (left) with those of a HMBXS (right)
(adapted from Bradt and McClintock 1983, and Bhattacharya and van den Heuvel 1991)

$$L_{opt}/L_x > 1$$

and they are concentrated on the Galactic plane (typical ages t $\lesssim 10^7$ yr). They all have
hard X–ray spectra (kT $\gtrsim$ 15 keV). They show regular X–ray pulsations and no X–ray
bursts, which indicates strong magnetic fields at the surface of the neutron star.

(ii) *Low–mas binary X–ray sources* (LMBXS) or *class II sources* (Her X–1 type sources).
The companion has a mass M $\lesssim$ 1 M$_\odot$. In those sources:

$$L_{opt}/L_x < 0.1$$

The optical emission is dominated by reprocessing of the X–rays by the accretion disk
around the neutron star, which makes the spectral type of the companion uncertain.
They are concentrated towards the Galactic center, spread around the Galactic plane,
and are abundant in globular clusters. Typical ages are thus t $\sim (5$–$15)\times 10^9$ yr. They
have soft X–ray spectra (kT $\lesssim$ 10 keV). Regular X–ray pulsations are rare, and they often
show X–ray bursts (thermonuclear flashes on the neutron star surface), which indicates
weak magnetic fields (B $\lesssim 10^9$ G).

A comparison of a typical LMBXS with an also typical HMBXS is shown in Figure 1.

Other strong Galactic X–ray sources are:

— Young supernova remnants (like the Crab Nebula and its pulsar).
— Black hole X–ray binaries.
— Peculiar X–ray binaries (SS433, Cyg X–3).

1.2. *Millisecond pulsars*

Radio pulsars provided the first evidence of neutron stars. Neutron stars had been
conceived (shortly after the discovery of the neutron) by Landau (1932) and by Baade and
Zwicky (1934), but the first clear sign of their existence only came almost 35 years later,
when Hewish et al. (1968) detected the first pulsar and those objects were interpreted
as being rapidly rotating, strongly magnetized neutron stars, powered by the loss of

rotational kinetic energy (Gold 1968). There are ~ 600 pulsars known, nowadays, and most of them are single neutron stars: only $\sim 3\%$ of them are known to be members of binary systems. The periods of "ordinary" pulsars are within the range:

$$P \sim 0.1 - 5 \ s$$

and their surface dipole magnetic fields:

$$B_s \sim 3 \times 10^{11} - 3 \times 10^{13} \ G$$

Millisecond radio pulsars have periods

$$P < 10 \ ms$$

and there seems to be a close relationship between *binary* and millisecond radio pulsars:

(i) Shorter pulse periods (about one half of all binary radio pulsars have periods P < 30 ms).

(ii) Weaker magnetic fields (most binary radio pulsars and all the millisecond pulsars have $B_s < 4 \times 10^{10}$ G).

(iii) About one half of all known millisecond pulsars are found in binaries.

(iv) Approximately one half of all binary radio pulsars, and more than one half of the millisecond pulsars are in *globular clusters* (which, in contrast, contain only $\sim 3\%$ of the single radio pulsars).

1.3. *X-ray binaries and millisecond pulsars in globular clusters*

As it should be expected, high–mass X–ray binaries are completely absent from globular clusters. On the contrary, both low–mass X–ray binaries and millisecond pulsars are overabundant there as compared with the rest of the Galaxy. This can be clearly seen from the following figures:

The number of globular clusters in the Galaxy is $N_{GC} \lesssim 200$, and they contain, on average, a number of stars $n_{stars} \sim 10^5 - 10^6$, with masses $M \lesssim 0.8 \ M_\odot$. Therefore, their total mass is $\lesssim 10^{-3}$ the total mass of the Galactic disk. But, in contrast with that, the globular clusters contain $\sim 20\%$ of all known LMBXS, $\sim 50\%$ of the binary pulsars, and more than 50% of the millisecond pulsars.

As we will see later, those numbers reflect the enhanced rate of production of binaries by stellar encounters, with formation of neutron star binaries in particular, in the dense stellar environments of globular clusters, as compared with the rest of the Galaxy.

2. Neutron star formation: core collapse of massive stars

If we just considered *single star evolution* without *mass loss*, the initial stellar mass range

$$0.08 \ M_\odot \lesssim M \lesssim 60 - 100 \ M_\odot \tag{2.2}$$

would be divided as follows:

$$M \gtrsim 0.08 \ M_\odot \tag{2.3}$$

sets the lower mass limit for ignition of H ($T \simeq 2 \times 10^7 \ K$).

$$M \gtrsim 0.5 \ M_\odot \tag{2.4}$$

is the lower mass limit for starting He burning ($T \simeq 2 \times 10^8$ K).

The stars with masses

$$0.5 \ M_\odot \lesssim M \lesssim 8 - 9 \ M_\odot \qquad (2.5)$$

develop electron–degenerate C–O cores prior to C ignition (due to $\nu\bar{\nu}$ cooling). At the same time, a mass

$$M \gtrsim 1.4 \ M_\odot \qquad (2.6)$$

is required, within this mass range, to ignite C. Therefore, all stars with initial masses

$$1.4 \ M_\odot \lesssim M \lesssim 8 - 9 \ M_\odot \qquad (2.7)$$

would explode upon igniting C at the center of a strongly electron–degenerate core with roughly the Chandrasekhar mass. However, mass loss (in particular during the thermally pulsing asymptotic giant branch phase) most likely stops core growth before reaching the point of C ignition, and those stars become C–O white dwarfs.

Stars with masses

$$M \gtrsim 8 - 9 \ M_\odot \qquad (2.8)$$

ignite C nonexplosively at $T \simeq 8 \times 10^8$ K (off–center and in mild flashes towards the lower mass end). C burning produces a mixture of Ne (mostly) with some Mg, plus the O left from He burning. Recent evolutionary calculations (García–Berro and Iben 1994; Gutiérrez et al. 1996; Ritossa, García–Berro, and Iben 1996; Iben, Ritossa, and García–Berro 1997) show that the mass range

$$8 - 9 \ M_\odot \lesssim M \lesssim 12 - 13 \ M_\odot \qquad (2.9)$$

is subdivided into two subintervals. For

$$8 - 9 \ M_\odot \lesssim M \lesssim 10 - 11 \ M_\odot \qquad (2.10)$$

the stars enter a thermally pulsing "super–AGB" phase with an electron–degenerate ONe(Mg) core still below the Chandrasekhar mass. Those stars likely become O–Ne white dwarfs. On the other hand, for

$$10 - 11 \ M_\odot \lesssim M \lesssim 12 - 13 \ M_\odot \qquad (2.11)$$

electron captures on Ne in the core trigger explosive Ne and O burning, but at densities so high ($\rho \gtrsim 10^{10}$ $g \ cm^{-3}$) that further (and very fast) electron captures on the incinerated (nuclear statistical equilibrium) material decrease the Chandrasekhar mass below the actual mass of the core, which leads to gravitational collapse.

Stars with $M \gtrsim 12 - 13 \ M_\odot$ ignite Ne nonexplosively (at $T \simeq 1.5 \times 10^9$ K). After burning Ne into Mg, O is ignited at $T \simeq 2 \times 10^9$ K. The gravitational contraction which follows O exhaustion in the core leads to the Si burning stage ($T \simeq 3 - 5 \times 10^9$ K). Si burning is, in fact, an increasing approach to nuclear statistical equilibrium (NSE), which is reached at $T \simeq 6 \times 10^9 - 10^{10}$ K. An electron–degenerate NSE ("Fe–Ni") core eventually forms, surrounded by a mantle (or "onion–like" structure) of thermonuclear burning shells.

The NSE core grows in mass from the material being processed through the Si–burning shell on top of it, until it reaches the Chandrasekhar mass. It then implodes in a hydro-dynamic time scale ($\sim 1\ s$) until reaching nuclear matter densities ($\rho \sim 10^{14}\ g\ cm^{-3}$).

What follows is still a matter of heated debate. Core collapse halts when the central regions overshoot ordinary nuclear matter density by a factor of a few. An inner core containig about 40% of the total core mass forms, in hydrostatic equilibrium, while a shock wave starts to sweep out the still infalling outer core. Due to energy losses (photodissociation of the heavy nuclei that still make most of the material ahead of the shock front) the shock wave stalls at some point, turning into a stationary accretion front before reaching the edge of the core. Material thus continues to fall in, which would eventually lead to growth of the core above the maximum mass for a neutron star ($\sim 2 - 3\ M_\odot$) and collapse into a black hole.

Neutrino energy deposition behind the stalled shock is generally thought to build up enough pressure (creation of a "hot bubble") to eventually revive the shock and launch it into the overlying mantle and envelope, blowing them apart and ejecting their material at high velocities into the interstellar medium. Indeed, when collapse first stops, the core is still trapping $\sim 10^{53}\ erg$ in $\nu\bar{\nu}$ pairs (electron, μ, and τ neutrinos). They are released (*deleptonization*) on a time scale $\sim 10\ s$. If just $\sim 1\%$ of this energy is deposited behind the shock front, that is enough to power a supernova explosion. Accurate physical modeling of neutrino release and deposition is, however, a difficult problem: there are three different *neutrinospheres* behind the shock front (one for each neutrino species); the layers above the neutrinospheres become convectively unstable (negative entropy gradient) and at the same time, due to differential deleptonization, layers below the neutrinospheres become Rayleigh–Taylor unstable (which should boost neutrino energy release). Thus, one–dimensional hydrodynamic modeling becomes inadequate, and whilst the general picture we have just sketched is supported by the detection of neutrinos from SN 1987A, a self–consistent physical model of the explosion mechanism is still lacking. As mentioned above, the total kinetic energy of the material ejected in a supernova explosion (several $M_\odot$ moving at velocities of up to $\sim 10,000\ km\ s^{-1}$) is only $\sim 10^{51}\ erg$, and the total energy emitted as electromagnetic radiation is much less: $\sim 10^{49}\ erg$. The fact that the explosion itself involves only a tiny fraction of the total energy balance encourages most supernova theorists in the search for a robust, self–consistent mechanism based only on neutrino energy deposition on an approximately spherical configuration, without resorting to the effects of rotation and/or magnetic fields, but at the same time it means that the explosion is just a marginal consequence of core collapse, which mainly results in the emission of huge $\nu\bar{\nu}$ fluxes by the proto–neutron star. Therefore, the physical modeling has to be very accurate to adequately deal with it (see, for instance, Burrows, Hayes, and Fryxell 1995; Janka and Müller 1996; Mezzacappa et al. 1998).

Independently from the knowledge of the exact explosion mechanism, the association of neutron star formation with supernova explosions has been established by the presence of radio pulsars in a number of supernova remnants, starting with the Crab Nebula. The ejecta of gravitational collapse supernovae contain the nucleosynthesis products of all the thermonuclear burning stages undergone by the massive star progenitors (modified by explosive nucleosynthesis during the supernova event itself), and they are the main agents of chemical evolution of the Universe after the big bang. That occurs in spite of the fact that only a minute fraction of all stars evolve up to the stage of gravitational collapse ($\sim 10^{-3}$ for a Salpeter IMF with $x = 1.35$).

The nature of the compact remnant left behind by the supernova explosion is thought to depend on the mass of the stellar progenitor: it would be a neutron star for $M \leq M_{crit}$ and a black hole for $M > M_{crit}$. The value of M_{crit} is highly uncertain and it might be

$$M_{crit} \sim 25 - 30 \ M_\odot \qquad (2.12)$$

The shock wave sent into the overlying stellar structure from the collapsed core should induce homologous expansion, the velocities of the different layers being proportional to their distances to the center of the explosion. That opens up the possibility that the ejected material would reach escape velocities only from some mass fraction outwards, the more internal layers falling back into the newly formed neutron star, which would then form a black hole. Mass ejection could thus occur simultaneously with production of a black hole (Brown and Bethe 1994).

Supernova explosions due to core collapse of massive stars often occur in binary systems. If the system survives the explosion and the two stars remain close enough after it, matter can be accreted by the neutron star (or the black hole) and that result in X-ray emission. If more than half the total mass of the system prior to the explosion is ejected, however, the binary will be disrupted and a single neutron star plus its previous stellar companion will be sent flying apart. In HMBXS, that can be avoided if the progenitor of the neutron star (the initially more massive member of the binary system, or *primary* star), when leaving the main sequence and thus increasing in radius, transfers enough mass to the companion (the initially less massive or *secondary* star). Then, the primary will be left with just its He core and when gravitational collapse occurs the amount of mass ejected in the supernova explosion will be small enough to leave the system still bound (van den Heuvel 1978). In contrast, it is harder to explain how a binary initially made of a primary with $M_1 \gtrsim 10 \ M_\odot$ plus a secondary with $M_2 \lesssim 1 \ M_\odot$ (as in a LMBXS) could survive a supernova explosion. A possible alternative mechanism of neutron star formation, via *nonexplosive* or *accretion–induced collapse* (AIC) of a white dwarf formed in a first stage of the evolution of the binary system, will be discussed in the next Section.

3. Neutron star formation: accretion–induced collapse of white dwarfs

The problem of whether a white dwarf could grow by accretion up to the Chandrasekhar mass and collapse to a neutron star without exploding along the way, was first addressed by Canal and Schatzman (1976). More than 20 years later, the answer still is that it is possible but not easy (see Canal, Isern and Labay 1990; Canal 1994, 1997, for reviews). In order to achieve gravitational collapse, the white dwarf has to overcome two types of obstacles:

(i) The material accreted from the companion must become a part of the electron–degenerate core. That mainly depends on the chemical composition of the accreted material and on the mass–accretion rate $\dot{M}$.

(ii) Explosive ignition of the core followed by its disruption (in a Type Ia or *thermonuclear* supernova event) must be avoided.

Problem (i) is also encountered for production of a Type Ia supernova (SNeIa) by a mass–accreting white dwarf.

3.1. *Mass growth: the outer layers*

As stated above, retention of the accreted material depends on its composition (H, He, C–O). LMBXS, however, show H_α, H_β emission, which means that the material responsible for the growth in mass of a possible white dwarf progenitor of the neutron star should also have been H–rich. Accretion of H–rich material has different pitfalls, depending on the mass–accretion rate. The following accretion regimes have to be avoided:

$$\dot{M}_H \lesssim 10^{-9} - 10^{-8} \ M_\odot \ yr^{-1} \qquad (3.13)$$

would lead to nova–like outbursts (explosive H ignition), or at least to thermonuclear flashes followed by large expansion of the accreted layers and their loss by interaction with the companion.

On the other hand:

$$\dot{M}_H \gtrsim 10^{-6} \ M_\odot \ yr^{-1} \qquad (3.14)$$

would lead to formation of a red–giant envelope on top of a H–burning shell. This envelope would engulf the companion star (*common–envelope* stage) and would eventually be ejected.

For even higher rates:

$$\dot{M}_H \gtrsim \dot{M}_{Edd} \simeq 10^{-5} \ M_\odot \ yr^{-1} \qquad (3.15)$$

a common envelope would directly form and the material shed by the companion would also be lost.

Recent results from Hachisu, Kato and Nomoto (1996) seem to indicate, however, that the companion star could lose mass at a much faster rate than the limits (3.14) or (3.15) without formation of a common envelope and subsequent stop of mass transfer, the excess material over (3.14) being carried away by a strong stellar wind. When applying this result to binaries in globular clusters, nonetheless, one must bear in mind that generation of the strong wind would depend on metallicity, the wind being suppressed at low metallicities (Kobayashi et al. 1998).

It must be stressed that the preceding limits are mostly based on calculations in which spherically–symmetric mass accretion has been assumed. More realistic modeling of the process might significantly change the results.

Thus, it would seem that for

$$10^{-9} - 10^{-8} \ M_\odot \ yr^{-1} \lesssim \dot{M}_H \lesssim 10^{-6} \ M_\odot \ yr^{-1} \qquad (3.16)$$

H can burn steadily into He. Those systems should be very luminous. There have been suggestions (Hertz and Grindlay 1983; Grindlay 1987) that low–luminosity X–ray sources in globular clusters might correspond to white dwarfs accreting mass in this regime.

The preceding mass range has to be further narrowed because accumulation of He from H burning at rates

$$10^{-9} \ M_\odot \ yr^{-1} \lesssim \dot{M}_{He} \lesssim 5 \times 10^{-8} \ M_\odot \ yr^{-1} \qquad (3.17)$$

would lead to detonation of the He layer. That might be a mechanism for triggering SNeIa, but it would, of course, drastically cut the path towards AIC. Thus, the mass–accretion range

$$5 \times 10^{-8} \ M_\odot \ yr^{-1} \lesssim \dot{M}_H \lesssim 10^{-6} \ M_\odot \ yr^{-1} \qquad (3.18)$$

would be the only possible "window" to mass growth up to the Chandrasekhar mass.

3.2. *Mass growth: the core*

The outcome of mass growth also depends on the chemical composition of the white dwarf core. In principle, there are He, C–O, and O–Ne white dwarfs.

He white dwarfs can directly be excluded as candidates to AIC: explosive ignition of He should occur when the central density becomes

$$\rho_c \gtrsim 4 \times 10^8 \ g \ cm^{-3}$$

even at zero temperature. The corresponding mass is $M \simeq 1.3 \ M_\odot$, well below the Chandrasekhar mass. That leaves only C–O and O–Ne white dwarfs. As we will see, the cooling stage preceding the start of mass accretion plays a crucial role in C–O white dwarfs while chemical composition (the Mg abundance, possible C leftovers), electron captures, and semiconvection determine the fate of mass–growing O–Ne white dwarfs.

3.3. *C–O white dwarfs*

A first important point is the maximum mass that a C–O white dwarf can reach without igniting C nonexplosively and starting to change the chemical composition into ONe(Mg). Current results indicate

$$M_{CO} \lesssim 1.1 - 1.2 \ M_\odot \tag{3.19}$$

As we will see, the wider the gap between the initial mass of the white dwarf and the Chandrasekhar mass, the harder it is to induce gravitational collapse by mass accretion.

We consider first the *cooling stage*. In binaries, its duration will depend on the mass of the secondary, and for LMBXS it should typically be of the order of several Gyr ($1 \ Gyr \equiv 10^9 \ yr$). An effect of cooling will be *crystallization* of the white dwarf core. The cooling times for the start of crystallization at the center of a C–O white dwarf depend on the white dwarf mass and they range from $t_{cool} \simeq 1.9 \ Gyr$ for a 0.6 $M_\odot$ white dwarf to $t_{cool} \simeq 1.1 \ Gyr$ for a 1.2 $M_\odot$ white dwarf (cooling times are shorter for higher masses because central densities are also higher).

In the case of the one–component plasma (OCP), phase transition from the fluid (Coulomb liquid) to the solid (crystal) phase corresponds to a critical value of the *plasma–coupling parameter*

$$\Gamma = \frac{Z^2 e^2}{r_s kT} \tag{3.20}$$

where Z is the atomic number of the ion, e the electron charge, k the Boltzmann constant, T the absolute temperature, and r_s the *ion–sphere radius*, that is the radius of the sphere containing, on average, one ion (it is proportional to $n^{-1/3}$, n being the ion number density). Crystallization takes place for

$$\Gamma_{crit} \simeq 170 - 180 \tag{3.21}$$

(Slattery, Doolen, and DeWitt 1982; Ogata and Ichimaru 1987). The resulting crystal is a *body–centered cubic lattice* (BCC): one ion at each of the eight vertices of a cube, plus another ion at the center of the cube.

The case of a *binary ion mixture* (BIM) such as C–O is less clear. There was a suggestion by Stevenson (1980) that C and O were not miscible in the solid phase: this would imply crystallization of pure O at the center of the white dwarf, the solid O core growing until O exhaustion in the fluid phase. That would be followed by crystallization of pure C in the more external layers. It this case, explosive ignition of O would take place at very high densities: $\rho_{ign} \simeq 2 \times 10^{10} \ g \ cm^{-3}$. Gravitational collapse should follow. More recent calculations (Barrat, Hansen, and Mochkovitch 1988; Ichimaru, Iyetomi, and Ogata 1988) have shown, however, that the solid phase is more O–rich than the fluid phase, but chemical separation is only partial and there is still enough C in the central layers to determine the ignition density (much lower than for O).

There are still open questions which might be relevant to determination of the ignition density of C–O white dwarfs. One is whether the C and O ions form a *random alloy* in the solid phase (the probability of a C ion being next to another C ion in the lattice being only a function of the C concentration) or, on the contrary, they form an *ordered alloy* (the C ions being preferentially surrounded by O ions). In the latter case, the ignition density could be as high as in the case of complete chemical separation of C and O.

Nevertheless, even in the most unfavourable case (random alloy with the same chemical composition as the fluid phase), C ignition can still be delayed up to very high densities (Hernanz et al. 1988), that depending on the initial white dwarf mass and on the accretion rate. This double dependence mainly arises from the fact that the outer layers are heated up both by compression and by the ignition of thermonuclear burning shells (H, He). Heat propagates inwards by conduction and that progressively melts the solid core formed during the cooling stage. If the central layers melt before C ignition, the latter takes place in the *strong screening regime* of the thermonuclear reactions whereas if the layers remain solid it happens in the *pycnonuclear regime*. Fast accretion and high mass favors reaching the point of ignition when the central layers are still in the solid phase, which means higher densities. For slower accretion and/or smaller initial mass, the white dwarf melts completely before C ignition and densities are lower. Depending on the $(M_{init}, \dot{M})$ combination, when the core melts completely:

$$(2-3) \times 10^9 \ g \ cm^{-3} \lesssim \rho_{ign} \lesssim 9.5 \times 10^9 \ g \ cm^{-3} \tag{3.22}$$

while if ignition takes place in the solid phase:

$$9.5 \times 10^9 \ g \ cm^{-3} \lesssim \rho_{ign} \lesssim 1.5 \times 10^{10} \ g \ cm^{-3} \tag{3.23}$$

Explosive ignition at the center initiates the last, dynamical stage of the evolution of the mass–accreting white dwarf. Its outcome depends on the competition between burning propagation, which releases energy and produces expansion of the white dwarf, and electron captures on the incinerated material (whose rate depends on density) which decrease the Chandrasekhar mass. If burning propagates fast, expansion quenches the electron captures and further burning induces accelerated expansion of the white dwarf, which explodes as a thermonuclear supernova (SNeIa). It burning propagation is slow, the Chandrasekhar mass decreases below the actual mass of the white dwarf core when still little nuclear energy has been released, and the star collapses. Gravitational collapse of the C–O white dwarf is most likely to occur when the center of the star has remained solid until C ignition: we then have the combination of the highest ignition densities (and thus the fastest electron captures) and the slowest propagation of burning. The central layers being still solid, hydrodynamic instabilities are inhibited and burning can only propagate by *conduction*. Conductive burning velocities can be approximated by:

$$v_{burn} \simeq 25 \left(\frac{\rho}{2 \times 10^9 \ g \ cm^{-3}} \right)^{0.8} \left(\frac{X_C}{0.5} \right) \ km \ s^{-1} \tag{3.24}$$

where X_C is the C mass fraction (Timmes and Woosley 1992). If we parameterize v_{burn} in terms of the local spound speed c_s, we have:

$$v_{burn} = \alpha \ c_s \quad (0.001 \lesssim \alpha \lesssim 0.01) \tag{3.25}$$

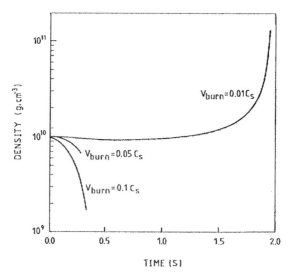

FIGURE 2. Central density *vs* time after explosive carbon ignition of a C–O white dwarf core at $\rho_{ign} = 10^{10}$ g cm^{-3}. Burning front velocity is parameterized as a fixed fraction of the local sound speed c_s. Bifurcation between explosion and collapse takes place between $v_{burn} = 0.01c_s$ and $v_{burn} = 0.05c_s$ (from Canal et al. 1990)

and comparison with the parameterized hydrodynamic calculations of Canal et al. (1990) (see Figure 2) shows that for $\alpha \lesssim 0.01$ collapse ensues.

3.4. *O–Ne white dwarfs*

O–Ne white dwarfs were first suggested as candidates to form neutron stars via AIC by Miyaji et al. (1980) (see also Nomoto 1984, 1987; Miyaji and Nomoto 1987). As we have seen above, O-Ne white dwarfs should form as the end products of the evolution of single stars in the initial mass range:

$$8 - 9 \ M_\odot \lesssim M \lesssim 10 - 11 \ M_\odot \tag{3.26}$$

and therefore they should also form when stars in this mass range are members of close binary systems. C burning will have transformed most C into Ne, with a small admixture of Mg. The O coming from He burning has remained intact. Mass growth from accretion will progressively increase the central density and with it the Fermi energy of the electrons. When the central density becomes $\rho_c \simeq 4 \times 10^9$ g cm^{-3}, the threshold for electron capture on ^{24}Mg is reached and we will have:

$$^{24}Mg + e^- \rightarrow \ ^{24}Na + \nu_e \tag{3.27a}$$

followed by (when the density becomes just slightly higher):

$$^{24}Na + e^- \rightarrow \ ^{24}Ne + \nu_e \tag{3.27b}$$

This sequence of electron captures sharply rises the temperature, but the ^{24}Mg abundance being small, this nuclide is exhausted before the temperature becomes high enough to trigger Ne and O burning. After ^{24}Mg and ^{24}Na exhaustion, the temperature drops again due to thermal $\nu\bar{\nu}$ emission. Compression of the central layers continues and when $\rho_c \simeq 9.5 \times 10^9$ g cm^{-3}, the threshold for electron capture on ^{20}Ne is reached:

$$^{20}Ne + e^- \rightarrow {}^{20}F + \nu_e \qquad (3.28a)$$

immediately followed by:

$$^{20}F + e^- \rightarrow {}^{20}O + \nu_e \qquad (3.28b)$$

The threshold for the electron capture on ^{20}F being lower than that for ^{20}Ne, each double capture now produces substantial heating. In addition, ^{20}Ne is much more abundant than ^{24}Mg ($X_{Ne} \simeq 0.25$ *vs* $X_{Mg} \simeq 0.03$). Thus, now the temperature rises above $T \sim 2 \times 10^9 \ K$ and that induces explosive Ne–O ignition. The exact ignition density depends on whether or not convection develops following the heating by the electron captures on ^{20}Ne: convection efficiently spreads the heat generated at the center over a larger region and thus delays explosive ignition up to $\rho_{ign} \simeq 2 \times 10^{10} \ g \ cm^{-3}$. The gradient of electron mole number Y_e, however, stabilizes the fluid against convective motions. Instead of the Schwarzschild criterion for the start of convection:

$$\nabla_T > \nabla_{ad} \qquad (3.29)$$

the Ledoux criterion:

$$\nabla_T > \nabla_L = \nabla_{ad} - \left(\frac{\partial lnT}{\partial lnY_e} \right)_{P,\rho} \nabla_{Y_e} \qquad (3.30)$$

with

$$\nabla_{Y_e} \equiv \frac{dlnY_e}{dlnP} \qquad (3.31)$$

appears to be more adequate. Adopting the Ledoux criterion, the ignition density is $\rho_{ign} \simeq 9.5 \times 10^{10} \ g \ cm^{-3}$. Semiconvection is likely to develop, however, and the ignition density will be somewhere in the interval:

$$9.5 \times 10^9 \ g \ cm^{-3} \lesssim \rho_{ign} \lesssim 2 \times 10^{10} \ g \ cm^{-3} \qquad (3.32)$$

(Gutiérrez et al. 1996; see Figure 3). The estimates of the burning propagation velocities in a O–Ne mixture by Timmes and Woosley (1992) place the critical ignition density for gravitational collapse to occur at:

$$\rho_{crit} \simeq 8.5 \times 10^9 \ g \ cm^{-3} \qquad (3.33)$$

which means that mass–accreting O–Ne white dwarfs should eventually collapse when reaching a mass $M \simeq 1.38 \ M_\odot$. The only possibility for those white dwarfs to explode would be that a C mass fraction $X_C \gtrsim 0.05$ were left unburned at the end of C burning (Gutiérrez et al. 1997). Some C is left indeed (Domínguez, Tornambè, and Isern 1993; Ritossa, García–Berro, and Iben 1996), but its amount seems to be too low to produce explosive ignition after further compression, with the possible exception of masses not much above the lower limit for nondegenerate C ignition ($M \simeq 8 - 9 \ M_\odot$).

Summarizing, C–O white dwarfs could produce neutron stars by AIC, provided that their initial mass were high ($M \gtrsim 1 \ M_\odot$), as well as the mass accretion rate ($\dot{M} \sim 10^{-7} - 10^{-6} \ M_\odot \ yr^{-1}$). The initial mass range would thus be narrow and the mass–accretion rate not much below the values which were thought to induce formation a red–giant envelope on top of the white dwarf. For O–Ne white dwarfs, there is no particular restriction on their initial mass (it should be high, anyway: $M \simeq 1.2 - 1.4 \ M_\odot$ for white

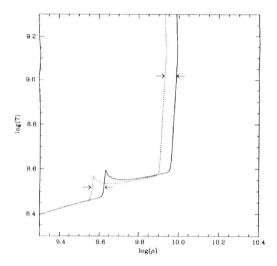

FIGURE 3. Evolution of the center of a O–Ne–Mg electron–degenerate core in the $log \, \rho - log \, T$ plane as it grows in mass, either from a C–burning shell ("super–AGB" case) or by accretion from a close binary companion (white dwarf case), when the Ledoux stability criterion is adopted (see text). Note the first temperature rise due to electron captures on Mg, which stops when this element is exhausted. The second temperature rise corresponds to electron captures on Ne: the rest of it and all the O are then ignited explosively, but the density is already so high that electron captures on the resulting material (in nuclear statistical equilibrium, NSE) promptly lower enough the Chandrasekhar mass to induce gravitational collapse. Dotted line is just for comparison with earlier results (from Gutiérrez et al. 1996)

dwarfs with this chemical composition), and the mass–accretion rate has to be just high enough to avoid He detonation ($\dot{M} \gtrsim 5 \times 10^{-8} \, M_\odot \, yr^{-1}$).

At present, it is still difficult to establish the relative importance of the contributions of C–O white dwarfs and O–Ne white dwarfs to the neutron star population in binaries via AIC. In any case, even if AIC were to explain the whole population of LMBXS, its rate should be much lower than the SNeIa rate (by a factor $\sim 10^{-4}$). There are also upper bounds to the frequency of this possible mechanism of neutron star formation, based on the ejection of very neutron–rich material coming from the outermost layers of the proto–neutron star (Woosley & Baron 1992; Fryer et al. 1998), but they are still compatible with an AIC origin of the neutron stars in LMBXS.

4. Binary X–ray sources and millisecond pulsars: evolutionary scenarios

4.1. *Basic concepts*

Before dealing with the evolutionary scenarios proposed to explain the origins of binary X–ray sources and millisecond pulsars in different environments, it will be useful to define different evolutionary time scales.

One is the *nuclear time scale*:

$$\tau_n \simeq 10^{10} \frac{M}{M_\odot} \left(\frac{L}{L_\odot} \right)^{-1} \, yr \qquad (4.34)$$

From which, using the mass–luminosity relations:

$$L \propto M^{3.5} \ (M > 1 \ M_\odot) \tag{4.35a}$$

and

$$L \propto M^{3} \ (M \le 1 \ M_\odot) \tag{4.35b}$$

we have:

$$\tau_n \simeq 10^{10} (M/M_\odot)^{-2.5} \ yr \ (M > 1 \ M_\odot) \tag{4.36a}$$

and

$$\tau_n \simeq 10^{10} (M/M_\odot)^{-2} \ yr \ (M \le 1 \ M_\odot) \tag{4.36b}$$

There is the *thermal time scale*:

$$\tau_{th} = \frac{GM^2}{RL} \tag{4.37}$$

Using the preceding mass–luminosity relations, plus the mass–radius relations:

$$R \propto M^{0.5} \ (M > 1 \ M_\odot) \tag{4.38a}$$

and

$$R \propto M \ (M \le 1 \ M_\odot) \tag{4.38b}$$

we obtain:

$$\tau_{th} \simeq 5 \times 10^{7} (M/M_\odot)^{-2} \ yr \tag{4.39}$$

The *dynamical time scale*:

$$\tau_d \simeq 50 \left(\frac{\bar{\rho}_\odot}{\bar{\rho}} \right)^{1/2} \ min \tag{4.40a}$$

or

$$\tau_d \simeq 0.04 \left(\frac{M_\odot}{M} \right)^{1/2} \left(\frac{R}{R_\odot} \right)^{1/2} \ days \tag{4.40b}$$

where $\bar{\rho}_\odot = 1.4 \ g \ cm^{-3}$.

The nuclear time scale corresponds to evolution governed by the rate of consumption of the nuclear fuel by thermonuclear reactions in the stellar interiors, when the components of a binary system basically evolve as if they were single stars. The thermal and the dynamic time scales are relevant to mass transfer: a mass–losing star will readjust its radius on a dynamical time scale and restore thermal equilibrium on a thermal time scale.

We will also very often refer to the *Roche geometry* of the equipotential surfaces of the gravitational field around a binary system (see Figure 4). Mass will flow from a star to its companion when the former, by increasing in radius, overfills its *Roche lobe*. The critical radius R_L (for the star with mass $M = M_1$) can be approximated by (Paczyński 1971):

$$\frac{R_L}{a} = 0.38 + 0.2 \ log \left(\frac{M_1}{M_2} \right) \quad (0.8 \le \frac{M_1}{M_2} < 2.0) \tag{4.41a}$$

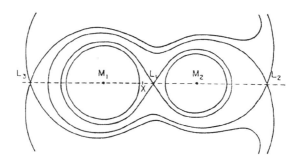

FIGURE 4. Schematic representation of the Roche geometry

and

$$\frac{R_L}{a} = 0.462 \left(\frac{M_1}{M_1 + M_2} \right)^{1/3} \quad (\frac{M_1}{M_2} < 0.8) \tag{4.41b}$$

where a is the separation between the two stars. Or by (Eggleton 1983):

$$\frac{R_L}{a} = \frac{0.49}{0.6 + q^{2/3} \ln(1 + q^{-1/3})} \tag{4.42}$$

with $q \equiv M_2/M_1$, which is accurate up to a few percent for all mass ratios.

4.2. *Mass transfer and mass loss*

We deal now with the effects of mass transfer and mass loss on the binary orbits (see, for instance, Bhattacharya and van den Heuvel 1991 for a more complete presentation). The angular momentum of the orbital motion (for a circular orbit) can be written as:

$$J_{orb}^2 = \frac{M_1 M_2 \Omega a^2}{M_1 + M_2} \tag{4.43a}$$

with

$$\Omega^2 = \frac{G M_{tot}}{a^3} \tag{4.43b}$$

where $M_{tot} \equiv M_1 + M_2$. If, during mass transfer, a fraction α of the exchanged matter leaves the system:

$$\frac{\dot{a}}{a} = -2 \left[1 + (\alpha - 1) \frac{M_1}{M_2} - \frac{\alpha}{2} \frac{M_1}{M_{tot}} \right] \frac{\dot{M}_1}{M_1} + 2 \frac{\dot{J}_{orb}}{J_{orb}} \tag{4.44}$$

If mass transfer is conservative, then $\alpha = \dot{J}_{orb} = 0$ and

$$\frac{\dot{a}}{a} = -2 \left(1 - \frac{M_1}{M_2} \right) \frac{\dot{M}_1}{M_1} \tag{4.45}$$

For mass transfer from the star with mass M_1 to the star with mass M_2 (then $\dot{M}_1 < 0$) we will thus have:

$$a \text{ decreases if } \frac{M_1}{M_2} > 1$$

and

$$a \text{ increases if } \frac{M_1}{M_2} < 1$$

Also, the orbital periods before and after mass transfer are related by:

$$\frac{P}{P_0} = \left[\frac{M_{1,0}(M_{tot} - M_{1,0})}{M_{1,0}(M_{tot} - M_1)} \right]^3 = \left(\frac{a}{a_0} \right)^{3/2} \tag{4.46}$$

If there is sudden mass loss, as in a supernova explosion, and the orbit was circular before mass ejection, the final orbit will have an eccentricity e and the initial and final separations are related through:

$$a_f = (1 - e)\, a_i \tag{4.47}$$

Assuming instantaneous mass ejection:

$$\frac{G(M_1 + M_2)}{a_i} = \frac{G(M_1 + M_2 - \Delta M)}{a_f} \frac{1 + e}{1 - e} \tag{4.48a}$$

and

$$e = \frac{\Delta M}{M_1 + M_2 - \Delta M} \tag{4.48b}$$

From the last expression, it becomes clear that the system will be disrupted if

$$\Delta M \geq \left(\frac{1}{2} \right)(M_1 + M_2)$$

When the system remains bound, there is a change in the velocity of the center of mass:

$$v_s = \frac{M_2 v_2 - (M_1 - \Delta M)v_1}{M_1 + M_2 - \Delta M} = e\, v_1 \tag{4.49}$$

where v_1, v_2 are the respective orbital velocities of stars 1 and 2 before explosion.

For eccentric orbits, tidal forces will eventually produce circularization. After that, the separation between the two stars will be:

$$a_c = (1 - e^2)\, a_f = (1 + e)\, a_i \tag{4.50}$$

In the preceding we have assumed that mass ejection by the exploding star is spherically symmetric. However, *kick velocities* $\sim 100 - 200\ km\ s^{-1}$ at birth, seem to be required for single neutron stars in order to explain the observed velocities of single radio pulsars (Lyne, Anderson, and Salter 1982; Bailes 1989; van den Heuvel and van Paradijs 1997). That would significantly enhance the chances of disruption when neutron stars are formed in binary systems via a supernova explosion.

4.3. *Spiral–In*

We have seen that conservative mass transfer from the more massive component of a binary system to the less massive one reduces the separation between the two stars. However, the main processes which make the binary orbits to shrink are *gravitational wave emission, magnetic braking*, and *common envelope evolution*.

Angular momentum loss by emission of gravitational waves becomes important when the orbital periods are very short. The rate of angular momentum loss is given by (Landau and Lifshitz 1962):

$$\frac{\dot{J}_{GWR}}{J_{orb}} = - \frac{32}{5} \frac{G^3}{c^5} \frac{M_1 M_2 (M_1 + M_2)}{a^4} \, s^{-1} \tag{4.51}$$

Magnetic braking can be important when the companion of a compact star is still in the main sequence and has a convective envelope. Coupling of the stellar wind with the magnetic field of the star out to distances of several stellar radii produces a quite effective way of slowing down rotation. That would make the starrotate asynchronously with the orbital revolution, but tidal forces tend to restore corotation. As a result, orbital angular momentum is tranferred to the star's spin. The corresponding expression is:

$$\frac{\dot{J}_{mb}}{J_{orb}} = - 3.8 \times 10^{-30} \frac{R_1^{4-\gamma}(M_1 + M_2)R_2^\gamma \, \Omega^2}{M_1 \, a^2} \tag{4.52}$$

where the parameter γ accounts for the dependence of braking on the radius of the mass–losing star (Verbunt & Zwaan 1981; Rappaport, Verbunt, and Joss 1983).

Finally, when a star overflows its Roche lobe so fast that its companion can only accrete a minor fraction of the material being shed, an envelope extending much farther than the orbital separation forms. The binary system then becomes a double core spinning at the center of a large, loosely bound envelope. Friction of the two stars with their common envelope transfers energy and angular momentum to it, resulting in its ejection and a significant decrease of the orbital separation a. We have:

$$\frac{GM_1 M_e}{\lambda R_{L,1}} = \alpha \left(\frac{GM_c M_2}{2a_f} - \frac{GM_1 M_2}{2a_i} \right) \tag{4.53a}$$

and

$$\frac{a_f}{a_i} = \frac{M_c}{M_1} \left(1 + \frac{2a_i}{\alpha \lambda R_{L,1}} \frac{M_e}{M_2} \right)^{-1} \tag{4.53b}$$

where $R_{L,1}$ is the Roche lobe radius of star 1 at the onset of mass transfer, M_c is the mass of the same star (its core mass) after loss of the common envelope of mass M_e ($M_1 = M_c + M_e$), λ is a weighting factor (< 1) for the gravitational binding energy of the envelope to the core, and α is a parameter accounting for the efficiency in the transfer of the gravitational binding energy, released by shrinking of the binary orbit, into the envelope (Webbink 1984; Bhattacharya and van den Heuvel 1991; Verbunt 1993).

4.4. *Formation of HMBXS*

The "classical" scenario for formation of HMBXS was first introduced by Paczyńsky (1967) and later completed by van den Heuvel and Heise (1972). It is based on conservative mass transfer between a pair of initially massive stars (see Figure 5).

First both components evolve independently, as if they were single stars, on a nuclear time scale. Since the more massive component evolves faster (see eq. 4.34), it is first to expand and reach the critical radius (4.41a). Then Roche lobe overflow starts. When most of the H–rich envelope has been transferred to the companion, the mass–losing star contracts and mass transfer stops. The system now consists of a H–deficient star (CHeB) plus an OB–type star that has accreted all the mass lost by its companion. The He core of the first star continues to evolve fast and reaches the point of gravitational collapse. By combining the initial masses of the two stars, it can be arranged that the mass ejected in the supernova explosion does not exceed half the total mass of the binary,

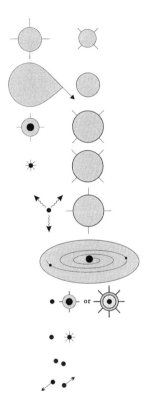

FIGURE 5. The "classical" scenario for the formation of HMBXS (adapted from Vanberen, De Loore and Van Rensbergen 1998)

and so disruption is avoided. The OB star and the newly formed neutron star are in a wide orbit. The neutron star now accretes mass from the wind of its companion and that powers X–ray emission. This evolutionary stage would correspond to the observed HMBXS. The OB star continues to evolve and eventually overfills its Roche lobe. As we have seen, the orbital separation a then decreases, and due to the extreme mass ratio of the system a common envelope forms. The neutron star spirals–in (see eq. 4.53b) and a very close binary (P $\sim$ few hr) results, consisting of a neutron star plus a He star (there is also the possibility of complete merging, with formation of a Thorne–Zytkov object). The He star evolves further and a second supernova explosion occurs. In most cases the system will be disrupted, thus producing two runaway pulsars. There is, however, a

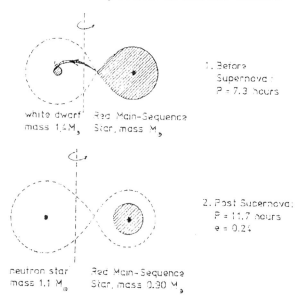

FIGURE 6. Formation of a LMBXS by AIC of a white dwarf (from Taam and van den Heuvel 1986)

small probability that the binary remains bound. A binary pulsar similar to PSR1855–09 might then result.

More realistically, mass transfer should be nonconservative, and depending on the binary parameters a common envelope can form in the first episode of mass transfer already. Besides, the first supernova explosion not only changes the system parameters but it can also strip the companion of a fraction of its outer layers (see Vanbeveren, de Loore and Van Rensbergen 1998, for a recent review).

4.5. *Formation of LMBXS*

A possible way to form a LMBXS is by AIC of a massive white dwarf in a cataclysmic–like binary. A specific scenario has been proposed by Taam and van den Heuvel (1986): the presupernova system would be made of a red main–sequence star with $M \simeq 1 \; M_\odot$ plus a white dwarf with $M_{WD} \simeq 1.4 \; M_\odot$, in a circular orbit, and would have a period $P = 7.3 \; hr$. After transfer of a small amount of mass, the white dwarf collapses into a neutron star. A mass $\Delta M \simeq 0.2 \; M_\odot$ is ejected, and the final system consists of a neutron star with $M_{NS} = 1.1 \; M_\odot$ plus a red main–sequence star with $M = 0.9 \; M_\odot$, with a period which has increased to $P = 11.7 \; hr$ and an eccentricity $e = 0.24$ (see Figure 6). Evidence for an AIC origin of a particular LMBXS (GRO J1744–28) has recently been gathered by van Paradijs et al. (1997). Arguments in favor of a similar origin for the neutron stars in binary radio pulsars with low–mass ($M \simeq 0.2 - 0.4 \; M_\odot$) companions are given by van den Heuvel and Bitzaraki (1995).

There is another possible way (in addition to the tidal capture and exchange encounter mechanisms, that we will consider in the next Section and which can only operate in globular clusters) to form a LMBXS, from the evolution of massive binaries with large mass ratios. Such a model (Sutantyo 1975; see also Verbunt, Wijers and Burm 1990) was proposed to explain the Her X–1 system, which consists of a neutron star ($M_{NS} \simeq 1.4 \; M_\odot$) plus an A–type companion ($M \simeq 2 \; M_\odot$). Her X–1 is located at $z = 3 \; kpc$ above

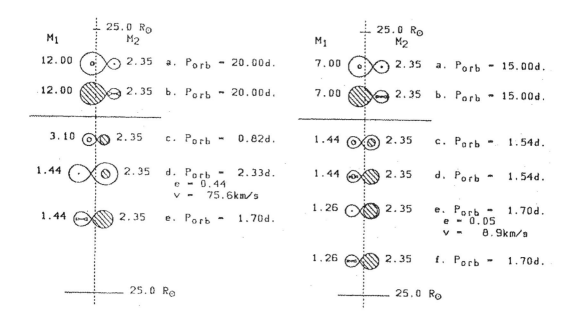

FIGURE 7. Two variations of an evolutionary scenario proposed in order to explain the origin of the LMXBS Her–X1 (from Verbunt, Wijers and Burm 1990)

the galactic plane, and since a 2 $M_\odot$ star has an age $\simeq 8 \times 10^8$ *yr*, the z–component of its velocity after the explosion must have been $v_z \gtrsim 120$ *km s*$^{-1}$ to reach its present height.

The initial system consists of two stars with masses $M_1 \simeq 15$ $M_\odot$ and $M_2 \simeq 2$ $M_\odot$ at large separation, with a period $P \gtrsim 1$ *yr*. The system evolves through a common–envelope phase and, at the end of it, consists of a ~ 4 $M_\odot$ He star plus a 2 $M_\odot$ A star, with a period $P \sim 1$ *day*. The He star explodes, leaving a neutron star with $M_{NS} \simeq 1.4$ $M_\odot$ in an eccentric orbit around the A–type star. The center of mass of the system acquires a velocity $v \sim 100$ *km s*$^{-1}$ (see eq. 4.49). When the companion of the neutron star evolves to fill its Roche lobe, the ensuing mass transfer powers the X–ray source (see Figure 7).

It has been suggested by Verbunt (1993) that LMBXS where the companion of the neutron star has a mass $M \simeq 1$ $M_\odot$ might also form in a similar scenario. The specific example presented, however, assumes that a star with initial mass $M \simeq 5$ $M_\odot$ evolves to form a neutron star, in clear contradiction with the much larger minimum values quoted in (2.11). There is also neglect of the problem of the kick velocities adquired by the neutron stars at birth, all of which makes the whole argument quite unconvincing.

4.6. *Formation of single millisecond pulsars by coalescence?*

It has been suggested that, in addition to descend from LMBXS through evaporation of the companion of the neutron star (see next Section), some single millisecond pulsars could result from coalescence of a neutron star and a white dwarf in systems descendant from HMBXS (Bonsema and van den Heuvel 1985). In systems where the companion of the neutron star has a mass $M \lesssim 10 - 11$ $M_\odot$, when the common envelope forms and is ejected, the core left behind would not have enough mass to evolve up to the stage of gravitational collapse and a massive white dwarf would form, as in PSR0655+64.

PSR0655+64 has an orbital period $P \sim 1$ *day*. If a similar system were formed with a shorter period $(P \lesssim 3 \ hr)$, then the orbit would decay by emission of gravitational radiation in $\sim 10^9 \ yr$ and the neutron star and the white dwarf would coalesce, that leading to the formation of a single, rapidly rotating neutron star. The system would resemble PSR1937+214, which has P = 1.5 ms.

5. Neutron star binaries in globular clusters

The high densities of stars in globular clusters make close stellar encounters much more frequent than in the rest of the Galaxy. However, in order to form a bound system of two stars, starting from a initially unbound configuration, a sink of orbital energy is required. There are three possiblities:

(i) A collision involving three stars: one star takes up the excess energy and escapes, leaving the other two stars in a bound system. The small probability of simultaneous convergence of three trajectories makes it irrelevant

(ii) Interaction of a neutron star with an existing binary leading to exchange: the neutron star replaces one of the components of the binary system.

(iii) Tidal capture: deformation of a normal star by close passage of a compact star takes relative kinetic energy of the orbit. The energy is then dissipated through oscillations finally resulting in heating. If the change in kinetic energy ΔE_k exceeds the total positive energy of the initial orbit, a bound system forms.

5.1. *Tidal capture*

We will give here just a simple description of the process (see Fabian, Pringle, and Rees 1975; Verbunt and Hut 1987; Verbunt 1988; Bhattacharya and van den Heuvel 1991). Let M be the mass of the "target star" and m that of the compact star. The kinetic energy of the relative motion is, at any time:

$$E_k = \frac{1}{2} \left(\frac{m \ M}{m + M} \right) v^2 \qquad (5.54)$$

and E_0 will correspond to v_∞ (when the separation is very large). We further call d the distance of closest approach, h the height of the tidal bulge, and m_t its mass. We then have

$$h \simeq \frac{m}{M} \frac{R^4}{d^3} \qquad (5.55a)$$

and

$$m_t \simeq k \ \frac{h}{R} \ M \qquad (5.55b)$$

R is the radius of the "target star" and k depends on the density distribution within the star: for instance, $k \sim 0.14$ for deeply convective stars.

The potential energy of the tidal bulge is:

$$E_t \simeq m_t \ \frac{GM}{R^2} \ h \simeq k \ \frac{G \ m^2}{R} \left(\frac{R}{d} \right)^6 \qquad (5.56)$$

If a fraction η of this energy is dissipated, then the condition for tidal capture is:

$$\eta E_t > E_0 \qquad (5.57)$$

which translates into:

$$d \lesssim 3R \left(\eta \frac{k}{0.14} \frac{m}{M} \frac{m+M}{2M_\odot} \frac{R_\odot}{R} \right)^{1/6} \left(\frac{10 \ km \ s^{-1}}{v_\infty} \right)^{1/3} \tag{5.58}$$

In a rough approximation, there is tidal capture when

$$d \lesssim 3R \tag{5.59}$$

We have assumed that the escape velocity from the stellar surface is $v_{esc} >> v_\infty$, which is true but for very distended giants.

Since tidal effects are negligible until very near closest approach, the impact parameter b is:

$$b = d \left(1 + \frac{2 \ G \ (m + M)}{v_\infty^2 \ d} \right)^{1/2} \tag{5.60}$$

The second term under the square root is typically $>> 1$. Therefore, the cross–section σ for tidal capture becomes:

$$\sigma = \pi \ b^2 \simeq \pi \ d \ \frac{2G \ (m + M)}{v_\infty^2} \tag{5.61}$$

and the rate of tidal capture per unit volume:

$$\Gamma_{bin} = n_c \ n \ v_\infty \ \sigma \simeq 6 \times 10^{-11} \ \frac{n_c}{10^2 \ pc^{-3}} \ \frac{n}{10^4 \ pc^{-3}} \ \frac{(m + M)}{M_\odot} \ \frac{3R}{R_\odot} \ \frac{10 \ km \ s^{-1}}{v_\infty} \ yr^{-1} \ pc^{-3} \tag{5.62}$$

n_c and n being, respectively, the number densities of compact objects and of "target stars".

The net tidal capture rate γ is obtained by integrating Γ over the cluster volume:

$$\gamma = \int_{V_{clust}} \Gamma_{bin} \ dV \tag{5.63}$$

5.2. *Exchange encounters*

In close encounters between a compact star and a preexisting binary, one of the binary components is ejected and it is substituted by the compact object. For this process to work, the distance of closest approach d of the compact object to the binary should be only a few times the initial binary separation a at most. For a bound system to be left, the binding energy of the binary typically has to be much larger than the initial kinetic energy of the incoming star E_0. There are two possiblities:

(i) Direct exchange

(ii) Resonance encounter: temporary formation of a triple system followed by ejection of one of the components.

In typical globular cluster environments, resonant encounters are roughly twice more frequent than direct exchanges. For three equal masses m, the rate of resonance exchange is (Hut and Bahcall 1983):

$$\Gamma_{bin} \simeq 5 \times 10^{-10} \ \frac{n_c}{10^2 \ pc^{-3}} \ \frac{n_{bin}}{10^2 \ pc^{-3}} \ \frac{m}{M_\odot} \ \frac{a}{1 \ AU} \ \frac{10 \ km \ s^{-1}}{v_\infty} \ yr^{-1} \ pc^{-3} \tag{5.64}$$

where n_{bin} is the number density of preexisting binaries.

Comparing this last expression with that for tidal captures (5.62), we see that the cross section for resonance exchange is larger (a vs $3R$), but on the other hand n_{bin} is smaller than n. Concerning the latter, it must be noted that the binary frequency seems to be much lower in globular clusters than in the Galactic disk. That, in turn, would be explained by the fact that globular clusters should have lost most of their primordial binaries (see King, this volume), together with the low probability of producing binary systems by tidal encounters between ordinary stars.

5.3. *Direct collisions*

From (5.59) we see that $\simeq 1/3$ of tidal captures involves physical collision between the two stars. Their outcome is uncertain and it depends on the evolutionary state of the target star. For very short distances of closest approach ($d \lesssim R/4$), a polytropic "target star" would merge with the compact star and jet–like ejection of matter would probably occur. We can distinguish two evolutionary stages:

(i) Direct hit of a compact star with a main–sequence star: it would produce the disruption of the main–sequence star, and part of its matter could form a massive disk around the compact star.

(ii) Collision of a compact star with a red–giant star. In this case, the compact star and the core of the red giant could form a binary orbiting inside the giant envelope. Friction would eject the envelope and the orbit would shrink, leaving a very compact binary.

Stellar mergers can be frequent in resonance encounters between compact objects and normal star binaries.

6. Origin and evolution of neutron star binaries in globular clusters

Tidal captures will result in two distinct classes of binaries, according to the characteristics of the companions of the compact objects:

(i) Binaries with extended main–sequence or giant companions, resulting from relatively distant encounters.

(ii) Binaries with degenerate companions, resulting from direct collisions with red giant stars.

In *class (i) binaries*, the initial eccentricity e will be close to unity, but quick circularization due to tidal effects should produce a circular orbit with $a \simeq 2d$. In a fraction of cases ($\sim 25\%$), the secondary overfills its Roche lobe immediately after capture, which induces rapid mass exchange. The mass–losing star being less massive than the compact object, that will *increase* the orbital separation a (see eq. 4.45), which in turn leads to slower evolution.

There will be two different evolutionary tracks, depending on the secondary being a main–sequence star or a giant:

In the case of *giant secondaries*, the mass transfer driven by the increase in radius due to the thermonuclear evolution of the interior *increases* a.

In the case of *main–sequence secondaries*, gravitational radiation and/or magnetic braking *decrease* a. That shrinks the Roche lobe radius and eventually drives mass transfer. There can be, however, a change of regime if the secondary is close to the turn–off mass.

In both cases, accretion of matter by the compact object produces a X–ray source, and $\dot{M}$ is larger in systems with evolving giant secondaries. Is is estimated that $\sim 70 - 90\%$ of all tidal captures are on main–sequence stars.

Further evolution will produce:

In the case of a neutron star plus a red giant, a system made of a rapidly spinning pulsar in a wide orbit plus a low–mass white dwarf. The orbit being loosely bound, there is the possibility that the neutron star would be released in subsequent stellar encounters.

In the case of a neutron star plus a main–sequence star, a contact binary when the period becomes $P \lesssim 6$ *hr*. When the period is further reduced to $P \sim 80$ *min*, the secondary becomes degenerate and that brings down mass transfer. The system will then consist of a slowly rotating neutron star plus a very small degenerate object.

In *class (ii) binaries*, evolution via emission of gravitational wave radiation by the neutron star/white dwarf pairs formed by direct collision will reduce the orbital separation a until forming a contact system (the period will be $P \sim$ a few min at contact). Then, mass transfer from the white dwarf to the more massive neutron star will increase a and $\dot{M}$ will eventually drop to zero when the white dwarf has already been reduced to a very small mass or even been evaporated by the pulsar wind.

There is the additional possibility that a single millisecond pulsar might result from direct collision of a neutron star with a main–sequence star, through formation of a massive disk in orbit around the neutron star. Accretion of matter form the disk would then spin up the pulsar. There are difficulties with this picture, however, since the mass flow into the neutron star might be highly super–Eddington and radiation could blow away most of the disk before significant accretion. A possible way out might be ejection of matter in jets along the spin axis but still allowing enough deposition of angular momentum into the neutron star.

A summary of the possible different outcomes of tidal capture is presented in Figure 8.

6.1. *Final evolution of LMBXS: evaporation of companion stars*

Interaction of the magnetic field of the neutron star with the accretion disk can produce a γ–ray flux that, in turn, would drive a wind from the companion star by heating its outer layers. This mechanism was proposed by Ruderman, Shaham, and Tavani (1989) to explain PSR1937+21, a single millisecond radio pulsar with $P = 1.55$ *ms*. Partial evaporation of the companion can stop accretion and the spun–up neutron star would appear as a *binary millisecond pulsar*. But the γ–ray flux from the pulsar can induce further evaporation of the companion until its total dissipation, and a single, rapidly rotating, "recycled" pulsar such as PSR1937+21 would then result.

That picture was confirmed by the later discovery of PSR1957+20, an eclipsing millisecond radio pulsar ($P = 1.6$ *ms*), with a companion of mass $M_2 \lesssim 0.02 \ M_\odot$ which was clearly evaporating. Further confirmation came from the discovery of PSR1744–24 ($P \simeq 11$ *ms*), a binary radio pulsar with $P_{orb} \simeq 1.9$ *hr*, in the globular cluster Terzan 5. Its companion has a mass $M_2 \sim 0.1 \ M_\odot$ and it is also evaporating.

There is still debate on the actual evaporation mechanism. Some authors disagree with the view that evaporation of the companion could already start during the accretion stage. There are also diverging views as to the heating mechanism of the outer layers of the companion: for Ruderman, Shaham, and Tavani (1989) it is the γ–rays produced in the accretion disk by the e^-e^+ flux coming from the magnetosphere of the pulsar, while for Krolik and Sincell (1990) it is the direct hit of this e^-e^+ flux on the companion which heats its outer layers via e^+ annihilation.

We will now outline a simplified evaporation model (Bhattacharya and van den Heuvel 1991). The primary energy source is the rotational energy of the pulsar:

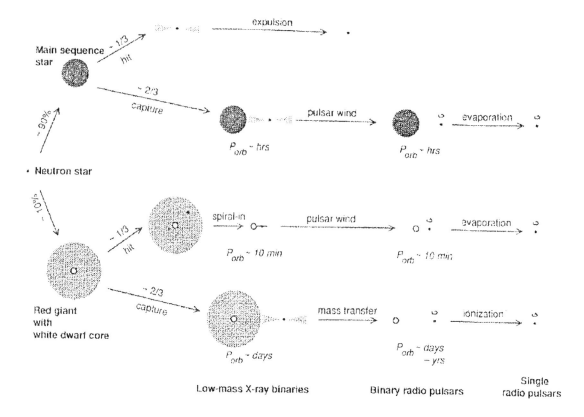

FIGURE 8. Summary of the possible outcomes of tidal captures in globular clusters (from Verbunt 1990; Bhattacharya and van den Heuvel 1991)

$$E_{rot} = \frac{1}{2} I \, \Omega^2 \simeq 2 \times 10^{52} \, P_{ms}^{-2} \; erg \qquad (6.65)$$

where I is the moment of inertia of the pulsar and P_{ms} its period in milliseconds. Such energy has to be compared with the binding energy of the companion star:

$$E_b \simeq \frac{GM^2}{R} \simeq 10^{48} \left(\frac{M}{M_\odot} \right)^2 \left(\frac{R}{R_\odot} \right)^{-1} \; erg \qquad (6.66)$$

Whether complete evaporation is possible or not depends on the efficiency of the evaporation process, the condition for complete evaporation being

$$g \, E_{rot}/E_b \geq 1 \qquad (6.67)$$

g can be expressed as

$$g = (R/2a)^2 \, f \qquad (6.68)$$

that is a geometric factor times the efficiency f of the evaporation process.

The energy emission by the pulsar is:

$$L_p = \frac{dE_{rot}}{dt} = \left(\frac{2R_n^6}{3c^3}\right) B_s^2 \left(\frac{2\pi}{P}\right)^4 \tag{6.69}$$

where R_n is the radius of the neutron star.

Then, for spherically symmetric wind emission ($\dot{M}$ is now the rate of mass loss through the wind):

$$\frac{1}{2} \dot{M} v^2 = f \left(\frac{R}{2a}\right)^2 \left(\frac{2R_n^6}{3c^3}\right) B_s^2 \left(\frac{2\pi}{P}\right)^4 \tag{6.70}$$

The wind velocity v has to be of the order of the escape velocity $v_e = (2GM/R)^{1/2}$, which gives an evaporation time scale

$$\tau_{evap} = \frac{M}{\dot{M}} = \left(\frac{3c^3}{2R_n^6}\right) \left(\frac{2GM^3}{R^3}\right) \left(\frac{a^2}{8\pi^4 f}\right) \frac{P^4}{B_s^2} \tag{6.71}$$

The pulsar spin–down time scale is:

$$\tau_{sd} = \frac{P}{2\dot{P}} = \left(\frac{P^2}{B_s^2}\right) \left(\frac{3c^3}{16\pi^2 R_n^6}\right) I \tag{6.72}$$

Hence:

$$\frac{\tau_{evap}}{\tau_{sd}} = \frac{1}{f(R/2a)^2} \frac{E_b}{E_{rot}} \tag{6.73}$$

and complete evaporation is possible if

$$\frac{\tau_{evap}}{\tau_{sd}} < 1 \tag{6.74}$$

For low values of M the companion star is degenerate and we have:

$$R/R_\odot \simeq 0.013 \, (1+X)^{5/3} \, (M/M_\odot)^{-1/3} \tag{6.75}$$

where X is the H mass fraction. In this case, $\tau_{evap} \propto M^3$. Then,

$$\frac{\tau_{evap}}{\tau_{sd}} = 10^{-2} \left(\frac{0.1}{f}\right) \left(\frac{a}{2R}\right)^2 \left(\frac{P}{1.55 \; ms}\right)^2 \left(\frac{M}{M_\odot}\right)^2 \left(\frac{R_\odot}{R}\right) \tag{6.76}$$

And, in the case of PSR1857+20 (the binary pulsar with $P \simeq 1.6 \; ms$), we have:

$$M = 0.02 \; M_\odot, \quad a = 2.5 \; R_\odot, \quad R = 0.166 \; R_\odot$$

which gives

$$\tau_{evap}/\tau_{sd} = 4 \times 10^{-3} \quad \text{for} \quad f = 0.1$$

and since

$$\tau_{sd} \sim 10^9 \; yr$$

then

$$\tau_{evap} \lesssim 10^7 \; yr$$

PSR1857+20, therefore, should become a single millisecond pulsar in a few million years.

7. Summary

We can now address a number of questions. The first one will be: how have the neutron stars in globular clusters formed?

In the *standard picture*, as we have seen, those neutron stars have formed from massive stars in the initial mass range

$$M \gtrsim 10 - 11 \ M_\odot$$

That immediately poses the problem of the velocity distribution of neutron stars at birth (the "kick velocities"). In the Galactic disk, the average velocity of pulsars is

$$\langle v \rangle \sim 200 \ km \ s^{-1}$$

while typical escape velocities from globular clusters are

$$v_{esc} \sim 25 \ km \ s^{-1}$$

This means that only neutron stars in the low–velocity tail of the distribution can be retained by the cluster. If we call this fraction f_{ret} and we use eq. (5.62) for the tidal capture rate, we obtain that in order to have a large enough number density of compact objects to account for all the LMBXS and recycled pulsars in globular clusters, a fraction

$$f_{ret} \sim 0.15$$

is required.

In the *AIC scenario*, white dwarf binaries would first form by tidal capture. An initial advantage over the standard scenario is that the number density of white dwarfs should be much higher than that of neutron stars ($n_{WD} \sim 10^3 \ n_{NS}$). Later, accretion of matter by the white dwarf would make it grow to the Chandrasekhar mass and eventually collapse into a neutron star without disrupting the binary.

If the mass–accreting white dwarf were a *C–O white dwarf*, the requirements would be:

$$M_{WD} \gtrsim 1 \ M_\odot \tag{7.77a}$$

and

$$\dot{M} \gtrsim 5 \times 10^{-8} - 10^{-7} \ M_\odot \ yr^{-1} \tag{7.77b}$$

If the white dwarf were a *O–Ne white dwarf*, there is no requirement on its mass (it would be $M_{WD} \gtrsim 1.2 \ M_\odot$, anyway), and the same lower limit on $\dot{M}$ as for C–O white dwarfs applies.

There is also the possibility of forming *single neutron stars* from merging of two white dwarfs in a binary (through emission of gravitational wave radiation) when the sum of their masses is larger than the Chandrasekhar mass. In the case of two C–O white dwarfs, the more massive star could ignite carbon nonexplosively in its outer layers, progressively changing its chemical composition into O–Ne as thermonuclear burning propagates towards the center. At the end, electron captures on Ne would trigger gravitational collapse.

In order to have a high accretion rate $\dot{M}$ after the tidal capture, the target star should be a red giant. In this case, $\dot{M}$ is driven by nuclear evolution in a wide binary. On the oher hand, double degenerate systems could form by direct collision of a massive white dwarf with a red giant star.

The AIC scenario has some advantage to explain the observed ratio of the number of radio pulsars to that of LMBXS in globular clusters. If both binary radio pulsars and single millisecond pulsars formed by evolution of LMBXS, then the life time of LMBXS should be

$$\tau_{LMBXS} \sim 10^7 \ yr$$

while evolutionary scenarios predict

$$\tau_{LMBXS} \gtrsim 10^8 \ yr$$

However, if neutron stars came from white dwarfs by AIC, binary pulsars would form directly. After gravitational collapse of the white dwarf, the system would be detached for $\sim 10^7 \ yr$ and the pulsar would start evaporating its companion (a white dwarf by then). At the end, a single, rapidly rotating radio pulsar would only remain.

Nonetheless, the AIC scenario would face two more problems:

(i) The neutron stars formed by AIC should have low surface magnetic fields B_s, since otherwise the pulsar would spin down too soon.

(ii) They should spin rapidly enough to be energetic pulsars.

7.1. *Primordial neutron stars*

The numbers of LMBXS and recycled pulsars should be computed from the basic characteristics of globular clusters. That involves the following steps:

(i) Compute the number of neutron star progenitors from the total mass of the globular cluster.

(ii) Compute f_{ret} from the escape velocity of the cluster and from the velocity distribution of newly born neutron stars.

(iii) Estimate f_{capt} of retained neutron stars which would undergo capture over the globular cluster life time, using the densities and the structure parameters of the clusters

The preceding poses a number of problems. One is to know the IMF appropriate for globular clusters when they formed, the main difficulty being the uncertainty in the mass function for the upper main sequence. Current observations of young globular clusters in the Large Magellanic Cloud (see Elson, this volume) should, however, greatly improve this point. Some constraints on the IMF can also be set from the effects of mass loss due to supernova explosion. Anyway, current estimates of the total number N_{prog} of neutron star progenitors for the whole globular cluster system of our Galaxy (with $M_{tot} \sim 5 \times 10^7 \ M_\odot$) are in the range

$$\sim 700 \ \text{to} \quad \sim 6 \times 10^5$$

A further problem is that both f_{ret} and f_{capt} depend sensitively on the structural parameters of the globular clusters.

In general, clusters predicted to contain many radio pulsars due to high f_{capt} also have large v_{esc} and thus also high f_{ret}.

In clusters with strong mass segregation, f_{capt} will be strongly enhanced. In this case, $f_{ret}f_{capt} \sim 1$. Not every tidal capture may lead to production of an active radio pulsar, however. If the neutron stars in most neutron star/main sequence pairs can not evaporate their companions in order to become observable radio pulsars, a large pulsar population would still be hard to account for.

7.2. *Accretion–induced collapse*

Although, in principle, it should be easy to form white dwarf/main sequence pairs by tidal capture in globular clusters, for a long while few cataclysmic variables had been detected. It was suggested that tidal capture might not produce many cataclysmic variables, the mass transfer subsequent to tidal capture being dynamically unstable in most cases and leading to merger of the two stars (Bailyn and Grindlay 1990). Recently, however, Cool et al. (1995) have discovered a candidate population of $\sim 5 - 20$ cataclysmic variables in NGC 6397. In addition, as we pointed out above, there is a population of low–luminosity X–ray sources ($L_x \sim 10^{32} - 10^{34}$ *erg* s^{-1}) which might also be evidence of white dwarf binaries.

The same as the population of primordial neutron stars, the number of AIC precursors should be computed from the total mass of the globular cluster system and the assumed IMF. Most candidates should come from tidal captures of massive white dwarfs by red giant stars. From eq. (5.62) one can estimate the corresponding fraction of all tidal captures. The cross section being proportional to the stellar radius:

$$f_{gia} = \frac{\sum_{gia} N_* R_*}{\sum_{tot} N_* R_*} \tag{7.78}$$

(Bhattacharya and van den Heuvel 1991). Typical estimates (Verbunt and Hut 1987; Verbunt 1988) give

$$f_{gia} \sim 0.1$$

but it can be larger if there is strong mass segregation (Verbunt and Meylan 1988).

There is considerable uncertainty on the mass range of the progenitors of the white dwarfs with masses $M_{WD} \gtrsim 1\ M_\odot$. Taking into account the most recent results on the evolution leading to O–Ne white dwarfs (see above), it could be

$$\sim 4 - (10 - 11)\ M_\odot$$

Then, for the whole globular cluster system of our Galaxy ($M_{tot} \sim 5 \times 10^7\ M_\odot$), the number of heavy white dwarfs would be

$$\sim 5 \times 10^3 - 7 \times 10^5$$

depending on the IMF and on the mass range of the progenitors. Out of those, a fraction f'_{capt} (smaller than f_{capt} for neutron stars, due to mass segregation in the clusters) will be tidally captured, and a fraction $f'_{capt} f_{gia}$ might undergo AIC.

It must be noted that, for $f_{gia} \lesssim 0.1$, AIC would be unable to generate $\sim 10^4$ neutron stars if $x \gtrsim 2$ in the exponent of the IMF. That is about the same constraint on the steepness of the IMF that can be derived from the alternative assumption that the binary and millisecond radio pulsars in globular clusters come from recycled primordial neutron stars.

7.3. *LMBXS in globular clusters and in the Galactic bulge*

LMBXS are comparatively abundant in the Galactic bulge. There, formation of neutron star/low–mass star pairs by tidal capture is unlikely, due to

(i) The high relative velocities of the stars ($\langle v \rangle \gtrsim 100 \ km \ s^{-1}$: more than one order of magnitude higher than in globular clusters).

(ii) Much lower n and n_c than in globular clusters.

(iii) The kinetic energy to be dissipated in the tidal capture would typically be larger than the binding energy of the dwarf star to be captured.

As an alternative to AIC, a globular cluster origin for those sources has been proposed: the source would form there by tidal capture, to be later released in the bulge when the cluster evaporates or is disrupted. The clusters with smaller masses should have evaporated by now, and a fraction of those with higher masses would have been tidally disrupted by repeated passages through the Galactic disk (Fall and Rees 1977).

A problem with the globular cluster origin is that the radial distribution of LMBXS in the Galactic bulge does not seem to fit with the disruption hypothesis. Besides, the luminosity functions of LMBXS in globular clusters and in the Galactic bulge appear to be different. However, Long and van Speybroeck (1983) pointed out that in M31 the sources within $\sim 400 \ pc$ from the galactic center resemble globular cluster sources by being more than twice as bright as the rest of the bulge sources in that galaxy.

7.4. *Conclusions*

From all the preceding, a few conclusions can be drawn:

(i) LMBXS in globular clusters could be formed either by tidal captures involving primordial neutron stars and/or from AIC of white dwarfs in binaries previously formed by tidal capture.

(ii) The relative importance of AIC and primordial massive star evolution in producing neutron stars in globular clusters remains an open question. Accurate modeling of the numbers of primordial neutron stars and AIC progenitors in globular clusters is still required.

(iii) The total number of radio pulsars in globular clusters, when properly estimated, might give constraints on the mass function of the upper main sequence of the clusters.

(iv) The high incidence of single radio pulsars in globular clusters may be explained by efficient evaporation of secondaries by the pulsar wind, plus maybe also exchange collisions of neutron stars and cluster binaries in which there is direct collision of the neutron star with another star.

(v) The evolution and thus the effective life times of LMBXS should be further investigated (for instance: does the pulsar wind stop mass accretion?).

(vi) The problem of the origin of neutron star binaries and radio pulsars in globular clusters is also linked to that of the origin and evolution of the magnetic fields of neutron stars.

(vii) Further radio and X–ray observations are needed, in order to overcome current biases (for instance: only ~ 600 radio pulsars are known at present, out of an estimated population of $\sim 10^4$ neutron stars in the Galaxy).

REFERENCES

BAADE, W. & ZWICKY, F. 1934 *Proc. Natl. Acad. Sci.* **20**, 254.

BAILES, M. 1989 *ApJ* **342**, 917.

BAILYN, C.D. & GRINDLAY, J.E. 1990 *ApJ* **353**, 159.

BARRAT, J.L., HANSEN, J.P. & MOCHKOVITCH, R. 1988 *A&A* **199**, L15.

BHATTACHARYA, D. & VAN DEN HEUVEL, E.P.J. 1991 *Phys. Reps.* **203**, 1.

BONSEMA, P.F.J. & VAN DEN HEUVEL, E.P.J. 1985 *A&A* **146**, L3.

BRADT, H.V. & MCCLINTOCK, J.E. 1983 *ARA&A* **21**, 13.

BROWN, G.E. & BETHE, H.A. 1994 *ApJ* **423**, 659.

BURROWS, A., HAYES, J. & FRYXELL, B.A. 1995 *ApJ* **450**, 830.

CANAL, R. 1994. In *Supernovae* (ed. S.A. Bludman, R. Mochkovitch & J. Zinn–Justin), p. 155. North–Holland.

CANAL, R. 1997. In *Thermonuclear Supernovae* (ed. P. Ruiz–Lapuente, R. Canal & J. Isern), p. 257. Kluwer.

CANAL, R. & SCHATZMAN, E. 1976 *A&A* **46**, 229.

CANAL, R., ISERN, J. & LABAY, J. 1990 *ARA&A* **28**, 183.

CANAL, R., GARCÍA, D., ISERN, J. & LABAY, J. 1990 *ApJ* **356**, L51.

COOL, A.M., GRINDLAY, J.E., COHN, H.N., LUGGER, P.M. & SLAVIN, S.D. 1995 *ApJ* **439**, 695.

DOMÍNGUEZ, I., TORNAMBÈ, A. & ISERN, J. 1993 *ApJ* **419**, 268.

EGGLETON, P.P. 1983 *ApJ* **268**, 368.

FABIAN, A.C., PRINGLE, J.E. & REES, M.J. 1975 *MNRAS* **172**, 15P.

FALL, M. & REES, M.J. 1977 *MNRAS* **181**, 37P.

FRYER, C.L., BENZ, W., HERANT, M. & COLGATE, S.A. 1998 *ApJ*, in press (preprint *astro-ph/9812058*).

GARCÍA–BERRO, E. & IBEN, I., JR. 1994 *ApJ* **434**, 306.

GIACCONI, R., GURSKY, H., KELLOGG, E., SCHREIER, E. & TANANBAUM, H. 1971 *ApJ* **167**, L67.

GOLD, T 1968 *Nature* **218**, 731.

GRINDLAY, J.E. 1987. In *The Origin and Evolution of Neutron Stars* (ed. D.J. Helfand & J.H. Huang), p. 173. Reidel.

GUTIÉRREZ, J., GARCÍA–BERRO, E., IBEN, I., JR., ISERN, J., LABAY, J. & CANAL, R. 1996 *ApJ* **459**, 701.

GUTIÉRREZ, J., CANAL, R., LABAY, J., ISERN, J. & GARCÍA–BERRO, E. 1997. In *Thermonuclear Supernovae* (ed. P. Ruiz–Lapuente, R. Canal & J. Isern), p. 303. Kluwer.

HACHISU, I., KATO, M. & NOMOTO, K. 1996 *ApJ* **470**, L97.

HERNANZ, M., ISERN, J., CANAL, R., LABAY, J. & MOCHKOVITCH, R. 1988 *ApJ* **324**, 331.

HERTZ, P. & GRINDLAY, J.E. 1983 *ApJ* **275**, 105.

HEWISH, A., BELL, S.J., PILKINGTON, J.D.H., SCOTT, P.F. & COLLINS, R.A. 1968 *Nature* **217**, 709.

HUT, P. & BAHCALL, J.N. 1983 *ApJ* **268**, 319.

IBEN, I., JR., RITOSSA, C. & GARCÍA–BERRO, E. 1997 *ApJ* **489**, 772.

ICHIMARU, S., IYETOMI, H. & OGATA, S. 1988 *ApJ* **334**, L17.

JANKA, H.-TH. & MÜLLER, E. 1996 *A&A* **306**, 167.

KOBAYASHI, C., TSUJIMOTO, T., NOMOTO, K., HACHISU, I. & KATO, M. 1998 *ApJ* **503**, L155.

KROLIK, J.H. & SINCELL, M.W. 1990 *ApJ* **357**, 208.

LANDAU, L.D. 1932 *Phys. Zeits. Sowjetunion* **1**, 285.

LANDAU, L.D. & LIFSHITZ, E.M. 1962 *The Classical Theory of Fields*. Pergamon.

LEWIN, W.H.G., RICKER, G.R. & MCCLINTOCK, J.E. 1971 *ApJ* **169**, L17.

LONG, K.S. & VAN SPEYBROECK, L.P. 1983. In *Accretion-Driven X-ray Sources* (ed. W.H.G. Lewin & E.P.J. van den Heuvel), p. 177. Cambridge.

LYNE, A.G., ANDERSON, B. & SALTER, M.J. 1982 *MNRAS* **201**, 503.

MEZZACAPPA, A., CALDER, A.C., BRUENN, S.W., BLONDIN, J.M., GUIDRY, M.W., STRAYER, M.R. & UMAR, A.S. 1998 *ApJ* **493**, 848.

MIYAJI, S. & NOMOTO, K. 1987 *ApJ* **318**, 307.

MIYAJI, S., NOMOTO, K., YOKOI, K. & SUGIMOTO, D. 1980 *Publ. Astron. Soc. Japan* **32**, 303.

NOMOTO, K. 1984 *ApJ* **277**, 791.

NOMOTO, K. 1987 *ApJ* **322**, 206.

PACZYŃSKI, B. 1967 *Acta Astronomica* **17**, 355.

PACZYŃSKI, B. 1971 *ARA&A* **9**, 183.

RAPPAPORT, S.A., VERBUNT, F. & JOSS, P.C. 1983 *ApJ* **275**, 713.

RITOSSA, C., GARCÍA–BERRO, E. & IBEN, I., JR. 1996 *ApJ* **460**, 489.

RUDERMAN, M., SHAHAM, J. & TAVANI, M. 1989 *ApJ* **336**, 507.

SCHREIER, E., LEWINSON, R., GURSKY, H., KELLOGG, E., TANANBAUM, H. & GIACCONI, R. 1972 *ApJ* **172**, L79.

STEVENSON, D.J. 1980 *J. Phys. Suppl. N° 3* **41**, C2–61.

SUTANTYO, W. 1975 *A&A* **41**, 47.

TAAM, R.E. & VAN DEN HEUVEL, E.P.J. 1986 *ApJ* *305*, 235.

TIMMES, F. & WOOSLEY, S.E. 1992 *ApJ* **396**, 649.

VAN DEN HEUVEL, E.P.J. 1978. In *Physics and Astrophysics of Neutron Stars and Black Holes* (ed. R. Giacconi & R. Ruffini), p. 828. North–Holland.

VAN DEN HEUVEL, E.P.J. & HEISE, J.G. 1972 *Nature Phys. Sci.* **239**, 67.

VAN DEN HEUVEL, E.P.J. & BITZARAKI, O. 1995 *A&A* **297**, L41.

VAN DEN HEUVEL, E.P.J. & VAN PARADIJS, J. 1997 *ApJ* **483**, 399.

VANBEVEREN, D., DE LOORE, C. & VAN RENSBERGEN, W. 1998 *A&A Rev.* **9**, 63.

VAN PARADIJS, J., VAN DEN HEUVEL, E.P.J., KOUVELIOTOU, C., FISHMAN, G.J., FINGER, M.H. & LEWIN, W.H.G. 1997 *A&A* **317**, L9.

VERBUNT, F. 1988. In *The Physics of Compact Objects: Theory vs Observation* (ed. N.E. White & L. Filipov), p. 529. Pergamon.

VERBUNT, F. 1993 *ARA&A* **31**, 93.

VERBUNT, F. & ZWAAN, C. 1981 *A&A* **100**, L7.

VERBUNT, F. & HUT, P. 1987. In *The Origin and Evolution of Neutron Stars* (ed. D.J. Helfand & J.H. Huang), p. 187. Reidel.

VERBUNT, F. & MEYLAN, G. 1988 *A&A* **203**, 297.

VERBUNT, F., WIJERS, R.A.M.J. & BURM, H.M.G. 1990 *A&A* **234**, 195.

WEBBINK, R.F. 1984 *ApJ* **277**, 355.

WOOSLEY, S.E. & BARON, E. 1992 *ApJ* **391**, 228.

WILLIAM E. HARRIS was born November 28, 1947 in Edmonton, Canada. As an undergraduate, he studied mathematics at the University of Alberta and graduated there in 1969. Graduate school followed at the University of Toronto, with a Master's degree in theoretical astrophysics in 1970 and PhD in astronomy in 1974, on "Globular Clusters in the Local Group Galaxies". After spending two years at Yale as a postdoctoral fellow, he moved to a faculty position at McMaster University in Hamilton, where he has been happy to stay conducting his teaching and research. As a graduate student, he developed a fascination with globular clusters and their place in galactic structure and galactic history which has lasted ever since. "No other subject in modern astrophysics has such a long history, or has re-invented itself so many times through technological advances and unexpected discoveries." Most of his current research is aimed at understanding the characteristics of globular cluster systems in giant elliptical galaxies, and at the ages and formation histories of globular clusters and their role in the earliest stages of galaxies.

Harris has found it a "terrific experience" to use telescopes all around the world during the period of rapid growth of astronomy beginning in the early 1970's and continuing into today's large-telescope era. He has chaired time allocation panels for the CFHT and HST, and was recently President of the Canadian Astronomical Society. In his free time, he enjoys choir singing (definitely as a bass), tennis in the summer, and in the winter the ice sport of curling a strange and unique game invented long ago in Scotland but now played more in Canada than anywhere else.

Globular Cluster Systems:
Formation Models and Case Studies

By WILLIAM E. HARRIS

Department of Physics & Astronomy, McMaster University, Hamilton ON L8S 4M1 Canada

Globular cluster systems represent only a small fraction of the total stellar mass of galaxy halos, but provide unique tracers which can be used to address models of galaxy formation. Several "case studies" of individually important galaxies are presented, in which we look at the characteristics of their globular clusters including the metallicity distributions, specific frequencies, luminosity (mass) distributions, and kinematics. Among these galaxies are the Milky Way, the nearby giant elliptical NGC 5128, the Virgo ellipticals NGC 4472 and M87, and the supergiant cD galaxies at the centers of rich clusters. In each case the possible roles of mergers, small-satellite accretions, and *in situ* formation in the growth of the galaxy are discussed. We also briefly touch on the connection between the globular clusters and the much more numerous field-halo stars. We conclude that in all formation scenarios, the presence or absence of gas at any stage of the galaxy's evolution plays a crucial role in determining the total cluster population, the number of distinguishable subpopulations, and the metallicity distribution of the clusters.

The analysis of globular cluster systems (GCSs) in other galaxies is starting to fulfil its long-held promise of informing us about galaxy formation in ways that are unique. The more we learn about GCSs, the more we realize that their role in galaxy formation is an intricate and varied process – yet with common themes that apply particularly to the old-halo population that is found in every type of galaxy.

A decade ago, it was possible to write a single review article (Harris 1991) which encompassed virtually all the themes in GCS research. Today, the field has proliferated in so many directions that such a job is impossible. Instead, we will take the approach that many of the key findings about globular cluster systems have relied on the thorough study of *a few critical individual galaxies* of representative types. These "case studies" form the outline of this chapter. We have space only to touch on a few highlights: to gain a fuller idea of the richness and diversity of this field, the reader is urged to see the more comprehensive recent discussions of Ashman & Zepf (1998) and Harris (1999, hereafter denoted H99).

1. Case Studies: The Milky Way GCS

Our logical starting point is the globular cluster system in the galaxy that we know best: the Milky Way. It is in many respects a typical large spiral, and current evidence suggests that the GCSs in other spirals like it have basically similar properties. In global perspective, the globular clusters of the Milky Way define a roughly spherical spatial distribution which has traditionally marked the paradigmatic "old halo" of our galaxy extending out to $R_{gc} \simeq 40$ kpc (Figure 1).

At distances beyond 40 kpc, the few remaining outermost-halo clusters join with several of the dwarf satellites of the Milky Way (Figure 2) to delineate a much larger-scale asymmetric planar distribution that may have had a distinct origin and history outside the Milky Way proper (e.g., Harris 1976; Zinn 1985; Majewski 1994; or H99).

At present, 147 globular clusters are known in the Milky Way (Harris 1996). Estimates of the true total population have differed widely over the years, but most of the "missing" undetected ones, if any, must lie at very low Galactic latitude behind extremely large

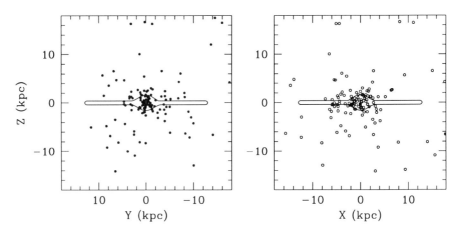

FIGURE 1. In these two projections (the YZ and XZ planes) the space distribution of the Milky Way globular clusters is seen to be approximately spherical except in the central bulge. In the XZ plane (where the Sun is at $(-8, 0)$), note the elongation of the inner points along the X-axis, which is largely a residual of random distance errors along the line of sight from the Sun. Note also the lack of points at large X-values along the disk beyond the Galactic center, where any "missing" clusters would likely be. Data for the Milky Way clusters in this and subsequent figures are taken from the catalog of Harris (1996).

amounts of foreground absorption. An occasional few may lie at extremely large distances and/or small luminosities, though only a handful of such objects have been found (usually accidentally) over the past two decades. An analysis of these factors by H99 suggests that the true total is $N \simeq 160 \pm 10$.

The existence of two major and remarkably distinct subpopulations in the GCS, based on metallicity and kinematics, was suspected long ago (see especially Kinman 1959, with even earlier hints in the literature) and was firmly established by the classic study of Zinn (1985). The *metal-poor component* (which I will abbreviate as MPC) contains about 3/4 of all the clusters and is spread throughout the halo; the *metal-rich component* (abbreviated MRC) contains the remaining 1/4 of the clusters and is almost entirely restricted to within the Solar circle, $R_{gc} \lesssim 8$ kpc. The total metallicity distribution function (MDF) can be well described by two Gaussians in [Fe/H] (Figure 3), and the clear $\gtrsim 1$ dex separation between the MRC and MPC groups already suggests distinct evolutionary histories for them.

Zinn (1985) and Armandroff (1989) interpreted the MRC as a "disk" population of clusters because of their high metallicity and large systemic rotation (see below), and since then it has become conventional to associate them with the old stellar thick disk or something analogous to it. Such an association, however, does not stand up entirely comfortably to the complete array of contemporary evidence. Minniti (1995) has argued persuasively that the sharply increasing space density of the MRC all the way in to the Galactic center resembles a bulge-like population far more than a disk-like one, while the net rotation of the innermost MRC clusters (see also Zinn 1996) is also like that of the old field giant stars in the Galactic bulge. The spatial distribution of the MRC is distinctly more flattened toward the Galactic plane than is the near-spherical MPC population (Zinn 1985; H99), and overall, this inner metal-rich population thus seems best associated with a *flattened bulge population*.

The *kinematical* analysis of the cluster population has drawn considerable attention

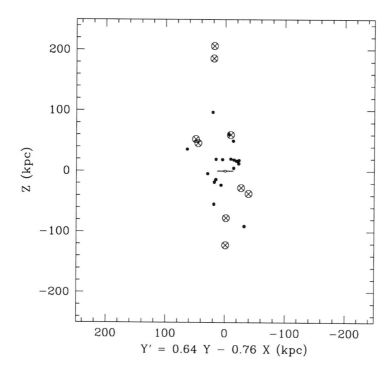

FIGURE 2. Spatial distribution of the Milky Way globular clusters (dots) more distant than 20 kpc from the Galactic center, and dwarf satellite galaxies (crosses). Most of these objects fall within a planar configuration rather than a spherically symmetric outer halo; the coordinate $Y' = 0.64Y - 0.76X$ (where X, Y are measured relative to the Galactic center; from Majewski 1994) is chosen to show the plane "edge-on".

in recent years, as a way of separating out possible subcomponents and thus formation histories; a group of clusters accreted, for example, from a single captured satellite galaxy such as we see happening now with the Sagittarius dwarf (Ibata et al. 1997; Da Costa & Armandroff 1995), could have distinctly different orbital properties than halo clusters formed *in situ* within the original potential well of the larger Galaxy. The discovery of the Sagittarius system has, in fact, stimulated numerous conjectures that the entire (metal-poor) halo might have accumulated by accretion of small satellites that were the original hosts of cluster formation.

Most of the kinematical analysis to date has necessarily relied on just the *radial velocities* of the clusters, although measurements of absolute proper motions and thus true three-dimensional space motions are steadily growing in reliability and importance (see below). The standard equation of condition for the radial velocity v_r of a cluster relative to the Solar Local Standard of Rest is (Frenk & White 1980)

$$v_r = V\cos\psi - V_0\cos\lambda$$

where $\mathbf{V}$, $\mathbf{V_0}$ are the Galactic rotation velocity vectors at the locations of the cluster and the Solar LSR; ψ is the angle between $\mathbf{V}$ and the vector $\mathbf{r}$ from the Sun to the cluster; and λ is the angle between $\mathbf{V_0}$ and $\mathbf{r}$ (Frenk & White 1980; Zinn 1985). To use this equation, we select a group of clusters (for example, those within a narrow range in metallicity or Galactocentric distance) and ask what the net rotation speed V is for the group. We

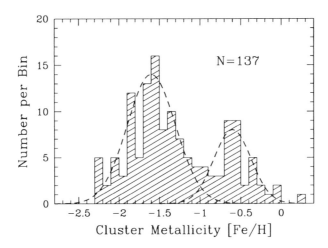

FIGURE 3. Metallicity distribution function for the globular clusters in the Milky Way. The two Gaussian curves have means and standard deviations of $(-1.6, 0.30)$ and $(-0.6, 0.23)$ and define the metal-poor (MPC) and metal-rich (MRC) components.

assume $V_0 = 220$ km s^{-1}, while r, v_r, λ are directly measured for each cluster, and ψ is deduced knowing $\mathbf{r}$. For an ensemble of n clusters we then plot $v_{LSR} = (v_r + V_0 \cos\lambda)$ against $\cos\psi$ and derive the mean slope $\langle V \rangle$ and line-of-sight dispersion σ_{los} about the mean.

Estimates for the mean rotation speeds and dispersions of various subsets of clusters, broken down by metallicity and Galactocentric distance, are summarized in Table 1 and shown in Figures 4 and 5 (taken from the recent compilation of H99). Let us look first at the metal-rich component: the *inner* MRC clusters (0 to 4 kpc) show a modest but clear rotation signal, consistent with bulge rotation and with the velocity pattern of bulge red-giant stars (see Minniti 1995; Zinn 1996). On the other hand, the 4-to-9 kpc MRC group has a much stronger rotation and lower dispersion, though still less than the ~ 180 km s^{-1} characteristic of the stellar thick disk (e.g. Majewski 1993). Is the MRC thus a composite of bulge and thick disk? If so, we might expect to see net age differences between them (with the disklike ones belonging, perhaps, to a later formation epoch). Unfortunately, no reliable differential age comparisons for these objects are yet available, primarily because of the very high reddenings and crowded fields in which the inner MRC clusters are found. A clear age comparison would, however, be of paramount importance and is very much worth pursuing.

The MPC (halo) clusters were, until this decade, generally thought to be a rather homogeneous group kinematically (except possibly for the few outermost Palomar-type clusters, which we will not discuss in detail here). They were characterized systemically by a very modest overall rotation ($V \sim 40$ km s^{-1}), high random motion ($\sigma_{los} \sim 120$ km s^{-1}), and a nearly isotropic orbit distribution ($\sigma_R \sim \sigma_\theta \sim \sigma_\phi$). Although these blanket statements remain true to first order, we now suspect that the MPC contains identifiable subgroups and perhaps traces of separate protogalactic 'fragments' and accreted satellites. The search for these continues, in the form of moving groups of field-halo stars, or elongated distributions of tidal debris from individual clusters or dwarf galaxies.

In the MPC subgroups broken down by radial zones, few significant trends emerge,

Group	Subgroup	$V(\mathrm{rot})$	σ_{los}
MRP	$0-4$ kpc	86 ± 40	99 ± 15
	$4-9$ kpc	147 ± 27	66 ± 12
MPP	$0-4$ kpc	56 ± 37	122 ± 16
	$4-8$ kpc	12 ± 31	79 ± 12
	$8-12$ kpc	26 ± 63	148 ± 29
	$[\mathrm{Fe/H}] < -1.7$	80 ± 43	130 ± 16
	$R_{gc} > 8$, BHB	55 ± 59	118 ± 18
	$R_{gc} > 8$, RHB	-39 ± 83	158 ± 24
	RHB ex.N3201	32 ± 89	149 ± 24

TABLE 1. Kinematical solutions for subgroups of globular clusters in the Milky Way. The rotation velocity $V(\mathrm{rot})$ and line of sight dispersion σ_{los} about the mean V are in km s^{-1}.

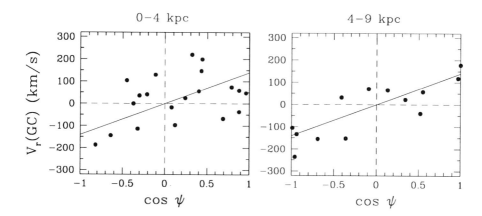

FIGURE 4. Kinematical solutions for the net rotation speeds of the metal-rich globular clusters in the Milky Way (adapted from Harris 1999). Here the radial velocity of the cluster relative to a stationary point at the LSR is plotted against $(\cos \psi)$ for the MRC clusters within $(0 < R_{gc} < 4)$ kpc (left panel) and $(4 < R_{gc} < 9)$ kpc (right panel).

all zones having small rotation and comparably high dispersion (see Table 1 and H99). However, when the sample is divided into metallicity subgroups – which may be more closely connected with cluster age – more significant trends do emerge. The clearest signal is that the *most metal-poor halo component* has a surprisingly significant net rotation: the clusters with $[\mathrm{Fe/H}] < -1.7$ have $V = (80 \pm 43)$ km s^{-1}, higher than any other MPC subgroup and quite comparable with the innermost MRC group (see H99). This significant rotation is driven by the metal-poor clusters in the *inner* halo $(R_{gc} \lesssim 10$ kpc), since the outer-halo objects necessarily have only a small range in $(\cos \psi)$ and therefore exert no leverage on V. Is this a fossil remnant of a formation process resembling the classic ELS model (Eggen, Lynden-Bell, & Sandage 1962), whereby the inner protogalaxy went through a monolithic collapse and spin-up?

The possibility that traces of entire accreted satellites might be detectable kinematically – for example, by systemic retrograde motion, if the satellite fell into the halo from

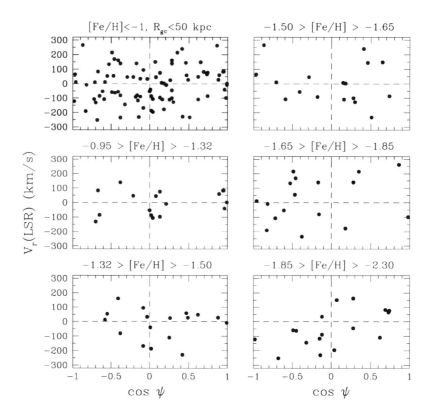

FIGURE 5. Kinematical solutions for the net rotation speeds of the metal-poor clusters in the Milky Way, excluding the four Sagittarius member clusters and any clusters more distant than 50 kpc from the Galactic center. The upper left panel shows all clusters combined, while the remaining panels show the distributions for five metallicity subgroups (see Table 1 for a similar breakdown by Galactocentric distance). NGC 3201, which strongly drives a "retrograde" solution for V in the bin for $(-1.50 > [Fe/H] > -1.65)$, is the point at uppermost left in the upper right panel (see text).

a direction counter to the normal rotation vector **V** – was raised by Rodgers & Paltoglou (1984), who suggested that a subgroup of clusters over a rather narrow range of [Fe/H] did indeed appear to have net retrograde motion. This approach was later given considerable impetus by Zinn (1993) and by the subsequent discovery of the Sagittarius dwarf with its retinue of four clusters (e.g., Da Costa & Armandroff 1995). These ideas are closely connected to one of the main underpinnings of the Searle & Zinn (1978) model in which the outer parts of the halo are envisaged as forming from many individual dwarf-sized pregalactic fragments. In this picture, the outer halo could reasonably be expected to harbor globular clusters of widely different ages or kinematics, condensing within these fragments at different times, then accreted on many different orbital directions.

A key piece of evidence connected with this argument is the morphological plot shown in Figure 6. *If* horizontal-branch type (red, blue, or intermediate) is a reliable indicator of cluster age, then it suggests that the outer-halo clusters would have younger ages on average (redder HB types at a given metallicity), as well as a larger spread in ages (Zinn 1985, 1993). This suggestion, however, relies strongly on the assumption that HB morphology is driven primarily by age at a given metallicity (Lee et al. 1994; Chaboyer

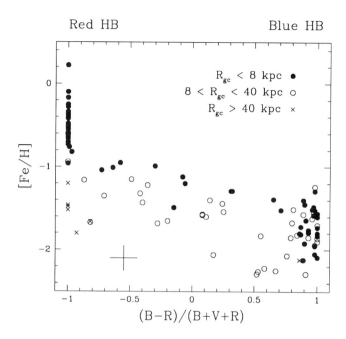

FIGURE 6. Globular cluster metallicity [Fe/H] plotted against horizontal-branch ratio (B-R)/(B+V+R) (the relative number of blue, red, or intermediate stars on the horizontal branch; from Lee et al. 1994 and Harris 1996). The HB morphology becomes progressively redder at large Galactocentric distances. The Zinn (1993) "RHB" clusters are the open circles with HB ratio less than 0.5 (see text).

et al. 1996). More direct tests must come from direct main-sequence photometry of these clusters, which now tend not to confirm the age hypothesis (see below).

To avoid prejudicing the discussion in terms of cluster age, I will refer to these morphological groups simply as the "red horizontal branch" (RHB) clusters and the "blue horizontal branch" (BHB) clusters. Zinn (1993) noted specifically that whereas the BHB group has the normal slightly prograde rotation characteristic of the overall halo, the RHB clusters by themselves seemed to possess a zero or slightly retrograde motion, perhaps indicative of their origin in an accretion or merger (see also Da Costa & Armandroff 1995 for lists of these clusters and another kinematic analysis). In a rediscussion of this issue with the most recent data (H99), I note that there are two serious problems with this interpretation: first and most important is that the RHB clusters are all drawn *by definition* from the outer halo ($R_{gc} \gtrsim 8$ kpc), whereas the BHB sample being compared with it was drawn *from all Galactocentric distances*, including large numbers from the inner halo. And, as we have seen above, the inner metal-poor halo has a distinct prograde rotation. We should, more correctly, compare the RHB and BHB subgroups from within the *same radial range*. When this is done (Table 1), no significant difference appears between them.

A second factor results from the unfortunate, and unavoidable, fact that both samples of objects are statistically small when we restrict ourselves to the outer halo. The nominal "retrograde" motion of the RHB group turns out to be entirely driven by just one cluster, NGC 3201. Because of its location on the sky ($\cos \psi \simeq -1$) and large radial velocity ($v_r = 494$ km s^{-1}), this single cluster has a uniquely powerful influence on the rotation solution: if we arbitrarily remove it from the sample, we find that V(rot) for the RHB group changes by $+70$ km s^{-1} and makes the result formally indistinguishable from the outer BHB group. In summary, the radial velocity data by themselves give no convincing evidence for net retrograde motion in any identifiable subgroup. Finding evidence for retrograde subgroups of stars in the halo must, instead, rely on the much larger samples of objects that can be accumulated from field stars (see Majewski's lectures in this volume).

Even though we cannot increase the *numbers* of clusters in our kinematical samples, we can eventually improve our knowledge of their *individual orbits* which may help us isolate potential kinematic differences. New data for absolute proper motions and thus true three-dimensional space velocities are steadily accumulating in both quantity and quality (see particularly the recent work of Dinescu et al. 1998 which also summarizes previous space motion studies). This material can be used, albeit for smaller numbers of clusters, to categorize the orbital types of each subsample discussed above. Inspection of their results shows that, when the RHB clusters are compared with the BHB clusters *in the same spatial region* ($8 - 40$ kpc), there is no distinguishable difference in mean orbital energy E or eccentricity e. Similar proportions of clusters are found in each group with individual prograde, retrograde, or plunging orbits. The traditional assumption of roughly isotropic orbits (with $\langle e \rangle \simeq 0.66$ in both RHB and BHB groups) is also confirmed.

Finally, differential age comparisons can now be performed for several clusters in each group, since recent precise photometry reaching well past the main-sequence turnoff has been obtained for a number of critical RHB/BHB "pairs" of similar metallicity, including the classic M3/M13 pair (Catelan & de Freitas Pacheco 1995; Johnson & Bolte 1998; Grundahl & Andersen 1998) and NGC 288/362/1851 (see Stetson et al. 1996; Sarajedini et al. 1997 for a representative range of views). Although individual analyses disagree in some details, the extremely precise comparisons in the papers listed above make it appear unlikely that the clusters in these comparison pairs differ strongly in age; for the M3/M13 pair in particular, such differences are well within the ~ 1 Gyr precision of the CMD matching methods. A view is emerging that the solution to the long-standing "second parameter problem" for horizontal-branch morphology lies in more than just age differences anmong these clusters, perhaps in a combination of factors involving detailed abundance differences, mixing, or even helium abundance (Johnson & Bolte 1998; Grundahl et al. 1998). In the outer Palomar-type clusters as well, the main sequence loci established from recent deep HST photometry (Stetson et al. 1999) translate into only very modest age differences ($\Delta\tau \lesssim 1.5$ Gyr) compared with M3 or M5, despite their much redder HB morphologies.

In summary, the most recent assemblage of data does not favor the view that there are major kinematic or age differences between the Zinn RHB and BHB clusters. We will have to continue exploring other avenues to understand the physical differences between them.

2. Interlude: Systemic Properties in Other Galaxies

Although we know about the Milky Way clusters in impressive detail, we must eventually look to other galaxies to see genuinely large samples of clusters over their full range of properties. Those in other large spiral galaxies generally appear very much like the Milky Way GCS (Harris 1991; Ashman & Zepf 1998), but the same is certainly not true for the many E galaxies, both giants and dwarfs, in which we now have GCS information. Working with the GCS in a distant galaxy has at least one very large advantage over observing the integrated field-star halo light: the globular clusters can be isolated one by one, and can thus be used to construct a full *distribution function* by metallicity, luminosity, or radial velocity. The lingering question is, of course, to ask whether or not the GCS is truly representative of the entire old halo. We will gain some additional clues to this important question as we go on.

2.1. Specific Frequencies

The simplest parameter describing a GCS is the total number of clusters it contains, quantified as the *specific frequency* S_N or number of clusters per unit galaxy luminosity (Harris & van den Bergh 1981; Harris 1991). In numerical terms it is defined simply as $S_N = N_{cl} \cdot 10^{0.4(M_V^T+15)}$ for a total cluster population N_{cl} and galaxy luminosity M_V^T. In the early years of this subject, it was a major suprise to find that individual galaxies differ in this ratio by factors of *twenty*! Why are some galaxies, apparently, vastly more efficient at creating – or holding – their globular clusters? This is the classic "specific frequency problem" which cuts across most types of galaxies, particularly the ellipticals. Understanding these differences between galaxies that are otherwise similar in structure has been a continual challenge to virtually all of our formation models.

As Figure 7 shows, the *average* S_N for ellipticals is $\simeq 3.5$, essentially constant over more than a 10^4 range in luminosity. But large scatter exists at all levels, and interesting trends affect both the the dE,N and cD-type galaxies (solid dots in Figure 7). These will be discussed below. The mean or "baseline" specific frequency $S_N^0 = 3.5$ can be translated into a typical *formation efficiency* e_0, which is essentially the number of clusters per unit stellar mass in the galaxy halo. Assuming $(M/L) \simeq 8$ for the old-halo stars in E galaxies, then we obtain directly from the definition of S_N above that $e_0 \simeq 1$ cluster per $2 \times 10^8 M_\odot$ of halo mass. The globular clusters evidently represent quite a tiny fraction of all the old stellar population in these galaxies, even for the higher$-S_N$ cases.

The S_N graph in Figure 7 falls conveniently into the dwarf and giant ellipticals, whose distributions are in some sense mirror images of each other (the largest specific frequencies occur at either the very lowest or very highest luminosities; as we will see below, these high$-S_N$ cases may have ultimately very similar causes). The giant ellipticals, as we discuss in more detail below, may have been built in a variety of ways (isolated or *in situ* from a single protogalaxy; by mergers; or by ongoing accretions of smaller satellites), and the supergiant brightest-cluster galaxies (BCGs) with their very high specific frequencies present a special problem which we will return to below.

The dE's are likely to present a fundamentally simpler story, but even here they fall into two natural subgroups according to the presence or absence of a central nucleus. The surveys of the GCSs in these small galaxies (Durrell et al. 1996; Miller et al. 1998) demonstrate that the non-nucleated dE's have rather similar specific frequencies in the range $S_N \sim 2 - 3$, like disk galaxies or ellipticals in sparse groups. These same dwarfs also tend to be more elongated in structure and, in cluster environments like Virgo or Fornax, form a spatially extended subsystem of galaxies resembling the spirals and irregulars rather than the giant ellipticals. Durrell et al. and Miller et al. suggest that most of the dE's may simply be original irregulars which formed relatively few clusters at early times and were then stripped of most of their gas in the cluster potential well, preventing further star formation.

The nucleated dE,N types exhibit a different trend, with the least luminous ones having the highest specific frequencies. These galaxies are also generally rounder in structure and occupy more centrally concentrated space distributions in Virgo and Fornax resembling the giant ellipticals. The authors cited above therefore interpret them as more likely to have been "genuine" small elliptical galaxies from the start, with their central nuclei forming either from residual gas infall or from individual globular clusters that were drawn in by dynamical friction. The interesting trend of S_N with luminosity can be

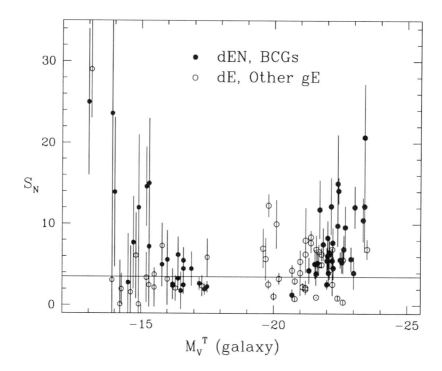

FIGURE 7. Specific frequency S_N against galaxy luminosity, for elliptical galaxies (with data from Harris et al. 1998 and Miller et al. 1998). For dwarfs (left side of figure), solid symbols denote nucleated dE,N types and open symbols are non-nucleated dE types. For giant ellipticals (right side of figure), solid symbols denote Brightest Cluster Galaxies (BCGs, most of which are cD galaxies at the centers of rich galaxy clusters). The horizontal line is at the "baseline" level of $S_N^0 = 3.5$ (see text for interpretation).

logically interpreted as the result of the main star formation stage. Presumably, in the smallest dE,N systems shown in Figure 7 (which have total stellar masses $\lesssim 10^8 M_\odot$), the globular clusters formed rather early, along with most of the field stars, but then the remaining gas (which would have been a high fraction of the original gas supply) was expelled from its small potential well by the first round of stellar winds and supernova (cf. the authors cited). The more massive ($\sim 10^9 M_\odot$) dE,N systems would have held on to more of their original gas supply and finished more star formation, leaving a lower $-S_N$ system in the end. A key part of this scenario is that the globular clusters need to form *early*, out of the densest initial clumps of gas. McLaughlin (1999) has used the Dekel & Silk (1986) model for gas loss from these small galaxies to show quantitatively that the expected trend of S_N with luminosity does indeed match the observations.

2.2. *Metallicity Distributions*

Information on the GCS *metallicity distribution function* (MDF) is now available for many galaxies. A rough general trend for mean cluster metallicity to increase with galaxy size was suspected to exist at a very early stage in the subject (Brodie & Huchra 1991; Harris 1991) and mimics the same trend for the halo light in the galaxies themselves. However, as more data have accumulated, it has become increasingly clear that large galaxy-to-galaxy differences in the MDF exist even between otherwise similar ellipticals. *Bimodal* MDFs are commonly, but not universally, found in large galaxies of both elliptical and spiral type (cf. Ashman & Zepf 1998; Ajhar et al. 1994; Kissler-Patig et al. 1997). When present, a bimodal MDF is usually interpreted as an indicator of at least two major phases of galaxy formation at early times, but the exact mechanism (whether by *in situ* formation, major mergers, or accretion) is still very much under debate. No one process (see below) can be said to conveniently answer all the observed situations equally well.

In an important recent study, Forbes et al. (1997) find that for bimodal MDFs in giant E galaxies, the redder, more metal-rich component (denoted MRC in parallel with the Milky Way above) is more closely correlated with galaxy size than the bluer, metal-poor component (MPC). The mean metallicity of the MPC differs considerably, and without obvious patterns, from one such galaxy to another, whereas in most gE galaxies the MRC consistently peaks at [Fe/H] ~ -0.2 or varies only slightly with M_V^T. By contrast, the MPC in most spiral galaxies and dwarfs peaks consistently at [Fe/H] $\simeq -1.5$, and the MRC (if it exists) at typically [Fe/H] $\simeq -0.5$, at lower levels than in the giant ellipticals. Forbes et al. use this, among other evidence, to argue against the hypothesis originally laid out by Schweizer (1987) and Ashman & Zepf (1992) that giant E galaxies are the result of major mergers between large disk galaxies (but see the discussion below).

2.3. *Luminosity and Mass Distributions*

By far the most robust and predictable feature of the GCS in different galaxies is the *luminosity distribution* of the clusters (LDF), which is the visible trace of the cluster mass spectrum. In its classic form plotted as number of clusters per unit magnitude M_V, the LDF has a roughly Gaussian-like shape with a characteristic "turnover" or peak point at $M_V^0 \simeq -7.4$, differing by little more than ± 0.2 magnitudes from any one large galaxy to another. The symmetric Gaussian-like shape has been consistently verified in every galaxy with sufficiently deep photometry (e.g., Harris et al. 1991; Whitmore et al. 1995; Kundu & Whitmore 1998), and the turnover luminosity is consistent enough to be an entirely respectable standard candle for distance determination (see H99; Whitmore 1997; Harris 1997 for comprehensive recent discussions).

The potential use of the LDF for standard-candle purposes was, in fact, the original stimulus for studying globular clusters in distant galaxies, beginning with the Virgo ellipticals (see Hanes 1977 and H99). But as more data have come in, and as the remarkable uniformity of the LDF has emerged more clearly, it has become more interesting for its strong astrophysical constraints on cluster formation and evolution. The LDF in its entirety must be the product of the *initial mass spectrum* of the clusters at formation, combined with the subsequent $\sim 10^{10}$ years of *dynamical evolution* in the tidal field of the parent galaxy. Most of the relevant destruction mechanisms (tidal shocking due to passage through the disk or bulge, evaporation coupled to the tidal field) are much more effective at lower cluster mass, and current models (e.g. Murali & Weinberg 1997; Gnedin & Ostriker 1997; Vesperini 1997, 1998; Vesperini & Heggie 1997; and Elson's lectures in this volume) indicate that the transition mass above which these effects are not critically important is the LDF turnover near $10^5 M_\odot$. Above this point, the LDF we observed today is therefore likely to be much closer to the original mass spectrum at the time of formation. Although a consensus is gradually growing in this direction, more comprehensive dynamical simulations incorporating the full range of dynamical effects, starting from a realistic formation mass spectrum, still need to be done.

Over the upper $\sim 90\%$ of the globular cluster mass range ($M \gtrsim 10^5 M_\odot$, where the very most massive clusters reach as high as $\sim 10^7 M_\odot$), the cluster mass spectrum is better plotted as *number of clusters per unit mass* or unit luminosity, rather than per unit magnitude. In this form, it is well described by a simple power-law form $dN/dM \sim M^{-1.8\pm0.2}$ in all galaxies (Harris & Pudritz 1994), gradually steepening to higher masses. This distribution is the form which needs to be reproduced by a quantitative formation model; it also needs to be virtually independent of other factors such as metallicity, total cluster population (S_N), or size and type of host galaxy! Whatever mechanism we settle on must be extremely robust.

McLaughlin & Pudritz (1996) have developed a quantitative theory for the LDF in which protocluster gas clouds build up within very large "supergiant" molecular clouds (SGMCs) by collisional agglomeration. The SGMC is visualized as supplying a large number of initial small-mass cloud "particles" (physically, these particles can be visualized as probably resembling the $\sim 100M_\odot$ cloud cores found in Galactic GMCs). These cloud particles then collide and amalgamate to form larger ones; after several crossing times, a power-law distribution of cloud masses results (e.g., Field & Saslaw 1965; Kwan 1979). The larger clouds become the "protoclusters" which eventually turn into full-fledged star clusters once their internal pressure support (due to turbulence and weak magnetic field) leaks away. Since the small clouds always vastly outnumber the large ones, the reservoir of gas contained in the entire SGMC needs to be very much larger than the masses of the individual protoclusters that build up inside it (observationally, the typical star cluster mass is $\sim 10^{-3}$ of the host GMC mass; see Harris & Pudritz 1994).

The emergent mass spectrum from this collisional growth process has the expected power-law form, but its detailed shape is controlled by two key input parameters: (a) the ratio of cloud lifetime against star formation relative to the cloud-cloud collision time; and (b) the dependence of cloud lifetime on mass. More massive clouds have higher mean densities and are expected to have shorter lifetimes. Thus, cloud growth is a stochastic race against time: large clouds continue to grow by absorbing smaller ones, but as they do, their survival time before turning into stars becomes shorter and shorter. Thus at the high-mass end, the slope of the mass spectrum dN/dM gradually steepens as it becomes more and more improbable that such massive clouds can survive before turning into stars. Encouragingly, this very feature is matched extremely well by the observations of LDFs in populous cluster systems such as in giant ellipticals (see McLaughlin & Pudritz 1996). Even more encouraging is the fact that this theory is also able to match the *entire* LDF of the newly formed star clusters in recent mergers such as NGC 4038/4039 (Whitmore & Schweizer 1995) and NGC 7252 (Miller et al. 1997). In these cases, we should be looking at something much closer to the initial mass spectrum, relatively unaffected even at low masses by dynamical evolution (see Figure 8).

The related new observations of globular clusters in *young and merging systems*, some of which were used above, represent a major stride forward in our understanding of globular cluster formation. Clusters are now seen to form in an amazingly wide range of situations: in protogalactic halos, in starburst dwarf galaxies with or without obvious external "triggers", in merging disk galaxies, or at the centers of giant ellipticals that are accreting gas. We see from the same observations that cluster formation is a *highly inefficient process*: the typical star cluster mass is $\sim 10^{-3}$ of the host GMC mass, and any one GMC appears to produce only a handful of star clusters that remain bound over many Gyr. In other words, the star clusters end up using typically less than 1% of the host GMC gas supply regardless of environment (see H99; Kissler-Patig et al. 1998; McLaughlin 1999). Although most *stars* may well form in "clustered" mode, the majority of these clumps and associations dissolve quickly away into the field, leaving the bound clusters as those rare sites in which the star formation efficiency was $\sim 50\%$ or more.

These observations show that it is no longer tenable to regard globular cluster for-

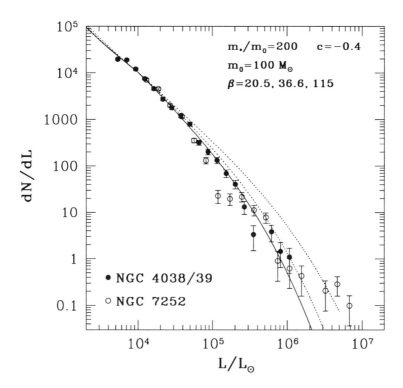

FIGURE 8. Luminosity distribution function (LDF, or number of clusters per unit luminosity) for the "young" globular clusters formed in the merger remnants NGC 4038/39 and NGC 7252. The model lines (from McLaughlin 1998, private communication) are computed from the collisional growth model of McLaughlin & Pudritz (1996). The original gas "particles" have masses $m_0 = 100 M_\odot$, the cloud lifetime varies as $\tau \sim m^{-0.4}$, and the parameter β defines the ratio of fiducial cloud lifetime to collisional crossing time. Larger values of β generate shallower mass functions that extend to higher mass. In this case a relatively small value $\beta \simeq 20$ fits the data.

mation as a "special" event which happened only in the early universe; though it is a rare mode of star formation, it is also clearly a robust process which can happen at any metallicity and at any time that a sufficient supply of gas is collected together. Exactly how the gas is accumulated into sufficiently large SGMCs during the protogalactic era is of interest on its own merit but is, apparently, not a critical issue for the mass spectrum of the globular clusters it produces.

At the present time, the collisional-growth model provides the only fully quantitative theoretical fit to the cluster mass spectrum at formation that we have available. Though it is obviously successful as far as it goes, many steps remain to be taken: for example, we would like to be able to predict from the gas dynamics of the host GMCs what the cloud lifetimes should be and exactly how the lifetime depends on mass. We also need to understand why the ratio of cluster mass to host GMC mass is typically $\sim 10^{-3}$, and what fraction of the total GMC mass can be expected to turn into clusters. (NB: In Elmegreen & Efremov (1997), a somewhat different scenario is adopted in which the mass distribution of clouds within a GMC is stated to resemble a fractal structure generated by turbulence. In this scheme, however, the expected slope $dN/dM \sim M^{-2}$ is slightly too steep to match most real galaxies, and does not clearly predict the progressive change in slope with mass that is embedded in the collisional growth model.)

3. Case Studies: NGC 5128

Globular clusters in the Milky Way and in other galaxies have generated a rich lore of data and theoretical interpretations for galaxy formation. Ultimately, however, we must return to the question that occupied even the earliest historical thinking about old stellar populations: how representative are the globular clusters of the *stellar* content of galactic halos? Are they a clearly distinguishable stellar population on their own?

For the Milky Way, we have the ability to bring in the full array of photometric, kinematic, and chemical composition properties of both kinds of stars, and the potential links of the clusters with such components as the metal-poor halo, the thick disk, the old disk, or accreted satellites, are numerous and have generated a continually growing literature. In progressively more distant galaxies, the information is inevitably more restricted until, at distances far beyond the Local Group, we can refer to the stellar halo component only by its integrated light, smoothing over all detail. For giant E galaxies, this problem has been especially unfortunate, since this is the one type of galaxy not represented in the Local Group, and the nearest rich collections of "typical" ellipticals are in the Virgo and Fornax clusters, some $\sim 15 - 18$ Mpc away.

There is, however, one giant E galaxy which is much closer: NGC 5128, the dominant member of the sparse Centaurus group at $d \simeq 4$ Mpc. Because of its well known inner gas and dust lane and active central regions where star formation is taking place, it has long been regarded as "peculiar" and was thus unduly neglected as a place where we might learn about the properties of *normal* ellipticals. However, much evidence accumulated in the past decade or more has revealed that major galaxies of all types undergo episodes of satellite accretion or stripping, gas infall, and even mergers with other large galaxies, and that what we see happening now in NGC 5128 is not an especially unusual event. Different theoretical channels exist for forming large elliptical galaxies (hierarchical merging of gas clouds in the early universe, or later merging of disk galaxies, to name two), and it is perhaps true that no single one will be found "typical" of all of them. But we now have no compelling reason to discard NGC 5128 as a basis for testing out many ideas about the way E galaxies are built, and eventually, we may find that it is just as representative of gE's as the Milky Way is of (say) large spirals.

NGC 5128 contains a healthy population of some ~ 1600 globular clusters, with a specific frequency $S_N = 2.6$ at a normal level for ellipticals in small groups (Harris 1991). Their spatial distribution, metallicities, and radial velocities have been studied in a series of papers by G. Harris and colleagues (1992 and references cited there). The GCS has been found to display the distinctly bimodal MDF that appears in many other gE's (Forbes et al. 1997), with roughly equal numbers of clusters in each component. In this respect, NGC 5128 so far reveals nothing fundamentally new. But of paramount importance for our purposes is that NGC 5128 is close enough for us to obtain (with HST imaging) color-magnitude photometry for its old-halo red giant stars, and thus *directly compare the metallicity distribution functions of the clusters and the halo stars.* At present, it is the only gE galaxy which is within reach in this way. With later and more advanced imaging tools (NGST and beyond), the stellar populations in the ellipticals of Virgo and Fornax will also come within our grasp; but NGC 5128 is an important prelude to what we can expect to glean from these systems.

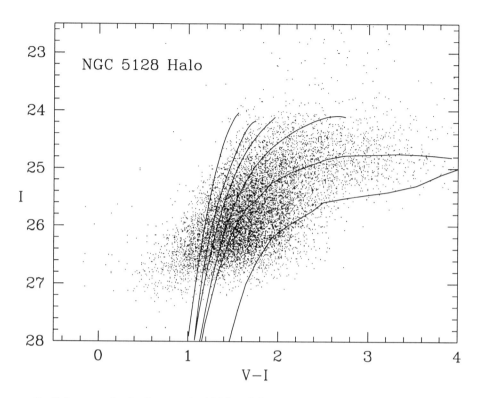

FIGURE 9. Color-magnitude diagram in $(I, V - I)$ for a halo field in NGC 5128, at a projected distance of 21 kpc south of the center of the galaxy (adapted from Harris et al. 1999). The limiting magnitude $I \simeq 27$ reaches to within about one magnitude above the expected level of the horizontal branch. The population of stars is dominated entirely by an old red-giant branch, with negligible contribution from any young or intermediate-age stars. Fiducial lines for different metallicities ([Fe/H] $= -2.2, -1.3, -0.75, -0.25, +0.1$ from left to right) are superimposed on the data to indicate the very large abundance range in the NGC 5128 halo.

Direct color-magnitude photometry of the halo stars in NGC 5128 was first obtained with HST by Soria et al. (1996) and then to a deeper level by G.Harris et al. (1999 [HHP99]). This latter study produced the color-magnitude diagram shown in Figure 9, showing that the halo of NGC 5128 is dominated by an old stellar population with an extremely broad range of color across the red giant branch. Only a tiny fraction of this range can be due to age differences (see the comparison of isochrones made in HHP99); most of it must be generated by a large spread of metallicity.

Interpolating within the fiducial lines, whose locations are a very nonlinear function of [Fe/H], we obtain the MDF shown in Figure 10. About two-thirds of the stars belong to a "metal-rich" component (MRC) starting at $Z \simeq 0.25 Z_\odot$ and trailing off to above-solar metallicity. The remaining one-third falls into the "metal-poor" component (MPC) starting at very low abundance ([Fe/H] < -2) and merging at its upper end with the onset of the MRC.

HHP99 show that these two components can be matched within the context of a classic one-zone chemical enrichment model (e.g., Pagel & Patchett 1975) in which an amount of gas G with initial abundance Z_0 is gradually converted to a mass of stars S, with each generation of stars ejecting back some enriched material to the interstellar medium. If the star formation is left to run to completion, then $G \rightarrow 0$ asymptotically, while the heavy-element abundance Z continually increases. Assuming complete mixing and instantaneous recycling, we find by solving the gains-and-losses equation for Z as a function of the remaining gas mass G that the relative number of stars at metallicity Z is described by a curve with a characteristic exponential-decay shape,

$$\frac{dN}{dZ} = \text{const} \cdot e^{-(Z-Z_0)/y} .$$

Here y is the "yield rate", the ratio of ejected mass in heavy elements to the mass locked up in dead stellar remnants for any given generation of stars.

This simplest of enrichment models, of course, has no direct timescale information in it, and does not take into account any of the interesting and important details in the enrichment process such as the different contributions of Type I and II supernovae. Nevertheless, its characteristic shape stands out remarkably well (Figure 10): the MPC started at primordial composition $Z_0 \simeq 0$ and was enriched at a rate $y \sim 0.002$, while the MRC began at $Z_0 \simeq 0.25 Z_\odot$ and was enriched with a clearly higher yield $y \simeq 0.006$.

The notably different yield ratios required by the data might, perhaps, represent different mixtures of Type I and II supernova or different IMFs in the two components. However, a more likely possibility may be connected with the fact that the MPC makes up only a minority of the stars (at most one-third of the halo, and perhaps much less if the MRC comprises a higher proportion of the inner halo; that is, if the halo has a metallicity gradient). Within the context of an *in situ* formation picture, this would mean that the first, earliest burst started from near-primordial composition and used up only a fraction of the gas, leaving it pre-enriched to $Z \sim 0.25 Z_\odot$ and waiting for the second, more major burst to start. The effective yield in the first burst would have been lowered if there had been substantial *gas loss* from the MPC star-forming clouds, as was noted long ago by Hartwick (1976) as a basic explanation for the low metallicity of the Milky Way halo stars.

An *in situ* synthesis of the formation of the NGC 5128 halo might then go something like this (HHP99; Forbes et al. 1997):

Round One: The protogalaxy consists of many pre-galactic fragments (SGMCs) of primordial composition, distributed throughout a still-lumpy larger potential well (Pudritz 1998). Star formation begins within these small clouds and consumes a fraction of the gas supply; but the ensuing stellar winds and supernovae interrupt the process, heating most of the (unused, but now partially enriched) gas and driving it out of these small local potential wells (Dekel & Silk 1986). The residue of this epoch is seen in the MPC stars, and in the metal-poor globular clusters.

Round Two: The gas filling the halo has now cooled and re-collapsed inward, and star formation begins again. This time, almost all the gas is used up, since it lies deeper within the fully formed potential well of the galaxy and does not escape. The heavy-element abundance drives on upward to Solar metallicity. This second stage, producing the MRC stars and clusters, is in fact what forms the bulk of the galaxy that we see today. The time interval between these two major phases is a matter of conjecture, but we can speculate that round Two must have happened at least $1 - 2$ Gy after Round One (that is, one or two free-fall times, enough for the ejected gas to cool, dissipate, and collapse back inward).

Round Three: The galaxy continues to build by later and more minor accretion of material, representing the ongoing phase of accumulation that we are witnessing now. Occa-

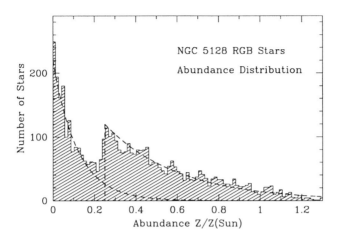

FIGURE 10. Metallicity distribution function for the red giant stars in the halo of NGC 5128, deduced from the color-magnitude data in the previous figure over the magnitude range $25 < I < 26$ (adapted from Harris et al. 1999). The MDF is shown as the number of stars per unit heavy-element abundance $Z/Z_\odot$. The *dashed lines* show the simple one-zone chemical enrichment model which has a characteristic exponential decay shape. The halo was enriched in two "bursts": the metal-poor component with a yield $y \simeq 0.002$ (left-hand curve), and the metal-richer component with $y \simeq 0.006$.

sional smaller satellites are captured, bringing their gas into the central ~ 5 kpc region that we see as the current site of star formation. The effects on the stellar content of the outer halo, however, appear to be minor.

This scenario can be checked against the MDF for the *globular clusters* (Figure 11, adapted from HHP99). The metal-rich clusters have a mean metallicity and MDF dispersion that are *indistinguishable* from the metal-richer stellar component – a strong consistency test that the two types of objects formed together in the major star-forming phase that produced most of the galaxy. Two interesting differences between the clusters and the field stars are, however, that (a) the clusters in the metal-poor "mode" make up about 60% of the halo GCS, whereas the MPC stars make up at most 1/3 of the stars; and (b) the MDF for the metal-poor *clusters* is sharply peaked at [Fe/H] = −1.2, and does not have the smooth ramp-up displayed by the MPC stars. An intriguing point connected with the comparison in Figure 11 is that the *specific frequencies* of the MPC and MRC must be quite different: the MPC contains one-third or less of the stars, but about two-thirds of the clusters. If we take these proportions literally, then we find (see HHP99) $S_N(\text{MPC}) \simeq 4.3$, rather like the Fornax and Virgo ellipticals, while $S_N(\text{MRC}) \simeq 1.5$ — a low value more like disk galaxies. This discrepancy would be consistent with the idea that globular clusters, as the densest units within the star-forming gas clouds, formed earliest; but that in "Round One", subsequent star formation was interrupted before it could run to completion, passing much of the gas on to the second round.

We must also ask whether or not other distinctly different models can produce the combination of features we observe. One alternate picture is the *accretion model* of Côté et al. (1998). The basis of their picture is that the MRC is regarded as the original E galaxy (formed in a single burst), while the MPC is accumulated later by accreted smaller galaxies, which have lower metallicity. The metal-poor halo and globular clusters are thus envisaged to build up over time. Large numbers of small galaxies are needed in this scheme: in the case of NGC 5128, the MPC stars make up as much as 1/3 of the

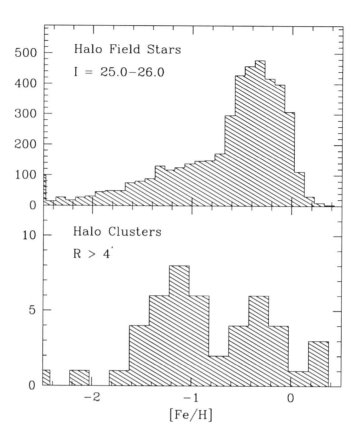

FIGURE 11. Metallicity distribution function (number of objects per [Fe/H] bin) for the halo stars in NGC 5128 (upper panel) and for the halo globular clusters (lower panel), with data from HHP99 and G. Harris et al. (1992).

halo, which would require the accretion of $\lesssim 200$ dwarf galaxies. For comparison, the present-day Centaurus group contains about 20 dwarfs, mostly irregulars (all of which have low S_N; it is the dE galaxies which have the higher S_N that we need for the MPC halo). These factors present difficulties for the halo accretion picture which may not be easy to overcome (see additional discussion in the next section).

An alternate and well known approach is the merger scenario outlined by Schweizer (1987) and Ashman & Zepf (1992), whereby a gE galaxy forms by the amalgamation of large disk galaxies (such as, indeed, we see happening today in cases like the Antennae). Here, the MRC is postulated to form during the merger, as long as the incoming progenitor galaxies are gas-rich. In the case of NGC 5128, we see at least 2/3 of the halo is metal-rich; thus, the merging "galaxies" would have had to be almost entirely gaseous in the first place. The distinction between this approach and a Searle-Zinn-like *in situ* model, in which a large galaxy builds up by hierarchical clustering of gas clouds, is therefore considerably blurred. We will return to these model comparisons in our later case studies.

4. Case Studies: NGC 4472 and M87

The Virgo cluster of galaxies is a testbed for globular cluster system studies which has proven fruitful over and over again, beginning with the classic survey photographic work of Hanes (1977). Among the Virgo ellipticals, the ones from which we have gathered the most informative data are the two supergiants, NGC 4472 (M49) and NGC 4486 (M87) – the latter galaxy being the prototypical "high−S_N" cD galaxy. Though their *stellar* populations are currently a bit beyond reach of our best imaging tools, we can still use their GCSs as extremely interesting touchstones for the various galaxy formation models mentioned above.

Both galaxies have strikingly similar bimodal MDFs, with roughly equal numbers of clusters in each "mode" (Geisler et al. 1996; Whitmore et al. 1995; Elson & Santiago 1996; Harris et al. 1998). Despite their factor-of-three differences in specific frequency, it has long been established that the luminosity distributions of their clusters have virtually identical forms, as they do in all giant ellipticals studied to date (Harris et al. 1991; Whitmore et al. 1995; H99).

4.1. *Velocities and Kinematics*

A new line of investigation which is being made increasingly possible by the 8- and 10-meter generation of telescopes is the kinematical analysis of these GCSs by large samples of radial velocities. Cohen & Ryzhov (1997), from accurate new velocity and abundance measurements of more than 200 clusters in M87, argue that virtually all the clusters (MRC or MPC) have traditionally "old" ages from their integrated line strengths, and use the radial velocity dispersion as a function of radius to trace out the mass profile of the halo. Kissler-Patig & Gebhardt (1998), using Cohen & Ryzhov's data, demonstrate that the *outer* part of the GCS ($R_{gc} \gtrsim 20$ kpc) shows substantial net rotation increasing past 200 km s^{-1}, along the major axis of the halo isophotes (dominated by the cD envelope). By contrast, the region inside ~ 20 kpc shows little significant rotation. Most of the rotation signal seems to arise from the *metal-poor* component, while the metal-richer clusters have a more modest and more nearly uniform $V_{rot} \sim 100$ km s^{-1} at all radii. This situation is in contrast to what we see in the Milky Way halo. Kissler-Patig & Gebhardt suggest that this angular momentum pattern was the result of a single merger between two already-large galaxies. Such a conclusion may not be the only possible one, since outward transport of angular momentum will result from a wide variety of formation scenarios involving either dissipational collapse or multiple mergers (or both), and even the merger product of several large galaxies can end up with a large outer-halo rotation (e.g., Weil & Hernquist 1996).

For NGC 4472, which has a total luminosity comparable to M87 but lacks a cD envelope or the "special" central position in the Virgo cluster potential well, similar velocity data have begun to accumulate. Sharples et al. (1998) have demonstrated from a new sample of cluster radial velocities that the MRC forms a dynamically "cooler" subsystem with distinctly lower velocity dispersion. In analogy with M87, there are hints that the MPC has a distinctly higher net rotation as well, though more data will be needed to confirm this.

The kinematical analysis of GCSs in other galaxies is just getting started. *Large* samples of velocities are essential to this type of study, and we can look forward in a few years to extremely informative new comparisons with theory.

4.2. *MDFs and Specific Frequencies: How Many Components?*

Integrated photometry of large samples of clusters in giant E galaxies like the Virgo members has made huge strides in the past few years. The especially detailed photometric study of NGC 4472 by Geisler et al. (1996) and Lee et al. (1998) is an excellent example of such work, and represents a major step in our understanding of the MDFs and their implications. Among the most important of their results, from a comprehensive and accurate metallicity-sensitive $(C - T_1)$ photometric survey of the GCS, is that the two components (MPC, MRC) have very different central concentrations (with the MRC closely following the spatial structure of the integrated halo light and the MPC following a much shallower distribution). Each of these components by itself displays little radial gradient in color. Perhaps most significant of all is the observation that the mean color of the MRC clusters as a function of radius *precisely matches the color of the halo light* – to well within the observational uncertainties of both. By inference, just as we found earlier for NGC 5128, the halo stellar population appears to be dominated by a metal-rich population whose MDF is very similar to the metal-rich globular cluster component. A reasonable conclusion is that the MRC clusters and the majority of the halo stars formed together.

Let us take this argument one step further (see Forbes et al. 1997 for similar reasoning). Suppose we assume that the halo has just two distinct components, whose mean metallicities are identical with those of the MRC and MPC clusters. Close inspection of the actual color indices and their internal uncertainties (see H99) then shows that $\lesssim 6\%$ of the halo stars can belong to the MPC, otherwise the integrated color of the entire halo would deviate from that of the MRC level beyond their error bars. Just as we did above for NGC 5128, we can translate this ratio into separate specific frequency estimates for the MPC and MRC, knowing that $S_N = 5.6$ for the whole galaxy (Harris et al. 1998). We obtain $S_N(\mathrm{MRC}) \simeq 2.4$ and a *lower limit* for the MPC of $S_N \gtrsim 50$! The specific frequency for the metal-rich component is similar to the average of many other normal ellipticals (see above), but the lower limit for the metal-poor component is already higher than the measured S_N for any known galaxy. If we interpret this result within a conventional *in situ* formation picture, it means that either the cluster formation in the first, metal-poor burst was outstandingly efficient, or that a very large fraction of the initial gas went unused, just as we speculated earlier for NGC 5128. The data for NGC 4472 suggest, apparently, a still more extreme difference between the relative amounts of metal-rich and metal-poor stellar components.

Very similar results emerge from a new study of the GCS in NGC 1399, the central giant galaxy in the Fornax cluster (Ostrov et al. 1998). Once again, a clear bimodal MDF emerges, in which the metal-richer clusters match the mean color of the galaxy halo light extremely well. The specific frequency of the MRC *alone* they estimate to be $S_N \simeq 3$, within just the same range that we found for NGC 4472.

4.3. *Merger Scenarios*

Let us now take a closer look at two alternate models for the construction of this two-component form of the MDF: mergers and accretions.

The idea that an elliptical galaxy could form from the direct merger of two colliding disk galaxies, thus forming more globular clusters out of the gas present within the progenitor galaxies, was forcefully presented by Schweizer (1987), followed by the more detailed modelling of Ashman & Zepf (1992 and subsequent papers). In this picture, most or all of the metal-poor clusters would have come from the progenitor galaxies while most of the metal-rich ones were formed out of the merger. Schweizer, Whitmore, and their colleagues have recently used the HST in a series of imaging studies of merger remnants of different ages (Whitmore & Schweizer 1995; Miller et al. 1997; Whitmore et al. 1997; Schweizer et al. 1996) to show in a very direct way that a large E galaxy indeed emerges out of the "train wreck" provided by the collision, and that a new, second population of globular clusters has formed in each merger.

It has been suggested in many papers (cf. the ones cited above) that the merger process is a way to understand the generic "high specific frequency problem" that ellipticals generally have larger values of S_N than disk galaxies by factors of 2 or 3. Many of these statements are wrong simply because they ignore the *field stars* that will also form during the merger (the specific frequency is the *ratio* of cluster numbers to field-star light, so S_N might stay the same, increase, or even decrease after the merger depending on the efficiency of cluster formation). In some other papers, this effect is recognized, but it is claimed that mergers ought to be an ideal place for super-efficient cluster formation because of the expected high degree of gas shocking and compression (though it is not obvious that these effects would be any more extreme than in, say, a more conventional Searle-Zinn formation process of merging gas clouds in a large potential well).

An excellent test of these arguments is to refer directly to the observations. The studies of merger remnants listed above show quantitatively that the newly born E galaxies have *low* specific frequencies, in the range $S_N \sim 1-3$ after the age-fading of the field-star light is taken properly into account (cf. the references cited above). That is, globular clusters are indeed formed in the merger, but in more or less the same proportions relative to field-star light as the original disk galaxies had, so there is little change in the net specific frequency. These studies indicate that many of the low$-S_N$ ellipticals found in sparse groups populated by spirals and irregulars might indeed be the simple results of earlier mergers of this type.

What about the much higher$-S_N$ ellipticals such as in Virgo and Fornax ($S_N \sim 5$), or even the M87-like objects with $S_N \gtrsim 10$? Are there ways in which the merger process can be adapted to allow all ellipticals to have formed this way? In fact, the problem of sheer numbers of clusters in many big ellipticals presents quite a severe challenge to the merger hypothesis, in at least two directions:

(a) Taking NGC 4472 as a typical case, we see that it contains $N(MP) \simeq 3660$ metal-poor clusters (Geisler et al. 1996). If all of them came from the progenitor spirals, then we would require the merger of not just two disk galaxies (at typically ~ 200 clusters per galaxy, like the Milky Way or M31, this would be out of the question) but rather more like 15 to 20 of them. With this many mergers, it is also less clear how such a sharply defined bimodal MDF would still exist in the final product, or how the outer halo could preserve the strong net rotation velocity mentioned above.

(b) In NGC 4472, the number of metal-rich clusters is $N(MR) \simeq 2440$. If all of them were formed during the mergers, at the normal efficiency $e_0 \simeq 1$ per $2 \times 10^8 M_\odot$ of gas, then the total input gas mass would have to be at least $5 \times 10^{11} M_\odot$. This enormous amount of gas does not resemble any merger happening today.

Both of these aspects of the problem put very extreme demands on the merger process: many mergers seem to be required, and the amount of input gas is so large that it is equivalent to an entire protogalaxy. Alternately, we must arbitrarily require that the cluster formation efficiency is vastly higher than normal.

Let us evaluate the problem more generally. If the progenitor galaxies have specific frequencies S_{N1}, S_{N2} and luminosities L_1, L_2 (appropriately age-faded to ~ 13 Gyr for an old E galaxy population), then it can quickly be shown (see H99) that the specific frequency S_N of the resulting elliptical will be

$$S_N = \frac{S_{N1}L_1 + S_{N2}L_2 + 3.5(e/e_0)L_3}{L_1 + L_2 + L_3}$$

where L_3 is the amount of field-star light (again, age-faded) formed in the merger, and e

is the efficiency of conversion of gas into new globular clusters relative to the "baseline" efficiency e_0 corresponding to $S_N^0 = 3.5$ (see above).

This equation can be used to predict the outcoming S_N for any pair of input galaxies and input gas mass $M_g = L_3/(M/L)$. Typically, in the present-day universe, even very "gas-rich" disk galaxies can supply only $10^9 - 10^{10} M_\odot$ of input gas, which is far short of the amount needed to make large numbers of clusters even if the efficiency e is an order of magnitude higher than normal. To change S_N from a typical disk-galaxy level of $S_N \sim 2$ up to the gE level of 6 or more requires *both* an extremely high $e-$factor (ten times e_0 or more!) *and extremely large amounts of gas* ($5 \times 10^{10} M_\odot$ or more). The one obvious era of the Universe's history when such large amounts of gas were available was, of course, in the protogalactic epoch. Such a merger picture differs significantly from the version originally presented by Schweizer (1987) or Ashman & Zepf (1992), and it is not clear how it could be distinguished from the hierarchical amalgamation of gas clouds as in Searle & Zinn.

We must emphasize that mergers, in the mode proposed by Ashman & Zepf, are a convincing way to interpret *some* E galaxies – the ones with low specific frequencies like their progenitor spirals. Such objects are quite clearly continuing to form in the low-redshift contemporary universe in sites like the Antennae galaxies and others. However, this model is not easily adapted to building the higher$-S_N$ ellipticals found in rich clusters. Other noteworthy difficulties with the merger hypothesis as a solution for all types of E formation are discussed at length by (e.g.) Geisler et al. (1996), Forbes et al. (1997), and Kissler-Patig et al. (1998).

4.4. Accretion Scenarios

An almost diametrically opposite approach to the problem is taken by Côté et al. (1998) through an accretion model. Here, the original E galaxy is imagined to form in a single major burst which generates the metal-rich stars and clusters. Then, over time it accretes many smaller galaxies (which each have lower metallicity) and builds up the metal-poor halo and cluster population. They show by numerical simulations that the accreted population can build up the right MDF shape for the clusters and, potentially, match many specific cases.

This model has the advantage of using a process that we are fairly certain is happening to real galaxies; Sagittarius and its disruption by the Milky Way is the best-known example. But this approach also is not without problems. A significant one is that for a typical gE, the number of metal-poor and metal-rich clusters is roughly equal, which implies that roughly half the present-day galaxy must have been accreted material. But then, why is the *color* of the halo light as red (metal-rich) as it is, and not halfway between the MRC and MPC clusters? In addition, the sheer numbers of accreted dwarf galaxies needed by the model are very large (many hundreds would be required, for example, to build up our test example, NGC 4472), and imply that the great majority of the original dwarf population must now be absorbed by the larger ones, despite their very small collision cross sections. Detailed dynamical simulations of the process are needed to determine whether or not this assumption is realistic. Finally, if many of the dwarfs are dE or irregular types, then their rather low specific frequencies ($S_N \sim 1 - 2$) would tend to lower the S_N of the final giant elliptical rather than keep it at its desired level of 5 or more.

The conditions of this accretion scenario would change noticeably if the incoming dwarfs were highly gaseous, so that new clusters could form in the process and also maintain most of the stars at the necessarily high metallicity that we see in giant ellipticals (see also Hilker 1998 for a similar view). If so, however, then the entire picture once again begins to resemble the standard Searle/Zinn galaxy formation model whereby a large galaxy builds up from many small dwarf-sized gas clouds. The presence or absence of gas is therefore a critical factor in evaluating the success of any formation picture.

5. Case Studies: The Brightest Cluster Ellipticals

For more than two decades, the class of cD-type galaxies that reside at the centers of rich galaxy clusters – the "brightest cluster members" or BCGs – have attracted considerable attention because of their exceptionally rich globular cluster populations. M87 in Virgo is the prototype of this class, and the GCSs in more than 30 such galaxies have now been studied (see Blakeslee et al. 1997; Harris et al. 1998 for summary lists). These galaxies are of special interest because they provide uniquely strong – and challenging – tests for our various galaxy formation models.

In M87, the basic shape of the MDF and the spatial distribution of the GCS are remarkably similar to what we see in NGC 4472. The special issue we are confronted with in M87 and the other BCGs is the classic "high specific frequency" problem: stated quite simply, why do they have so many globular clusters? In the 20 or more years we have been aware of the problem, no obvious answer has emerged. In any of the competing scenarios (*in situ*, mergers, accretions), appeals have been made to rather arbitrary conditions (such as an outstandingly high cluster formation efficiency in high density protogalactic gas; or especially violent compression shocks during mergers; or accretion of dwarf E galaxies which themselves had high S_N). All of these alternatives have tended to be uncompelling.

5.1. *The Role of the Halo Gas: Understanding High S_N*

New clues emerging from the recent surveys (Blakeslee 1997; Blakeslee et al. 1997; Harris et al. 1998) show rather clearly that S_N for these large galaxies displays a *continuum* of values from the "normal" $S_N^0 \simeq 3-4$ up to 15 and more. More importantly, S_N is clearly correlated with the luminosity of the BCG, and with the total mass of the surrounding galaxy cluster as measured by the velocity dispersion of the galaxies or the temperature and luminosity of the hot X-ray gas in the cluster potential. Larger BCGs are found in more massive potential wells, and have higher specific frequencies.

Let us go back to the definition of S_N. To obtain an unusually high ratio N_{cl}/L_{gal}, we really have only two ways to do it. (a) First, we can increase N_{cl} at a given L_{gal}. This route remains arbitrary, since we have no physical theory for predicting the formation efficiency e. (b) Second, we can decrease L_{gal} for the same number of clusters. The most obvious way to take this second path, following the clues mentioned above for NGC 5128 and NGC 4472, is to leave a lot of the original gas supply unused. To put this another way, we can assume that N_{cl} is proportional to the *original* protogalactic gas mass; but that in certain galaxies (the BCGs in particular), for whatever reason a large amount of the gas ended up not being converted to stars.

This latter option is raised by Blakeslee (1997) and Harris et al. (1998) and pursued extensively by McLaughlin (1999). Summarizing their arguments, suppose now that we redefine the globular cluster formation efficiency as a *mass ratio*,

$$\epsilon = \frac{M_{cl}}{M_\star + M_{gas}}$$

where $M_\star$ is the mass now in visible stars, while M_{gas} is the remaining mass in the galactic halo which was originally part of the protogalaxy. We now explicitly assume that ϵ *is constant*, and thus the total number of globular clusters is directly proportional to the

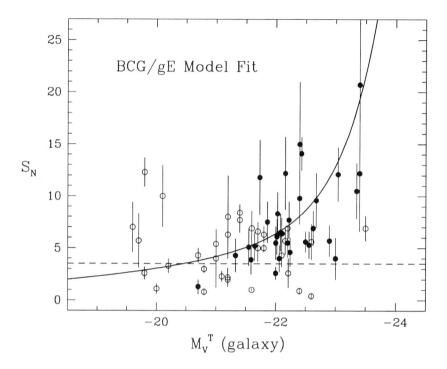

FIGURE 12. Relation between specific frequency and galaxy luminosity for giant E galaxies (BCGs are shown by filled circles, other giant ellipticals by open circles. The dashed line indicates $S_N = 3.5$ shown previously in Figure 7. The solid line is from the model of McLaughlin (1999), in which a constant globular cluster formation efficiency ϵ per unit *initial gas mass* is assumed. The curve ramps up steeply at high luminosity to account for the observation that more luminous BCGs have considerably more high-temperature halo gas (see text).

original protogalactic mass. (We do not expect, of course, that ϵ can be precisely the same in all environments, since cluster formation must at some level be a stochastic process. We are, however, using this hypothesis to reduce the order-of-magnitude differences in S_N down to much smaller levels that can more easily be accommodated within the observational scatter.) In most galaxies, M_{gas} is much smaller than $M_\star$: their halos do not contain much leftover gas. However, the BCGs usually have quite large amounts of high-temperature X-ray gas, and the amount goes up dramatically with luminosity ($M_X \sim L_{gal}^{2.5}$). For the very biggest BCGs, M_{gas} is larger than $M_\star$, which would require us to assume in this picture that *most* of their protogalactic gas is still unconverted.

McLaughlin (1999) tests this idea quantitatively by showing that in three well studied galaxies (M87 and NGC 4472 in Virgo and NGC 1399 in Fornax), the ratio of mass densities $\rho_{cl}/(\rho_\star + \rho_{gas})$ throughout the halo is equal to $\simeq 0.0025$ in all three galaxies. That is, the addition of the X-ray gas mass makes this ratio much more nearly consistent with the hypothesis that there is a quasi-universal cluster formation efficiency ϵ. By using the fundamental plane relations for E galaxies which relate the mass-to-light ratio, scale radius, and internal velocity dispersion to L_{gal}, McLaughlin (1999) also shows that the ratio $M_{gas}/M_\star$ scales as $L_{gal}^{1.5}$, leading to the following scaling for the specific frequency:

$$S_N \sim \frac{N_{cl}}{L_{gal}} \sim \epsilon \, (1 + \frac{M_{gas}}{M_\star}) L_{gal}^{0.3} \, .$$

There are two principal effects acting to define the shape of this relation: first is that the mass-to-light ratio increases gradually with luminosity ($M/L \sim L^{0.3}$), so that bigger ellipticals have more clusters per unit light. Second is that bigger BCGs have much more surrounding halo gas, thus (by hypothesis) they generated many more clusters at early times. Combining both of these effects, McLaughlin shows that the overall relation between N_{cl} and L_{gal} matches the BCG pattern remarkably well (Figure 12). In the graph, the solid line representing the $\epsilon = $ constant $= 0.0025$ model continues to steepen at higher luminosity because of the progressively increasing *relative* amounts of halo gas.

This model provides, for the first time, a plausible and quantitative view of the specific frequencies in BCGs, and as such, gives us a partial solution to the overall problem of understanding the global galaxy-to-galaxy scatter in specific frequency. Clearly, interesting anomalies still remain above and below the mean line. Do all the E galaxies with very low$-S_N$ values represent merger remnants? And do the few remaining normal ellipticals with anomalously high S_N represent cases of genuinely more efficient cluster formation? In addition, for the BCGs themselves, it is still not certain how the large amounts of halo gas were left in place: the gas might have been ejected and heated during the earliest star formation phase by the first round of supernovae, as we speculated earlier for NGC 5128 (this phase would have had to be particularly violent and short-lived to distinguish it from other, more normal ellipticals, perhaps involving a major galactic wind; see Harris et al. 1998). Alternately, the protogalactic SGMCs may have been stripped of much of their gas in the deep central BCG potential well before star formation could begin in earnest (Blakeslee 1997). Whatever the answer, these large central galaxies represent an extreme set of conditions which still needs to be modelled in detail.

5.2. *The Coma Ellipticals: New Anomalies and Challenges*

As a final example of particularly interesting case studies, we will turn to the E galaxies in the Coma cluster. Coma (at redshift $cz \simeq 7000$ km s^{-1}, about 4 magnitudes more distant than Virgo or Fornax) represents the nearest example of a genuinely rich Abell-cluster environment. Its central supergiant, NGC 4874, is an extremely luminous cD galaxy and the ~ 1 Mpc core of the cluster contains dozens of large ellipticals and hundreds of dwarfs.

At present, we have information for GCSs in three Coma giants: NGC 4874 itself (Harris 1987; Thompson & Valdes 1987; Blakeslee & Tonry 1995; Kavelaars et al. 1999), NGC 4881 (Baum et al. 1995), and IC 4051 (Baum et al. 1997; Woodworth & Harris 1999), primarily from HST imaging. The latter two galaxies have luminosities similar to the brighter Virgo ellipticals. These three cases, though they represent only a small fraction of the Coma population, are already showing us GCS characteristics that we have not seen before.

IC 4051: The halo of this galaxy is spatially quite centrally concentrated, possibly because it has been tidally trimmed by repeated passages through the Coma central potential (its radial velocity relative to the Coma center is -2100 km s^{-1}, suggesting that it is now passing through the cluster core; see Woodworth & Harris 1999). More notable, however, is the MDF for the globular clusters (Figure 13). It is almost entirely metal-rich, and unimodal (in fact, the observed dispersion of the histogram, σ[Fe/H] $= -0.65$, is scarcely larger than the $\simeq 0.5$ dex spread expected from observational scatter alone, suggesting that the intrinsic dispersion is closer to ~ 0.3 dex). If more than one

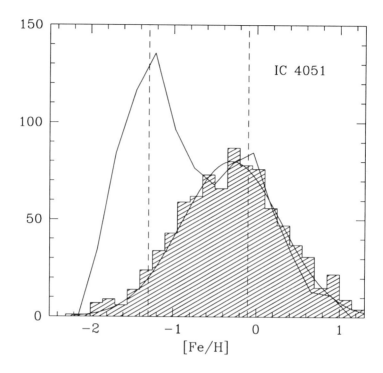

FIGURE 13. Metallicity distribution function for the globular clusters in the Coma giant elliptical IC 4051 (adapted from Woodworth & Harris 1999). The Gaussian fitted line to the histogram has a mean at [Fe/H] = −0.3 and a dispersion σ[Fe/H] = 0.65 dex. The broken solid line shows the MDF for the globular clusters in the Virgo elliptical NGC 4472; in IC 4051, the metal-poor component is almost entirely missing.

"mode" is indeed present in this MDF, the two peak locations must be close enough to be almost entirely obscured by the observational scatter. The only other galaxy we have found in which the metal-poor component is virtually absent is the cD in Hydra I, NGC 3311 (Secker et al. 1995). IC 4051 is, however, not a "central cluster" elliptical.

Still more remarkable is its specific frequency, $S_N = 12 \pm 3$, at a level comparable with some of the brightest BCGs despite the fact that it is not a cD, not exceptionally luminous, and not at the center of any identifiable group or subgroup of galaxies.

NGC 4881: This galaxy is similar in size and structure to IC 4051, but as shown by Baum et al. (1995), the globular cluster system has an extremely *low* specific frequency ($S_N \lesssim 2$) and an MDF which is entirely *metal-poor* (with a mean [Fe/H] = −1.2).

NGC 4874: This central cD in Coma has the expected high $S_N \simeq 10 - 15$ and a very broad MDF like M87 (Virgo) or NGC 1399 (Fornax). Notably, however, it has an extremely extended spatial structure, with a core radius approaching ~ 15 kpc, almost three times larger than the GCS in M87 and an order of magnitude larger than in the Milky Way. Are we looking at a globular cluster population which, at least partially, belongs to the extended potential well of the Coma cluster itself rather than the central galaxy?

All three of these interesting galaxies do, however, display the same GCLF shape that we have come to expect: a Gaussian-like distribution by magnitude (Figure 14) very

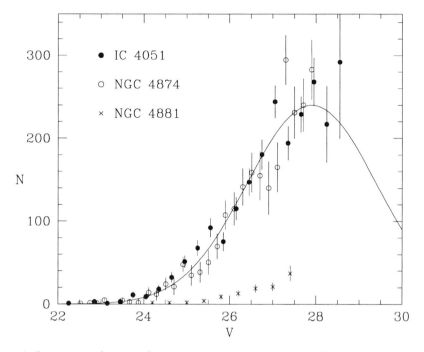

FIGURE 14. Luminosity function for the globular clusters in three Coma ellipticals. The Gaussian curve (solid line) has a turnover (peak point) at $V^0 = 27.8 \pm 0.2$ and a dispersion $\sigma = 1.5$ magnitudes.

similar to M87 and the Virgo and Fornax ellipticals, despite their extreme differences in metallicity distribution and specific frequency. If we *assume* that the peak luminosity is $M_V^0 = -7.3$, consistent with the distance calibrations to Virgo and Fornax through various stellar standard candles including Cepheids, planetary nebulae, and red-giant stars, then we obtain a Hubble constant $H_0 \simeq 70$ km s^{-1} Mpc^{-1} (see H99 for more detailed discussion).

Interpreting these results from the Coma galaxies presents new challenges to all the galaxy formation scenarios. For example, if the *in situ* approach is basically correct, then in IC 4051 we would have to ask how the galaxy was able to bypass the first, low-metallicity enrichment stage almost totally. If its high S_N is, as argued above, due to gas ejection during the major star formation stage, then we would expect rather low effective yield y during this stage and thus a low-metallicity GCS, contrary to what we see. On the other hand, if IC 4051 formed almost entirely from mergers, then where is the low-[Fe/H] component from the smaller progenitor galaxies? How, also, did the high specific frequency arise, since all the observed merger cases seem to produce low S_N?

In NGC 4881, we have more or less the opposite set of problems. It is, apparently, entirely missing the high-metallicity component that is present in all other gE's. Thus, a merger formation would have had to be almost entirely gas-free (though this would be consistent with its low S_N). If, instead, it formed through accretion, then where is the high-[Fe/H] cluster component from the original E galaxy? Finally, if most of the galaxy formed *in situ*, then it is difficult to understand the large (~ 1 dex) mean metallicity difference between the GCS and the field-star light. All the current models are left with significant problems – which means that we have much to learn!

6. A Brief Synthesis

Throughout this chapter, an attempt has been made to relate observable features of globular cluster systems (kinematics, metallicity distributions, luminosity and mass distributions, specific frequencies) to the competing hypotheses for galaxy formation: *in situ*, mergers, or accretions. Where do we stand, at the current stage of development of the subject?

Each of these models in its nominal form represents an extreme view of the formation process, and each has merits. The *in situ* model comes in two basic varieties: the Searle/Zinn (1978) picture whereby a large galaxy is assembled over a relatively long period of time (perhaps several Gyr) from many small, dwarf-sized gas clouds; and the monolithic collapse picture of Eggen, Lynden-Bell, & Sandage (1962). The arguments that are presented to contrast, and choose between, these two extremes now involve a far greater body of evidence than was available when these key papers were written. For example, the full complexity of behavior of the globular cluster systems in these galaxies – the large ranges in metallicity, the existence of two or more distinct subpopulations in most large galaxies, the kinematical subgroups, and the evidence for late infall and accretion – clearly favors the SZ scheme, especially for their outer halos. The ELS picture may still have relevance for describing simpler, smaller systems such as nucleated dwarf ellipticals or the central bulges of larger galaxies, in which the principal formation epoch may have been dominated by rapid dissipation and collapse. Nevertheless, for most large galaxies, elliptical or spiral, it seems necessary to invoke at least two major formation stages. Surprisingly, we may need *only* two stages for at least some gE galaxies if we interpret the two-component metallicity distributions in the simplest possbile way.

The basic *accretion* picture would start with a large initial galaxy which had already formed through an *in situ* process. As described above, we then add a sequence of smaller galaxies to it and thus build up the metal-poor halo component. Here also, we can invoke two basic varieties: If the accreted objects are gas-free, then we would build up a larger galaxy with a normal specific frequency and a broad MDF, but with a halo that should have an average metallicity that is intermediate or moderately low. If, however, the accreted smaller galaxies are very gas-rich, then new clusters and halo stars can form in the process and (possibly) change the GCS specific frequency and weight the MDF increasingly toward the metal-rich end. However, in a rather direct sense, this latter alternative could be regarded as a version of the generic SZ picture.

Lastly, the *merger* approach involves the amalgamation of pre-existing galaxies of roughly equal size. The result, almost invariably, should be an elliptical of low specific frequency if the merging objects are pre-existing disk galaxies. Once again, the presence or absence of gas will play an important role in the outcome. If the mergers are taking place at high redshift (that is, at very early times), then we can expect the progenitors to be largely gaseous, in which case the majority of stars in the merged product would actually form during the merger of gas clouds. This, too, can be viewed as an extension of the basic SZ galaxy formation picture. On the other hand, if the merging is happening at low redshift nearer to the present day, then the galaxies are much more likely to have smaller amounts of gas, much less star formation can happen during the merger, and the result will be a low$-S_N$ elliptical such as we see in small groups.

No one of these three approaches can be taken as the answer to all cases. Instead, we need all of them to tell the complete story. The evidence is plain that mergers and accretions are happening today and must be part of the ongoing construction of large galaxies. Similarly, *in situ* formation at very early times is clearly an essential part of the history of giant ellipticals and BCGs. In their full range of possibilities, these formation pictures make up a continuum of processes which can be seen to operate in the real world, and thus the competition among them which often appears in the literature is more than a little artificial. The challenge we face is, for any one galaxy, to identify the particular set of processes which has ended up dominating its present-day structure.

For practical help and many conversations on these topics, I am grateful to Dean McLaughlin, J. J. Kavelaars, and Gretchen Harris.

REFERENCES

Ajhar, E. A., Blakeslee, J. P., & Tonry, J. L. 1994, AJ, 108, 2087

Armandroff, T. E. 1989, AJ, 97, 375

Ashman, K. M., & Zepf, S. E. 1992, ApJ, 384, 50

Ashman, K. M., & Zepf, S. E. 1998, Globular Cluster Systems (Cambridge: Cambridge University Press)

Baum, W. A. et al. 1995, AJ, 110, 2537

Baum, W. A. et al. 1997, AJ, 113, 1483

Blakeslee, J. P. 1997, ApJ, 481, L59

Blakeslee, J. P., & Tonry, J. L. 1995, ApJ, 442, 579

Blakeslee, J. P., Tonry, J. L., & Metzger, M. R. 1997, AJ, 114, 482

Brodie, J. P., & Huchra, J. P. 1991, ApJ, 379, 157

Catelan, M., & de Freitas Pacheco, J. A. 1995, AAp, 297, 345

Chaboyer, B., Demarque, P., & Sarajedini, A. 1996, ApJ, 459, 558

Cohen, J. G., & Ryzhov, A. 1997, ApJ, 486, 230

Côté, P., Marzke, R. O., & West, M. J. 1998, ApJ, 501, 554

Da Costa, G. S., & Armandroff, T. E. 1995, AJ, 109, 2533

Dekel, A., & Silk, J. 1986, ApJ, 303, 39

Dinescu, D. I., Girard, T. M., & van Altena, W. F. 1998, preprint

Durrell, P. R., Harris, W. E., Geisler, D., & Pudritz, R. E. 1996, AJ, 112, 972

Eggen, O. J., Lynden-Bell, D., & Sandage, A. R. 1962, ApJ, 136, 735 (ELS)

Elmegreen, B. G., & Efremov, Y. N. 1997, ApJ, 480, 235

Elson, R. A. W., & Santiago, B. X. 1996, MNRAS, 280, 971

Field, G. B., & Saslaw, W. C. 1965, ApJ, 142, 568

Forbes, D. A., Brodie, J. P., & Grillmair, C. J. 1997, AJ, 113, 1652

Frenk, C. S., & White, S. D. M. 1980, MNRAS, 193, 295

Geisler, D., Lee, M. G., & Kim, E. 1996, AJ, 111, 1529

Gnedin, O. Y., & Ostriker, J. P. 1997, ApJ, 474, 225

Grundahl, F., & Andersen, M. I. 1998, in Galaxy Evolution: Connecting the Distant Universe with the Local Fossil Record, Observatoire de Meudon, Paris

Grundahl, F., VandenBerg, D. A., & Andersen, M. I. 1998, ApJ, 500, L179

Hanes, D. A. 1977, MNRAS, 180, 309

Harris, G. L. H., Geisler, D., Harris, H. C., & Hesser, J. E. 1992, AJ, 104, 613

Harris, G. L. H., Harris, W. E., & Poole, G. B. 1999, AJ, 117, in press (HHP99)

Harris, W. E. 1976, AJ, 81, 10953

Harris, W. E. 1987, ApJ, 315, L29

Harris, W. E. 1991, ARAA, 29, 543

Harris, W. E. 1996, AJ, 112, 1487

Harris, W. E. 1997, in The Extragalactic Distance Scale (Poster Papers), edited by M. Donahue & M. Livio (Baltimore: StScI), 29

Harris, W. E. 1999, in Star Clusters, 28th Saas-Fee Advanced Course for Astrophysics and Astronomy, in press (H99)

Harris, W. E., Allwright, J. W. B., Pritchet, C. J., & van den Bergh, S. 1991, ApJS, 76, 115

Harris, W. E., & van den Bergh, S. 1981, AJ, 86, 1627

Harris, W. E., Harris, G. L. H., & McLaughlin, D. E. 1998, AJ, 115, 1801

Harris, W.E., & Pudritz, R. E. 1994, ApJ, 429, 177

Hartwick, F. D. A. 1976, ApJ, 209, 418

Hilker, M. 1998, PhD thesis, University of Bonn

Ibata, R. A., Wyse, R. F. G., Gilmore, G., Irwin, M. J., & Suntzeff, N. B. 1997, AJ, 113, 634

Johnson, J. A., & Bolte, M. 1998, AJ, 115, 693

Kavelaars, J. J., Harris, W. E., Hanes, D. A., Pritchet, C. J., & Hesser, J. E. 1999, in preparation

Kinman, T. D. 1959, MNRAS, 119, 538

Kissler-Patig, M., Forbes, D. A., & Minniti, D. 1998, MNRAS, 298, 1123

Kissler-Patig, M., & Gebhardt, K. 1998, AJ, 116, 2237

Kissler-Patig, M., Kohle, S., Hilker, M., Richtler, T., Infante, L., & Quintana, H. 1997, A&A, 319, 470

Kundu, A., & Whitmore, B. C. 1998, AJ, 116, in press

Kwan, J. 1979, ApJ, 229, 567

Lee, M. G., Kim, E., & Geisler, D. 1998, AJ, 115, 947

Lee, Y. W., Demarque, P., & Zinn, R. 1994, ApJ, 423, 248

Majewski, S. R. 1993, ARAA, 31, 575

Majewski, S. R. 1994, ApJ, 431, L17

McLaughlin, D. E. 1999, AJ, in press

McLaughlin, D. E., & Pudritz, R. E. 1996, ApJ, 457, 578

Miller, B. W., Lotz, J. M., Ferguson, H. C., Stiavelli, M., & Whitmore, B. C. 1998, ApJ, in press

Miller, B. W., Whitmore, B. C., Schweizer, F., & Fall, S. M. 1997, AJ, 114, 2381

Minniti, D. 1995, AJ, 109, 1663

Murali, C., & Weinberg, M. D. 1997, MNRAS, 291, 717

Ostrov, P. G., Forte, J. C., & Geisler, D. 1998, preprint

Pagel, B. E. J., & Patchett, B. E. 1975, MNRAS, 172, 13

Pudritz, R. E. 1998, private communication

Rodgers, A. W., & Paltoglou, G. 1984, ApJ, 283, 5

Sarajedini, A., Chaboyer, B., & Demarque, P. 1997, PASP, 109, 1321

Schweizer, F. 1987, in Nearly Normal Galaxies, edited by S. M. Faber (New York: Springer), 18

Schweizer, F., Miller, B. W., Whitmore, B. C., & Fall, S. M. 1996, AJ, 112, 1839

Searle, L., & Zinn, R. 1978, ApJ, 225, 357 (SZ)

Secker, J., Geisler, D., McLaughlin, D. E., & Harris, W. E. 1995, AJ, 109, 1019

Sharples, R. M. et al. 1998, AJ, 115, 2337

Soria, R. et al. 1996, ApJ, 465, 79

Stetson, P. B., VandenBerg, D. A., & Bolte, M. 1996, PASP, 108, 560

Stetson, P. B. et al. 1999, AJ, 117, in press

Thompson, L. A., & Valdes, F. 1987, ApJ, 315, L35

Vesperini, E. 1997, MNRAS, 287, 915

Vesperini, E. 1998, MNRAS, 299, 1019

Vesperini, E., & Heggie, D. C. 1997, MNRAS, 289, 898

Weil, M. L., & Hernquist, L. 1996, ApJ, 460, 101

Whitmore, B. C. 1997, in The Extragalactic Distance Scale, edited by M. Livio, M. Donahue, & N. Panagia (Cambridge University Press), 254

Whitmore, B. C., Miller, B. W., Schweizer, F., & Fall, S. M. 1997, AJ, 114, 1797

Whitmore, B. C., & Schweizer, F. 1995, AJ, 109, 960

Whitmore, B. C., Sparks, W. B., Lucas, R.A., Macchetto, F. D., & Biretta, J. A. 1995, ApJ, 454, L73

Woodworth, S. C., & Harris, W. E. 1999, in preparation

Zinn, R. 1985, ApJ, 293, 424

Zinn, R. 1993, in The Globular Cluster – Galaxy Connection, ASP Conference Series, 48, edited by G. H. Smith & J. P. Brodie (San Francisco: ASP), 38

Zinn, R. 1996, in Formation of the Galactic Halo ... Inside and Out, ASP Conference Series, 92, edited by H. Morrison & A. Sarajedini (San Francisco: ASP), 211